Springer Complexity

Springer Complexity is an interdisciplinary program publishing the best research and academic-level teaching on both fundamental and applied aspects of complex systems – cutting across all traditional disciplines of the natural and life sciences, engineering, economics, medicine, neuroscience, social and computer science.

Complex Systems are systems that comprise many interacting parts with the ability to generate a new quality of macroscopic collective behavior the manifestations of which are the spontaneous formation of distinctive temporal, spatial or functional structures. Models of such systems can be successfully mapped onto quite diverse "real-life" situations like the climate, the coherent emission of light from lasers, chemical reaction–diffusion systems, biological cellular networks, the dynamics of stock markets and of the Internet, earthquake statistics and prediction, freeway traffic, the human brain, or the formation of opinions in social systems, to name just some of the popular applications.

Although their scope and methodologies overlap somewhat, one can distinguish the following main concepts and tools: self-organization, nonlinear dynamics, synergetics, turbulence, dynamical systems, catastrophes, instabilities, stochastic processes, chaos, graphs and networks, cellular automata, adaptive systems, genetic algorithms and computational intelligence.

The two major book publication platforms of the Springer Complexity program are the monograph series "Understanding Complex Systems" focusing on the various applications of complexity, and the "Springer Series in Synergetics", which is devoted to the quantitative theoretical and methodological foundations. In addition to the books in these two core series, the program also incorporates individual titles ranging from textbooks to major reference works.

Understanding Complex Systems

Founding Editor: J.A. Scott Kelso

Future scientific and technological developments in many fields will necessarily depend upon coming to grips with complex systems. Such systems are complex in both their composition – typically many different kinds of components interacting simultaneously and nonlinearly with each other and their environments on multiple levels – and in the rich diversity of behavior of which they are capable.

The Springer Series in Understanding Complex Systems series (UCS) promotes new strategies and paradigms for understanding and realizing applications of complex systems research in a wide variety of fields and endeavors. UCS is explicitly transdisciplinary. It has three main goals: First, to elaborate the concepts, methods and tools of complex systems at all levels of description and in all scientific fields, especially newly emerging areas within the life, social, behavioral, economic, neuro- and cognitive sciences (and derivatives thereof); second, to encourage novel applications of these ideas in various fields of engineering and computation such as robotics, nano-technology and informatics; third, to provide a single forum within which commonalities and differences in the workings of complex systems may be discerned, hence leading to deeper insight and understanding.

UCS will publish monographs, lecture notes and selected edited contributions aimed at communicating new findings to a large multidisciplinary audience.

R. Dahlhaus · J. Kurths · P. Maass · J. Timmer
(Eds.)

Mathematical Methods in Signal Processing and Digital Image Analysis

With 96 Figures and 20 Tables

Springer

Volume Editors

Rainer Dahlhaus
Universität Heidelberg
Inst. Angewandte Mathematik
Im Neuenheimer Feld 294
69120 Heidelberg
Germany
dahlhaus@statlab.uni-heidelberg.de

Jürgen Kurths
Universität Potsdam
Inst. Physik, LS Theoretische Physik
Am Neuen Palais 19
14469 Potsdam
Germany
jkurths@agnld.uni-potsdam.de

Peter Maass
Universität Bremen
FB 3 Mathematik/Informatik
Zentrum Technomathematik
28334 Bremen
Germany
pmaass@uni-bremen.de

Jens Timmer
Universität Freiburg
Zentrum Datenanalyse
Eckerstr. 1
79104 Freiburg
Germany
jens.timmer@fdm.uni-freiburg.de

ISBN: 978-3-642-09482-8 e-ISBN: 978-3-540-75632-3

Understanding Complex Systems ISSN: 1860-0832

Cover Design: WMXDesign GmbH, Heidelberg

Printed on acid-free paper

9 8 7 6 5 4 3 2 1

springer.com

Preface

Interest in time series analysis and image processing has been growing very rapidly in recent years. Input from different scientific disciplines and new theoretical advances are matched by an increasing demand from an expanding diversity of applications. Consequently, signal and image processing has been established as an independent research direction in such different areas as electrical engineering, theoretical physics, mathematics or computer science. This has lead to some rather unstructured developments of theories, methods and algorithms. The authors of this book aim at merging some of these diverging directions and to develop a consistent framework, which combines these heterogeneous developments. The common core of the different chapters is the endavour to develop and analyze mathematically justified methods and algorithms. This book should serve as an overview of the state of the art research in this field with a focus on nonlinear and nonparametric models for time series as well as of local, adaptive methods in image processing.

The presented results are in its majority the outcome of the DFG-priority program SPP 1114 "Mathematical methods for time series analysis and digital image processing". The starting point for this priority program was the consideration, that the next generation of algorithmic developments requires a close cooperation of researchers from different scientific backgrounds. Accordingly, this program, which was running for 6 years from 2001 to 2007, encompassed approximately 20 research teams from statistics, theoretical physics and mathematics. The intensive cooperation between teams from different specialized disciplines is mirrored by the different chapters of this book, which were jointly written by several research teams. The theoretical findings are always tested with applications of different complexity.

We do hope and expect that this book serves as a background reference to the present state of the art and that it sparks exciting and creative new research in this rapidly developing field.

This book, which concentrates on methodologies related to identification of dynamical systems, non- and semi-parametric models for time series, stochastic methods, wavelet or multiscale analysis, diffusion filters and mathematical morphology, is organized as follows.

The Chap. 1 describes recent developments on multivariate time series analysis. The results are obtained from combinig statistical methods with the theory of nonlinear dynamics in order to better understand time series measured from underlying complex network structures. The authors of this chapter emphasize the importance of analyzing the interrelations and causal influences between different processes and their application to real-world data such as EEG or MEG from neurological experiments. The concept of determining directed influences by investigating renormalized partial directed coherence is introduced and analyzed leading to estimators of the strength of the effect of a source process on a target process.

The development of surrogate methods has been one of the major driving forces in statistical data analysis in recent years. The Chap. 2 discusses the mathematical foundations of surrogate data testing and examines the statistical performance in extensive simulation studies. It is shown that the performance of the test heavily depends on the chosen combination of the test statistics, the resampling methods and the null hypothesis.

The Chap. 3 concentrates on multiscale approaches to image processing. It starts with construction principles for multivariate multiwavelets and includes some wavelet applications to inverse problems in image processing with sparsity constraints. The chapter includes the application of these methods to real life data from industrial partners.

The investigation of inverse problems is also at the center of Chap. 4. Inverse problems in image processing naturally appear as parameter identification problems for certain partial differential equations. The applications treated in this chapter include the determination of heterogeneous media in subsurface structures, surface matching and morphological image matching as well as a medically motivated image blending task. This chapter includes a survey of the analytic background theory as well as illustrations of these specific applications.

Recent results on nonlinear methods for analyzing bivariate coupled systems are summarized in Chap. 5. Instead of using classical linear methods based on correlation functions or spectral decompositions, the present chapter takes a look at nonlinear approaches based on investigating recurrence features. The recurrence properties of the underlying dynamical system are investigated on different time scales, which leads to a mathematically justified theory for analyzing nonlinear recurrence plots. The investigation includes an analysis of synchronization effects, which have been developed into one of the most powerfull methodologies for analyzing dynamical systems.

Chapter 6 takes a new look at strucutred smoothing procedures for denoising signals and images. Different techniques from stochastic kernel smoother to anisotropic variational approaches and wavelet based techniques are analyzed and compared. The common feature of these methods is their local and adaptive nature. A strong emphasize is given to the comparison with standard methods.

Chapter 7 presents a novel framework for the detection and accurate quantification of motion, orientation, and symmetry in images and image sequences. It focuses on those aspects of motion and orientation that cannot be handled successfully and reliably by existing methods, for example, motion superposition (due to transparency, reflection or occlusion), illumination changes, temporal and/or spatial motion discontinuities, and dispersive nonrigid motion. The performance of the presented algorithms is characterized and their applicability is demonstrated by several key application areas including environmental physics, botany, physiology, medical imaging, and technical applications.

The authors of this book as well as all participants of the SPP 1114 “Mathematical methods for time series analysis and digital image processing” would like to express their sincere thanks to the German Science Foundation for the generous support over the last 6 years. This support has generated and sparked exciting research and ongoing scientific discussions, it has lead to a large diversity of scientific publications and – most importantly- has allowed us to educate a generation of highly talented and ambitious young scientists, which are now spread all over the world. Furthermore, it is our great pleasure to acknowledge the impact of the referees, which accompangnied and shaped the developments of this priority program during its different phases. Finally, we want to express our gratitude to Mrs. Sabine Pfarr, who prepared this manuscript in an seemingly endless procedure of proof reading, adjusting images, tables, indices and bibliographies while still keeping a friendly level of communication with all authors concerning those nasty details scientist easily forget.

Bremen,
November 2007

Rainer Dahlhaus, Jürgen Kurths,
Peter Maass, Jens Timmer

Contents

List of Contributors

Til Aach
RWTH Aachen University, Aachen, Germany
Til.Aach@lfb.rwth-aachen.de

Jens F. Acker
University of Dortmund, Dortmund, Germany
jens.acker@math.uni-dortmund.de

Björn Andres
University of Heidelberg, Heidelberg, Germany
bjoern.andres@iwr.uni-heidelberg.de

Christoph Bandt
University of Greifswald, Greifswald, Germany
bandt@uni-greifswald.de

Erhardt Barth
University of Lübeck, Lübeck, Germany
barth@inb.uni-luebeck.de

Anatoly Berdychevski
Weierstraß-Institut Berlin, Berlin, Germany
berdichevski@wias-berlin.de

Benjamin Berkels
University of Bonn, Bonn, Germany
benjamin.berkels@ins.uni-bonn.de

Martin Böhme
University of Lübeck, Lübeck, Germany
boehme@inb.uni-luebeck.de

Kristian Bredies
University of Bremen, Bremen, Germany
kbredies@math.uni-bremen.de

Rainer Dahlhaus
University of Heidelberg, Heidelberg, Germany
dahlhaus@statlab.uni-heidelberg.de

Stephan Dahlke
University of Marburg, Marburg, Germany
dahlke@mathematik.uni-marburg.de

Mamadou S. Diallo
ExxonMobil, Houston, TX, USA
mamadou.s.diallo@exxonmobil.com

Stephan Didas
Saarland University, Saarland, Germany
didas@mia.uni-saarland.de

Marc Droske
University of Bonn, Bonn, Germany
droske@iam.uni-bonn.de

Jürgen Franke
University of Kaiserslautern, Kaiserslautern, Germany
franke@mathematik.uni-kl.de

Christoph S. Garbe
University of Heidelberg, Heidelberg, Germany
Christoph.Garbe@iwr.uni-heidelberg.de

Andreas Groth
University of Greifswald, Greifswald, Germany
groth@uni-greifswald.de

Siana Halim
Petra-Christian University, Surabaya, Indonesia
halim@peter.petra.ac.id

Martin Haker
University of Lübeck, Lübeck, Germany
haker@inb.uni-luebeck.de

Wolfram Hesse
University of Jena, Jena, Germany
wolfram.hesse@mti.uni-jena.de

Matthias Holschneider
University of Potsdam, Potsdam, Germany
hols@math.uni-potsdam.de

Jaroslav Hron
University of Dortmund, Dortmund, Germany
jaroslav.hron@math.uni-dortmund.de

Bernd Jähne
University of Heidelberg, Heidelberg, Germany
bernd.jaehne@iwr.uni-heidelberg.de

Karsten Koch
University of Marburg, Marburg, Germany
koch@mathematik.uni-marburg.de

Claudia Kondermann
University of Heidelberg, Heidelberg, Germany
Claudia.Nieuwenhuis@iwr.uni-heidelberg.de

Kai Krajsek
University of Frankfurt, Frankfurt, Germany
kai.krajsek@vsi.cs.uni-frankfurt.de

Michail Kulesh
University of Potsdam, Potsdam, Germany
mkulesh@math.uni-potsdam.de

Jürgen Kurths
University of Potsdam, Potsdam, Germany
jkurths@agnld.uni-potsdam.de

Lutz Leistritz
University of Jena, Jena, Germany
lutz.leistritz@mti.uni-jena.de

Dirk Lorenz
University of Bremen, Bremen, Germany
dlorenz@math.uni-bremen.de

Peter Maass
University of Bremen, Bremen, Germany
pmaass@math.uni-bremen.de

Thomas Maiwald
University of Freiburg, Freiburg, Germany
maiwald@fdm.uni-freiburg.de

Enno Mammen
University of Mannheim, Mannheim, Germany
emammen@rumms.uni-mannheim.de

Norbert Marwan
University of Potsdam, Potsdam, Germany
marwan@agnld.uni-potsdam.de

Rudolf Mester
University of Frankfurt, Frankfurt, Germany
mester@iap.uni-frankfurt.de

Cicero Mota
University of Frankfurt, Frankfurt, Germany; Federal University of Amazonas, Manaus, Brazil
cicmota@gmail.com

Pavel Mrázek
UPEK Prague R & D Center, Prague, Czech Republic
pavel.mrazek@upek.com

Matthias Mühlich
RWTH Aachen, Aachen, Germany
mm@lfb.rwth-aachen.de

Stephan Müller
Hoffmann-La Roche AG, Basel, Switzerland
stephan.mueller@roche.com

Swagata Nandi
Indian Statistical Institute, New Delhi, India
nandi@isid.ac.in

Nadine Olischläger
University of Bonn, Bonn, Germany
nadine.olischlaeger@ins.uni-bonn.de

Pavel Pavlov
University of Heidelberg, Heidelberg, Germany
pavel.pavlov@iwr.uni-heidelberg.de

Heinz-Otto Peitgen
Center for Complex Systems and Visualization, Bremen, Germany
heinz-otto.peitgen@cevis.uni-bremen.de

Jörg Polzehl
Weierstrass-Institute Berlin, Berlin, Germany
polzehl@wias-berlin.de

Tobias Preusser
Center for Complex Systems and Visualization, Bremen, Germany
preusser@cevis.uni-bremen.de

M. Carmen Romano
University of Potsdam, Potsdam, Germany
romano@agnld.uni-potsdam.de

Michael Rosenblum
University of Potsdam, Potsdam, Germany
mros@agnld.uni-potsdam.de

Martin Rumpf
University of Bonn, Bonn, Germany
Martin.Rumpf@ins.uni-bonn.de

Bärbel Schack
University of Jena, Jena, Germany

Karl Schaller
Hôpitaux Universitaires de Genève, Genève, Switzerland
karl.schaller@hcuge.ch

Hanno Scharr
Research Center Jülich GmbH, Jülich, Germany
h.scharr@fz-juelich.de

Björn Schelter
University of Freiburg, Freiburg, Germany
schelter@fdm.uni-freiburg.de

Frank Scherbaum
University of Potsdam, Potsdam, Germany
fs@geo.uni-potsdam.de

Stefan Schiffler
University of Bremen, Bremen, Germany
schiffi@math.uni-bremen.de

Tobias Schuchert
Research Center Jülich GmbH, Jülich, Germany
t.schuchert@fz-juelich.de

Vladimir Spokoiny
Weierstrass-Institute Berlin, Berlin, Germany
spokoiny@wias-berlin.de

Andreas Stämpfli
Hoffmann-La Roche AG, Basel, Switzerland
andreas.staempfli@roche.com

Gabriele Steidl
University of Mannheim, Mannheim, Germany
steidl@math.uni-mannheim.de

Ingo Stuke
University of Lübeck, Lübeck, Germany
Ingo.Stuke@t-online.de

Suhasini Subba Rao
University of Texas, Austin, TX, USA
suhasini@stat.tamu.edu

Joseph Tadjuidje
University of Kaiserslautern, Kaiserslautern, Germany
tadjuidj@mathematik.uni-kl.de

Gerd Teschke
Konrad-Zuse-Center Berlin, Berlin, Germany
teschke@zib.de

Marco Thiel
University of Potsdam, Potsdam, Germany
thiel@agnld.uni-potsdam.de

Herbert Thiele
Bruker Daltonics GmbH, Bremen, Germany
Herbert.Thiele@bdal.de

Jens Timmer
University of Freiburg, Freiburg, Germany
jeti@fdm.uni-freiburg.de

Stefan Turek
University of Dortmund, Dortmund, Germany
ture@featflow.de

Joachim Weickert
Saarland University, Saarbrücken, Germany
weickert@mia.uni-saarland.de

Manuel Werner
University of Marburg, Marburg, Germany
werner@mathematik.uni-marburg.de

Herbert Witte
University of Jena, Jena, Germany
herbert.witte@mti.uni-jena.de

1

Multivariate Time Series Analysis

Björn Schelter[1], Rainer Dahlhaus[2], Lutz Leistritz[3], Wolfram Hesse[3], Bärbel Schack[3], Jürgen Kurths[4], Jens Timmer[1], and Herbert Witte[3]

[1] Freiburg Center for Data Analysis and Modeling, University of Freiburg, Freiburg, Germany
{schelter,jeti}@fdm.uni-freiburg.de

[2] Institute for Applied Mathematics, University of Heidelberg, Heidelberg, Germany
dahlhaus@statlab.uni-heidelberg.de

[3] Institute for Medical Statistics, Informatics, and Documentation, University of Jena, Jena, Germany
{lutz.leistritz,wolfram.hesse,herbert.witte}@mti.uni-jena.de

[4] Institute for Physics, University of Potsdam, Potsdam, Germany
jkurths@agnld.uni-potsdam.de

In Memoriam

Bärbel Schack (1952–2003)

On July 24th, 2003, Bärbel Schack passed away. With her passing, the life sciences have lost one of their most brilliant, original, creative, and compassionate thinkers.

1.1 Motivation

Nowadays, modern measurement devices are capable to deliver signals with increasing data rates and higher spatial resolutions. When analyzing these data, particular interest is focused on disentangling the network structure underlying the recorded signals. Neither univariate nor bivariate analysis techniques are expected to describe the interactions between the processes sufficiently well. Moreover, the direction of the direct interactions is particularly important to understand the underlying network structure sufficiently well. Here, we present multivariate approaches to time series analysis being able to distinguish direct and indirect, in some cases the directions of interactions in linear as well as nonlinear systems.

1.2 Introduction

In this chapter the spectrum of methods developed in the fields ranging from linear stochastic systems to those in the field of nonlinear stochastic systems is discussed. Similarities and distinct conceptual properties in both fields are presented.

Of particular interest are examinations of interrelations and especially causal influences between different processes and their applications to real-world data, e.g. interdependencies between brain areas or between brain areas and the periphery in neuroscience. There, they present a primary step toward the overall aim: the determination of mechanisms underlying pathophysiological diseases, primarily in order to improve diagnosis and treatment strategies especially for severe diseases [70]. The investigations are based on considering the brain as a dynamic system and analyzing signals reflecting neural activity, e.g. electroencephalographic (EEG) or magnetoencephalographic (MEG) recordings. This approach has been used, for instance, in application to data sets recorded from patients suffering from neurological or other diseases, in order to increase the understanding of underlying mechanisms generating these dysfunctions [18, 20, 21, 22, 24, 51, 52, 65, 68]. However, there is a huge variety of applications not only in neuroscience where linear as well as nonlinear time series analysis techniques presented within this chapter can be applied successfully.

As far as the linear theory is considered, various time series analysis techniques have been proposed for the description of interdependencies between dynamic processes and for the detection of causal influences in multivariate systems [10, 12, 16, 24, 50, 67]. In the frequency domain the interdependencies between two dynamic processes are investigated by means of the cross-spectrum and the coherence. But an analysis based on correlation or coherence is often not sufficient to adequately describe interdependencies within a multivariate system. As an example, assume that three signals originate from distinct processes (Fig. 1.1). If interrelations were investigated by an application of a bivariate analysis technique to each pair of signals and if a relationship was detected between two signals, they would not necessarily be linked directly (Fig. 1.1). The interdependence between these signals might also be mediated by the third signal. To enable a differentiation between direct and indirect influences in multivariate systems, graphical models applying partial coherence have been introduced [8, 9, 10, 53, 57].

Besides detecting interdependencies between two signals in a multivariate network of processes, an uncovering of directed interactions enables deeper insights into the basic mechanisms underlying such networks. In the above example, it would be possible to decide whether or not certain processes project their information onto others or vice versa. In some cases both directions might be present, possibly in distinct frequency bands. The concept of Granger-causality [17] is usually utilized for the determination of causal influences. This probabilistic concept of causality is based on the common sense conception that causes precede their effects in time and is formulated

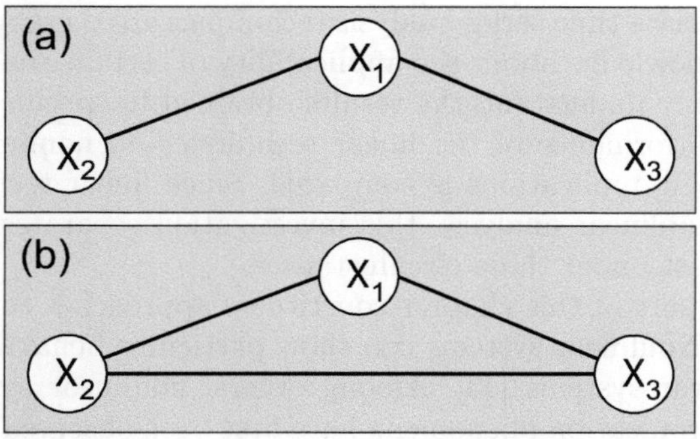

Fig. 1.1 (**a**) Graph representing the true interaction structure. Direct interactions are only present between signals X_1 and X_2 and X_1 and X_3; the direct interaction between X_2 and X_3 is absent. (**b**) Graph resulting from bivariate analysis, like cross-spectral analysis. From the bivariate analysis it is suggested that all nodes are interacting with one another. The spurious edge between signals X_2 and X_3 is mediated by the common influence of X_1

in terms of predictability. Empirically, Granger-causality is commonly evaluated by fitting vector auto-regressive models. A graphical approach for modeling Granger-causal relationships in multivariate processes has been discussed [11, 14]. More generally, graphs provide a convenient framework for causal inference and allow, for example, the discussion of so-called spurious causalities due to confounding by unobserved processes [13].

Measures to detect directed influences in multivariate linear systems that are addressed in this manuscript are, firstly, the Granger-causality index [24], the directed transfer function [28], and, lastly, partial directed coherence [2]. While the Granger-causality index has been introduced for inference of linear Granger-causality in the time domain, partial directed coherence has been suggested to reveal Granger-causality in the frequency domain based on linear vector auto-regressive models [2, 24, 49, 56, 57, 70, 71]. Unlike coherence and partial coherence analysis, the statistical properties of partial directed coherence have only recently been addressed. In particular, significance levels for testing nonzero partial directed coherences at fixed frequencies are now available while they were usually determined by simulations before [2, 61]. On the one hand, without a significance level, detection of causal influences becomes more hazardous for increasing model order as the variability of estimated partial directed coherences increases leading to false positive detections. On the other hand, a high model order is often required to describe the dependencies of a multivariate process examined sufficiently well. The derivation of the statistics of partial directed coherence suggests a modification with superior properties to some extent that led to the concept of renormalized partial directed coherence.

A comparison of the above mentioned techniques is an indispensable prerequisite to reveal their specific abilities and limitations. Particular properties of these multivariate time series analysis techniques are thereby discussed [70]. This provides knowledge about the applicability of certain analysis techniques helping to reliably understand the results obtained in specific situations. For instance, the performance of the linear techniques on nonlinear data which are often faced in applications is compared. Since linear techniques are not developed for nonlinear analysis, this investigation separates the chaff from the wheat at least under these circumstances.

The second part of this chapter constitutes approaches to nonlinear time series analysis. Nonlinear systems can show particular behaviors that are impossible for linear systems [43]. Among others, nonlinear systems can synchronize. Synchronization phenomena have first been observed by Huygens for coupled self-sustained oscillators. The process of synchronization is an adaptation of certain characteristics of the two processes. Huygens has observed an unison between two pendulum clocks that were mounted to the same wall. The oscillations between the clocks showed a phase difference of 180° [4, 42]. A weaker form of synchronization has recently been observed between two coupled chaotic oscillators. These oscillators were able to synchronize their phases while their amplitudes stay almost uncorrelated [6, 38, 42, 43, 46]. Nowadays, several forms of synchronization have been described ranging from phase synchronization to lag synchronization to almost complete synchronization [7, 43, 47]. Generalized synchronization is characterized by some arbitrary function that relates processes to one another [30, 48, 60].

The process of synchronization is necessarily based on self-sustained oscillators. By construction linear systems are not self-sustained oscillators and therefore synchronization cannot be observed for those linear systems [58, 72]. However, as will be shown, techniques for the analysis of synchronization phenomena can be motivated and derived based on the linear analysis techniques [55].

As the mean phase coherence, a measure able to quantify synchronization, is originally also a bivariate technique, a multivariate extension was highly desired. This issue is related to the problem of disentangling direct and indirect interactions as discussed in the vicinity of linear time series analysis. Two synchronized oscillators are not necessarily directly coupled. One commonly influencing oscillator is sufficient to warrant a spurious coupling between the first two. Again similar to the linear case, interpretations of results are thus hampered if a disentangling was not possible. But a multivariate extension of phase synchronization analysis has been developed. A procedure based on the partial coherence analysis was employed and carried over to the multivariate nonlinear synchronizing systems [55]. By means of a simulation study it is shown that the multivariate extension is a powerful technique that allows disentangling interactions in multivariate synchronizing systems.

The chapter is structured as follows. First the linear techniques are introduced. Their abilities and limitations are discussed in an application to real-world data. The occurrence of burst suppression patterns is investigated by means of an animal model of anesthetized pigs. In the second part, nonlinear synchronization is discussed. First, the mean phase coherence is intuitively introduced and then mathematically derived from cross-spectral analysis. A multivariate extension of phase synchronization concludes the second part of this Chapter.

1.3 Mathematical Background

In this section, we summarize the theory of the multivariate linear time series analysis techniques under investigation, i.e. partial coherence and partial phase spectrum (Sect. 1.3.1), the Granger-causality index, the partial directed coherence, and the directed transfer function (Sect. 1.3.2). Finally, we briefly introduce the concept of directed graphical models (Sect. 1.3.3).

1.3.1 Non-Parametric Approaches

Partial Coherence and Partial Phase Spectrum

In multivariate dynamic systems, more than two processes are usually observed and a differentiation of direct and indirect interactions between the processes is desired. In the following we consider a multivariate system consisting of n stationary signals X_i, $i = 1, \ldots, n$.

Ordinary spectral analysis is based on the spectrum of the process X_k introduced as

$$S_{X_k X_k}(\omega) = \left\langle \mathcal{FT}\left\{X_k\right\}(\omega)\ \mathcal{FT}\left\{X_k\right\}^*(\omega)\right\rangle , \tag{1.1}$$

where $\langle \cdot \rangle$ denotes the expectation value of $(\cdot)$, and $\mathcal{FT}\left\{\cdot\right\}(\omega)$ the Fourier transform of $(\cdot)$, and $(\cdot)^*$ the complex conjugate of $(\cdot)$. Analogously, the cross-spectrum between two processes X_k and X_l

$$S_{X_k X_l}(\omega) = \left\langle \mathcal{FT}\left\{X_k\right\}(\omega)\ \mathcal{FT}\left\{X_l\right\}^*(\omega)\right\rangle , \tag{1.2}$$

and the normalized cross-spectrum, i.e. the coherence as a measure of interaction between two processes X_k and X_l

$$\mathrm{Coh}_{X_k X_l}(\omega) = \frac{\left|S_{X_k X_l}(\omega)\right|}{\sqrt{S_{X_k X_k}(\omega)\ S_{X_l X_l}(\omega)}} \tag{1.3}$$

are defined. The coherence is normalized to $[0, 1]$, whereby a value of one indicates the presence of a linear filter between X_k and X_l and a value of zero its absence.

To enable a differentiation in direct and indirect interactions bivariate coherence analysis is extended to partial coherence. The basic idea is to subtract linear influences from third processes under consideration in order to detect directly interacting processes. The partial cross-spectrum

$$S_{X_k X_l|Z}(\omega) = S_{X_k X_l}(\omega) - S_{X_k Z}(\omega) S_{ZZ}^{-1}(\omega) S_{Z X_l}(\omega) \tag{1.4}$$

is defined between process X_k and process X_l, given all the linear information of the remaining possibly more-dimensional processes $Z = \{X_i | i \neq k, l\}$. Using this procedure, the linear information of the remaining processes is subtracted optimally. Partial coherence

$$\mathrm{Coh}_{X_k X_l|Z}(\omega) = \frac{|S_{X_k X_l|Z}(\omega)|}{\sqrt{S_{X_k X_k|Z}(\omega)\, S_{X_l X_l|Z}(\omega)}} \tag{1.5}$$

is the normalized absolute value of the partial cross-spectrum while the partial phase spectrum

$$\Phi_{X_k X_l|Z}(\omega) = \arg\left\{S_{X_k X_l|Z}(\omega)\right\} \tag{1.6}$$

is its argument [8, 10]. To test the significance of coherence values, critical values

$$s = \sqrt{1 - \alpha^{\frac{2}{\nu - 2L - 2}}} \tag{1.7}$$

for a significance level α depending on the dimension L of Z are calculated [66]. The equivalent number of degrees of freedom ν depends on the estimation procedure for the auto- and cross-spectra. If for instance the spectra are estimated by smoothing the periodograms, the equivalent number of degrees of freedom [5]

$$\nu = \frac{2}{\sum_{i=-h}^{h} u_i^2}\,, \text{ with } \sum_{i=-h}^{h} u_i = 1 \tag{1.8}$$

is a function of the width $2h + 1$ of the normalized smoothing window u_i.

Time delays and therefore the direction of influences can be inferred by evaluating the phase spectrum. A linear phase relation $\Phi_{X_k X_l|Z}(\omega) = d\omega$ indicates a time delay d between processes X_k and X_l. The asymptotic variance

$$\mathrm{var}\left\{\Phi_{X_k X_l|Z}(\omega)\right\} = \frac{1}{\nu}\left[\frac{1}{\mathrm{Coh}^2_{X_k X_l|Z}(\omega)} - 1\right] \tag{1.9}$$

for the phase $\Phi_{X_k X_l|Z}(\omega)$ again depends on the equivalent number of degrees of freedom ν and the coherence value at frequency ω [5]. The variance and therefore the corresponding confidence interval increases with decreasing coherence values. Large errors for every single frequency prevent a reliable estimation of the phase spectrum for corresponding coherence values which are smaller than the critical value s. For signals in a narrow frequency band, a linear phase relationship is thus difficult to detect. Moreover, if the two processes considered were mutually influencing each other, no simple procedure exists to detect the mutual interaction by means of one single phase spectrum especially for influences in similar frequency bands.

Marrying Parents of a Joint Child

When analyzing multivariate systems by partial coherence analysis, an effect might occur, which might be astonishingly in the first place. While bivariate coherence is non-significant the partial coherence can be significantly different from zero. This effect is called *marrying parents of a joint child* and is explained as follows (compare Fig. 1.2):

Imagine that two processes X_2 and X_3 influence process X_1 but do not influence each other. This is correctly indicated by a zero bivariate coherence between oscillator X_2 and oscillator X_3. In contrast to bivariate coherence, partial coherence between X_2 and X_3 conditions on X_1. To explain the significant partial coherence between the processes X_2 and X_3, the specific case $X_1 = X_2 + X_3$ is considered. The optimal linear information of X_1 in X_2 is $1/2\,X_1 = 1/2\,(X_2+X_3)$. Subtracting this from X_2 gives $1/2\,(X_2-X_3)$. Analogously, a subtraction of the optimal linear information $1/2\,X_1 = 1/2\,(X_2+X_3)$ from X_3 leads to $-1/2\,(X_2 - X_3)$. As coherence between $1/2\,(X_2 - X_3)$ and $-1/2\,(X_2-X_3)$ is one, the partial coherence between X_2 and X_3 becomes significant. This effect is also observed for more complex functional relationships between stochastic processes X_1, X_2 and X_3. The "parents" X_2 and X_3 are connected and "married by the common child" X_1. The interrelation between X_2 and X_3 is still indirect, even if the partial coherence is significant. In conclusion, the *marrying parents of a joint child* effect should not be identified as a direct interrelation between the corresponding processes and is detected by simultaneous consideration of bivariate coherence and partial coherence.

Finally we mention that in practice the effect usually is much smaller than in the above example; e.g. if $X_1 = X_2 + X_3 + \varepsilon$ with independent random variables of equal variance, then it can be shown that the partial coherence is 0.5.

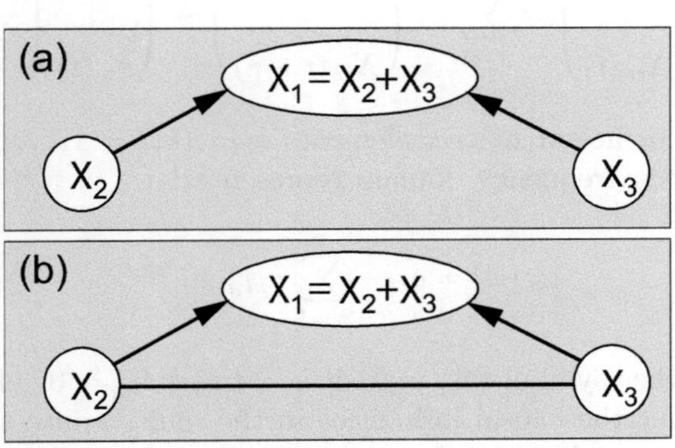

Fig. 1.2 (**a**) Graph representing the true interaction structure. Signal X_1 is the sum of two signals X_2 and X_3, which are independent processes, i.e. the direct interaction between X_2 and X_3 is absent. (**b**) Graph resulting from multivariate analysis. From the multivariate analysis it is suggested that all nodes are interacting with one another. The spurious edge between signal X_2 and X_3 is due to the so-called *marrying parents of a joint child* effect

1.3.2 Parametric Approaches

Besides the non-parametric spectral concept introduced in the previous section, we investigate three parametric approaches to detect the direction of interactions in multivariate systems. The general concept underlying these parametric methods is the notion of causality introduced by Granger [17]. This causality principle is based on the common sense idea, that a cause must precede its effect. A possible definition of Granger-causality based on the principle of predictibilty may be given by the following supposition. For dynamic systems a process X_l is said to Granger-cause a process X_k, if knowledge of the past of process X_l improves the prediction of the process X_k compared to the knowledge of the past of process X_k alone and several other variables under discussion. In the following we will speak of multivariate Granger-causality if additional variables are used or of bivariate Granger-causality if no additional variables are used. The former corresponds in some sense to partial coherence while the latter corresponds in some sense to ordinary coherence. A comparison of bivariate and multivariate Granger-causality can be found in Eichler, Sect. 9.4.4 [15].

Commonly, Granger-causality is estimated by means of vector autoregressive models. Since a vector autoregressive process is linear by construction, only linear Granger-causality can be inferred by this methodology. In the following, we will use the notion causality in terms of linear Granger-causality although not explicitly mentioned.

The parametric analysis techniques introduced in the following are based on modeling the multivariate system by stationary n-dimensional vector autoregressive processes of order p (VAR[p])

$$\begin{pmatrix} X_1(t) \\ \vdots \\ X_n(t) \end{pmatrix} = \sum_{r=1}^{p} a_r \begin{pmatrix} X_1(t-r) \\ \vdots \\ X_n(t-r) \end{pmatrix} + \begin{pmatrix} \varepsilon_1(t) \\ \vdots \\ \varepsilon_n(t) \end{pmatrix} . \tag{1.10}$$

The estimated coefficient matrix elements $\hat{a}_{kl,r}$ $(k,l = 1,\ldots,n;\ r = 1\ldots,p)$ themselves or their frequency domain representatives

$$\hat{A}_{kl}(\omega) = \delta_{kl} - \sum_{r=1}^{p} \hat{a}_{kl,r}\, \mathrm{e}^{-\mathrm{i}\omega r} \tag{1.11}$$

with the Kronecker symbol ($\delta_{kl} = 1$, if $k = l$ and $\delta_{kl} = 0$, else) contain the information about the causal influences in the multivariate system. The coefficient matrices weight the information of the past of the entire multivariate system. The causal interactions between processes are modeled by the off-diagonal elements of the matrices. The influence of the history of an individual process on the present value is modeled by the diagonal elements. For bivariate Granger-causality n is set to 2 and $X_1(t)$ and $X_2(t)$ are the two processes under investigation.

The estimated covariance matrix $\hat{\Sigma}$ of the noise $\varepsilon(t) = (\varepsilon_1(t), \ldots, \varepsilon_n(t))'$ contains information about linear instantaneous interactions and therefore, strictly speaking, non-causal influences between processes. But changes in the diagonal elements of the covariance matrix, when fitted to the entire systems as well as the sub-systems, can be utilized to investigate Granger-causal influences, since the estimated variance of the residuals $\varepsilon_i(t)$ reflects information that cannot be revealed by the past of the processes.

Following the principle of predictability, basically all multivariate process models, which provide a prediction error, may be used for a certain definition of a Granger-causality index. Such models are e.g. time-variant autoregressive models or self-exciting threshold autoregressive (SETAR) models. The first one results in a definition of a time-variant Granger-causality index, the second one provides the basis for a state-dependent Granger-causality index.

Time-Variant Granger-Causality Index

To introduce a Granger-causality index in the time-domain and to investigate directed influences from a component X_j to a component X_i of a n-dimensional system, n- and $(n-1)$-dimensional VAR-models for X_i are considered. Firstly, the entire n-dimensional VAR-model is fitted to the n-dimensional system, leading to the residual variance $\hat{\Sigma}_{i,n}(t) = \text{var}\,(\varepsilon_{i,n}(t))$. Secondly, a $(n-1)$-dimensional VAR-model is fitted to a $(n-1)$-dimensional subsystem $\{X_k, k = 1, \ldots, n | k \neq j\}$ of the n-dimensional system, leading to the residual variance $\hat{\Sigma}_{i,n-1}(t) = \text{var}\left(\varepsilon_{i,n-1|j}(t)\right)$.

A time-resolved Granger-causality index quantifying linear Granger-causality is defined by [24]

$$\hat{\gamma}_{i \leftarrow j}(t) = \ln \left(\frac{\hat{\Sigma}_{i,n-1|j}(t)}{\hat{\Sigma}_{i,n}(t)} \right) . \tag{1.12}$$

Since the residual variance of the n-dimensional model is expected to be smaller than the residual variance of the smaller $(n-1)$-dimensional model, $\gamma_{i \leftarrow j}(t)$ is larger than or equal to zero except for some biased estimation of parameters. For a time-resolved extension of the Granger-causality index, a time-variant VAR-parameter estimation technique is utilized by means of the recursive least square algorithm RLS which is a special approach of adaptive filtering [35]. Consequently, a detection of directed interactions between two processes X_i and X_j is possible in the time domain.

Here, the time-resolved Granger-causality index is the only analysis technique under investigation reflecting information about multivariate systems in the time-domain. The multivariate extensions of alternative time-domain analysis techniques, such as the widely used cross-correlation function, are usually also based on operations in the frequency-domain. Partial correlation functions are commonly estimated by means of estimating partial auto- and cross-spectra. Furthermore, complex covariance structures between time lags and processes prevent a decision about statistically significant time lags obtained by cross-correlation analysis. Moreover, high values of the cross-correlation function do not reflect any statistical significance.

State-Dependent Granger-Causality Index

Many investigations of interaction networks are based on event-related data. Independent of the used data source – EEG, MEG or functional MRI (fMRI) – this is combined with the processing of transient signals or signals with nonlinear properties. Thus, a modeling of the underlying processes by means of autoregressive processes is questionable and remains controversial. A possible extension of the linear Granger-causality is given by SETAR models which are suitable to model biomedical signals with transient components or with nonlinear signal properties [32].

Let $N > 1$ be the dimension of a process X, and let $R_1, \ldots, R_K$ be a partition of $\mathbb{R}^N$. Furthermore, let

$$X_{n,d}^{(k)} = \begin{cases} 1, & \text{if} \quad X(n-d) \in R_k \\ 0, & \text{if} \quad X(n-d) \notin R_k \end{cases} \tag{1.13}$$

be the indicator variable that determines the current regime of the SETAR process. Then any solution of

$$X(n) + \sum_{k=1}^{K} X_{n,d}^{(k)} \left[a_0^{(k)} + \sum_{i=1}^{p_k} A_i^{(k)} X(n-i) \right] = \sum_{k=1}^{K} X_{n,d}^{(k)} \omega^{(k)}(n) \tag{1.14}$$

is called (multivariate) SETAR process with delay d. The processes $\omega^{(k)}$ are zero mean uncorrelated noise processes. Thus, SETAR processes realize a regime state-depended autoregressive modeling. Usually, the partition $R_1, \ldots, R_K$ is defined by a thresholding of each underlying real axis of $\mathbb{R}^N$.

Let $\Psi_{-j} = (X_1, \ldots, X_{j-1}, X_{j+1}, \ldots, X_N)^T$ be the reduced vector of the observed process, where the j-th component of X is excluded. Then, two variances $\hat{\Sigma}_{i|\Psi_{-j}}(k)$ and $\hat{\Sigma}_{i|X}(k)$ of prediction errors $\omega_i^{(k)}|\Psi_{-j}$ with respect to the reduced process Ψ_{-j} and $\omega_i^{(k)}|X$ with respect to the full process X may be estimated for each regime $R_k, k = 1, \ldots, K$. Clearly, two different decompositions of $\mathbb{R}^N$ have to be considered using a SETAR modeling of Ψ_{-j} and X. If X is in the regime R_k for any arbitrary k, then the reduced process Ψ_{-j} is located in the regime defined by the projection of R_k to the hyper plane of $\mathbb{R}^N$, where the j-th component is omitted.

Let I_k be the index set, where the full process is located in the regime R_k. That is, it holds

$$I_k = \left\{ n : X_{n,d}^{(k)} = 1 \right\} . \tag{1.15}$$

Now the relation

$$I_k \subseteq \left\{n : \Psi_{n,d}^{(k_{-j})} = 1\right\} \tag{1.16}$$

is fulfilled for all j. Thus, the index set I_k may be transferred to Ψ_{-j}, and the variance of $\omega_i^{(k_{-j})}|\Psi_{-j}$ may be substituted by a conditional variance $\omega_i^{(k)}|\Psi_{-j}$, which is estimated by means of I_k. Now, the following definition of the regime or state dependent Granger-causality index considers alterations of prediction errors in each regime separately

$$\hat{\gamma}_{i\leftarrow j}^{(k)} = \ln\left(\frac{\hat{\Sigma}_{i|\Psi_{-j}}(k)}{\hat{\Sigma}_{i|X}(k)}\right), \ k = 1, \ldots, K\ . \tag{1.17}$$

Significance Thresholds for Granger-Causality Index

Basically, Granger-causality is a binary quantity. In order to define a binary state dependent or time-variant Ganger causality a significance threshold is needed that indicates $\gamma_{i\leftarrow j}^{(k)} > 0$ or $\gamma_{i\leftarrow j}(t) > 0$, respectively. Generally, thus far we do not have the exact distribution of the corresponding test statistics. A possible way out is provided by shuffle procedures. To estimate the distribution under the hypothesis $\gamma_{i\leftarrow j}^{(k)} = 0$ or $\gamma_{i\leftarrow j}(t) = 0$, respectively, shuffle procedures may be applied. In this case, only the j-th component is permitted to be shuffled; the temporal structure of all other components has to be preserved.

In the presence of multiple realizations of the process X which is often the case dealing with stimulus induced responses in EEG, MEG or fMRI investigations, Bootstrap methods may be applied e.g. to estimate confidence intervals. Thereby, the single stimulus responses (trials) are considered as i.i.d. random variables [23, 33].

Partial Directed Coherence

As a parametric approach in the frequency-domain, partial directed coherence has been introduced to detect causal relationships between processes in multivariate dynamic systems [2]. In addition, partial directed coherence accounts for the entire multivariate system and renders a differentiation between direct and indirect influences possible. Based on the Fourier transform of the coefficient matrices (cf. 1.11), partial directed coherence

$$\pi_{i\leftarrow j}(\omega) = \frac{|A_{ij}(\omega)|}{\sqrt{\sum_k |A_{kj}(\omega)|^2}} \tag{1.18}$$

between processes X_j and X_i is defined, where $|\cdot|$ is the absolute value of $(\cdot)$. Normalized between 0 and 1, a direct influence from process X_j to process X_i

is inferred by a non-zero partial directed coherence $\pi_{i \leftarrow j}(\omega)$. To test the statistical significance of non-zero partial directed coherence values in applications to finite time series, critical values should be used that are for instance introduced in [56]. Similarly to the Granger-causality index, a significant causal influence detected by partial directed coherence analysis has to be interpreted in terms of linear Granger-causality [17]. In the following investigations, parameter matrices have been estimated by means of multivariate Yule-Walker equations.

Renormalized Partial Directed Coherence

Above, partial directed coherence has been discussed and a pointwise significance level has been introduced in [56]. The pointwise significance level allows identifying those frequencies at which the partial directed coherence differs significantly from zero, which indicates the existence of a direct influence from the source to the target process. More generally, one is interested in comparing the strength of directed relationships at different frequencies or between different pairs of processes. Such a quantitative interpretation of the partial directed coherence and its estimates, however, is hampered by a number of problems.

(i) The partial directed coherence measures the strength of influences relative to a given signal source. This seems counter-intuitive since the true strength of coupling is not affected by the number of other processes that are influenced by the source process. In particular, adding further processes that are influenced by the source process decreases the partial directed coherence although the quality of the relationship between source and target process remains unchanged. This property prevents meaningful comparisons of influences between different source processes or even between different frequencies as the denominator in (1.18) varies over frequency.
In contrast it is expected that the influence of the source on the target process is diminished by an increasing number of other processes that affect the target process, which suggests to measure the strength relative to the target process. This leads to the alternative normalizing term

$$\left(\sum_k \left| \hat{A}_{ik}(\omega) \right|^2 \right)^{1/2} , \tag{1.19}$$

which may be derived from the factorization of the partial spectral coherence in the same way as the original normalization by [2]. Such a normalization with respect to the target process has been used in [28] in their definition of the directed transfer function (DTF). Either normalization may be favorable in some applications but not in others.

(ii) The partial directed coherence is not scale-invariant, that is, it depends on the units of measurement of the source and the target process. In particular, the partial directed coherence can take values arbitrarily close to either one or zero if the scale of the target process is changed accordingly. This problem becomes important especially if the involved processes are not measured on a common scale.

(iii) When the partial directed coherence is estimated, further problems arise from the fact that the significance level depends on the frequency unlike the significance level for the ordinary coherence derived in Sect. 1.3.1 [5]. In particular, the critical values derived from

$$|\pi_{i\leftarrow j}(\omega)| \stackrel{d}{=} \left(\frac{\hat{C}_{ij}(\omega)\,\chi_1^2}{N \sum_k \left|\hat{A}_{kj}(\omega)\right|^2} \right)^{1/2} \tag{1.20}$$

compensate for the effects of normalization by

$$\sqrt{\sum_k |\hat{A}_{kj}(\omega)|^2}\,,$$

that is, the significance of the partial directed coherence essentially depends on the absolute rather than the relative strength of the interaction. A naïve approach to correct for this would be to use the significance level and reformulate it such that

$$\frac{|\pi_{i\leftarrow j}(\omega)|^2}{\hat{C}_{ij}(\omega)} N \sum_k \left|\hat{A}_{kj}(\omega)\right|^2 \stackrel{d}{=} \chi_1^2 \tag{1.21}$$

holds. Thereby,

$$C_{ij}(\omega) = \Sigma_{ii} \left[\sum_{k,l=1}^{p} H_{jj}(k,l) \left(\cos(k\omega)\cos(l\omega) + \sin(k\omega)\sin(l\omega)\right) \right] \tag{1.22}$$

with

$$\lim_{N\to\infty} N \operatorname{cov}\left(\hat{a}_{ij}(k), \hat{a}_{ij}(l)\right) = \Sigma_{ii}\, H_{jj}(k,l)\,, \tag{1.23}$$

where Σ_{ii} is the variance of noise process ε_i in the autoregressive process. The elements $H_{jj}(k,l)$ are entries of the inverse $\mathbf{H} = \mathbf{R}^{-1}$ of the covariance matrix $\mathbf{R}$ of the vector auto-regressive process $\mathbf{X}$. However, as shown below this is not the optimal result that can be obtained. Moreover, it can be shown that a χ_2^2-distribution with two degrees of freedom is obtained.

(iv) Although the pointwise significance level adapts correctly to the varying uncertainty in the estimates of the partial directed coherence, this behavior shows clearly the need for measures of confidence in order to be able to compare estimates at different frequencies. Without such measures, it remains open how to interpret large peaks that exceed the significance level only slightly and how to compare them to smaller peaks that are clearly above the threshold.

In summary, this discussion has shown that partial directed coherence as a measure of the relative strength of directed interactions does not allow conclusions on the absolute strength of coupling nor does it suit for comparing the strength at different frequencies or between different pairs of processes. Moreover, the frequency dependence of the significance level shows that large values of the partial directed coherence are not necessarily more reliable than smaller values, which weakens the interpretability of the partial directed coherence further. In the following, it is shown that these problems may be overcome by a different normalization.

A New Definition of Partial Directed Coherence: Renormalized Partial Directed Coherence

For the derivation of an alternative normalization, recall that the partial directed coherence is defined in terms of the Fourier transform $A_{ij}(\omega)$ in (1.11). Since this quantity is complex-valued, it is convenient to consider the two-dimensional vector

$$\mathbf{P}_{ij}(\omega) = \begin{pmatrix} \operatorname{Re} A_{ij}(\omega) \\ \operatorname{Im} A_{ij}(\omega) \end{pmatrix} \tag{1.24}$$

with $\mathbf{P}_{ij}(\omega)'\mathbf{P}_{ij}(\omega) = |A_{ij}(\omega)|^2$. The corresponding estimator $\hat{\mathbf{P}}_{ij}(\omega)$ with $\hat{A}_{ij}(\omega)$ substituted for $A_{ij}(\omega)$ is asymptotically normally distributed with mean $\mathbf{P}_{ij}(\omega)$ and covariance matrix $\mathbf{V}_{ij}(\omega)/N$, where

$$\mathbf{V}_{ij}(\omega) = \sum_{k,l=1}^{p} H_{jj}(k,l)\, \Sigma_{ii} \begin{pmatrix} \cos(k\omega)\cos(l\omega) & -\cos(k\omega)\sin(l\omega) \\ -\sin(k\omega)\cos(l\omega) & \sin(k\omega)\sin(l\omega) \end{pmatrix} . \tag{1.25}$$

For $p \geq 2$ and $\omega \neq 0 \bmod \pi$, the matrix $\mathbf{V}_{ij}(\omega)$ is positive definite [56], and it follows that, for large N, the quantity

$$N\, \hat{\lambda}^{\circ}_{ij}(\omega) = N\, \hat{\mathbf{P}}_{ij}(\omega)'\mathbf{V}_{ij}(\omega)^{-1}\hat{\mathbf{P}}_{ij}(\omega)$$

has approximately a noncentral χ^2-distribution with two degrees of freedom and noncentrality parameter $N\,\lambda_{ij}(\omega)$, where

$$\lambda_{ij}(\omega) = \mathbf{P}_{ij}(\omega)'\mathbf{V}_{ij}(\omega)^{-1}\mathbf{P}_{ij}(\omega).$$

If $p = 1$ or $\omega = 0 \bmod \pi$, the matrix $\mathbf{V}_{ij}(\omega)$ has only rank one and thus is not invertible. However, it can be shown that in this case $N\,\hat{\lambda}^{\circ}_{ij}(\omega)$ with $\mathbf{V}_{ij}(\omega)^{-1}$ being a generalized inverse of $\mathbf{V}_{ij}(\omega)$ has approximately a noncentral χ^2-distribution with one degree of freedom and noncentrality parameter $N\,\lambda_{ij}(\omega)$ [56].

The parameter $\lambda_{ij}(\omega)$, which is nonnegative and equals zero if and only if $A_{ij}(\omega) = 0$, determines how much $\mathbf{P}_{ij}(\omega)$ and thus $A_{ij}(\omega)$ differ from zero. Consequently, it provides an alternative measure for the strength of the effect of the source process X_j on the target process X_i.

The most important consequence of the normalization by $\mathbf{V}_{ij}(\omega)$ is that the distribution of $\hat{\lambda}^{\circ}_{ij}(\omega)$ depends only on the parameter $\lambda_{ij}(\omega)$ and the sample size N. In particular, it follows that the α-significance level for $\hat{\lambda}^{\circ}_{ij}(\omega)$ is given by $\chi^2_{df,1-\alpha}/N$ and thus is constant unlike in the case of the partial directed coherence. Here, $\chi^2_{df,1-\alpha}$ denotes the $1-\alpha$ quantile of the χ^2-distribution with the corresponding degrees of freedom (2 or 1). More generally, confidence intervals for parameter $\lambda_{ij}(\omega)$ can be computed; algorithms for computing confidence intervals for the noncentrality parameter of a noncentral χ^2-distribution can be found, for instance, in [29]. The properties of noncentral χ^2-distributions (e.g. [26]) imply that such confidence intervals for $\lambda_{ij}(\omega)$ increase monotonically with $\hat{\lambda}^{\circ}_{ij}(\omega)$, that is, large values of the estimates are indeed likely to correspond to strong influences among the processes. Finally, the parameter $\lambda_{ij}(\omega)$ can be shown to be scale-invariant.

With these properties, $\hat{\lambda}^{\circ}_{ij}(\omega)$ seems an "ideal" estimator for $\lambda_{ij}(\omega)$. However, it cannot be computed from data since it depends on the unknown covariance matrix $\mathbf{V}_{ij}(\omega)$. In practice, $\mathbf{V}_{ij}(\omega)$ needs to be estimated by substituting estimates $\hat{\mathbf{H}}$ and $\hat{\Sigma}$ for $\mathbf{H}$ and Σ in (1.25). This leads to the alternative estimator

$$\hat{\lambda}_{ij}(\omega) = \hat{\mathbf{P}}_{ij}(\omega)'\hat{\mathbf{V}}_{ij}(\omega)^{-1}\hat{\mathbf{P}}_{ij}(\omega)\,.$$

It can be shown by Taylor expansion that under the null hypothesis of $\lambda_{ij}(\omega) = 0$ this statistic is still χ^2-distributed with two respectively one degrees of freedom, that is, the α-significance level remains unchanged when $\hat{\lambda}^{\circ}_{ij}(\omega)$ is replaced by $\hat{\lambda}_{ij}(\omega)$. In contrast, the exact asymptotic distribution of the new estimator under the alternative is not known. Nevertheless, extensive simulations have revealed that approximate confidence intervals can be obtained by applying the theoretical results yielded for the "ideal" estimator $\hat{\lambda}^{\circ}_{ij}(\omega)$ to the practical estimator $\hat{\lambda}_{ij}(\omega)$ [54].

Directed Transfer Function

The directed transfer function is an alternative frequency-domain analysis technique to detect directions of interactions and is again based on the Fourier transformation of the coefficient matrices (cf. (1.11)). The transfer function $H_{ij}(\omega) = A^{-1}_{ij}(\omega)$ leads to the definition of the directed transfer function [2, 28]

$$\delta_{i \leftarrow j}(\omega) = \frac{|H_{ij}(\omega)|^2}{\sum_l |H_{il}(\omega)|^2} \, . \tag{1.26}$$

The directed transfer function is again normalized to [0, 1]. An interaction from process X_j to process X_i is detected if $\delta_{i \leftarrow j}(\omega)$ is unequal to zero. The normalization in the definition of the directed transfer function and the partial directed coherence is a major difference between both analysis techniques [31].

A similar discussion compared to the discussion of partial directed coherence above is also possible for the directed transfer function [59].

We mention though that for the three parametric approaches under investigation, values quantifying the directed influences cannot be identified with the strength of the interactions directly. Only renormalized partial directed coherence is capable in quantifying the interaction strength.

Time-Resolved Extension of Parametric Approaches

In order to detect non-stationary effects in the interrelation structure of the multivariate system, an extension of the parametric approaches is introduced. To this aim a time-resolved parameter estimation technique is utilized. The Granger-causality index has already been introduced as a time resolved procedure applying the recursive least square algorithm [35].

An alternative way to estimate time-resolved parameters in VAR-models and to consider explicitly observation noise influence in the multivariate system is based on time-variant state space models (SSM) [19, 62]

$$\begin{aligned} B(t) &= \quad B(t-1) + \eta(t) \\ X(t) &= B(t-1)\, X(t-1) + \varepsilon(t) \\ Y(t) &= \quad C(t)\, X(t) + \varrho(t) \, . \end{aligned} \tag{1.27}$$

State space models consist of hidden state equations $B(t)$ and $X(t)$ as well as observation equations $Y(t)$. The hidden state equation for $B(t)$ includes the parameter matrices $a_r(t)$. The observation equation $Y(t)$ takes explicitly account for observation noise $\varrho(t)$. For $\eta(t) \neq 0$, the equation for $B(t)$ enables a detection of time-varying parameters. For a numerically efficient procedure to estimate the parameters in the state space model, the EM-algorithm based on the extended Kalman filter is used in the following [69].

1.3.3 Directed Graphical Models

Graphical models are a methodology to visualize and reveal relationships in multivariate systems [11]. Such a graph is shown in Fig. 1.3. The vertices reflect the processes and the arrows the significant results of the applied analysis technique. For example, if partial directed coherences are non-significant between process X_3 and process X_4, both processes are identified as not influencing each other and arrows between the processes in the corresponding

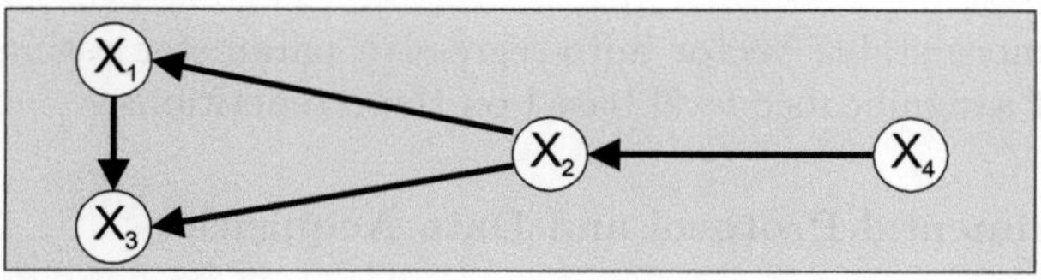

Fig. 1.3 Directed graph summarizing the interdependence structure for an exemplary multivariate system

graphical model are missing. In contrast, if partial directed coherence is only significant for one direction, for example from process X_4 to process X_2 but not in the opposite direction, an arrow is drawn from process X_4 to process X_2.

1.4 Application to Neural Data

In order to examine time-variant causal influences within distinct neural networks during defined functional states of brain activity, data obtained from an experimental approach of deep sedation were analyzed. Burst suppression patterns (BSP) in the brain electric activity were used for the analysis. This specific functional state was chosen because BSP represent a defined reference point within the stream of changes in electroencephalographic (EEG) properties during sedation [73] leading to secured unconsciousness. An analysis of well-described alternating functional states of assumed differences of signal transfer in a time frame of seconds is possible. It has been shown that a hyperpolarization block of thalamo-cortical neurons evoked mainly by facilitated inhibitory input of reticular thalamic nucleus (RTN) activity induces inhibition of thalamo-cortical volley activity which is reflected by cortical interburst activity [27, 63, 64]. This in turn is assumed to be responsible for disconnection of afferent sensory input leading to unconsciousness. The role of burst activity in terms of information transfer remains elusive. Therefore, BSP is studied in order to elaborate time and frequency dependent features of information transfer between intrathalamic, thalamo-cortical and cortico-thalamic networks. Patterns were induced by propofol infusion in juvenile pigs and derived from cortical and thalamic electrodes.

The analysis was performed to clarify a suggested time-dependent directed influence between the above mentioned brain structures known to be essentially involved in regulation of the physiological variation in consciousness during wakefulness and during sleep [25, 40] as well as responsible to induce unconsciousness during administration of various anesthetic and sedative compounds. In addition, the alternating occurrence pattern characteristic of burst activity allowed a triggered analysis of the Granger-causality index. Multiple trials enable to use a generalized recursive least square estimator [24, 35], providing a more stable vector auto-regressive parameter estimation and a calculation of a significance level based on these repetitions.

1.4.1 Experimental Protocol and Data Acquisition

The investigation was carried out on six female, domestic juvenile pigs (mixed breed, 7 weeks old, 15.1 ± 1.4 kg body weight (b.w.)) recorded at the University Hospital of Jena by the group of Dr. Reinhard Bauer. Deep sedation with burst suppression patterns was induced by continuous propofol infusion. Initially, 0.9 mg/kg b.w./min of propofol for approximately 7 min were administered until occurrence of burst suppression patterns (BSP) in occipital leads [37], followed by a maintenance dose of 0.36 mg/kg b.w./min. Ten screw electrodes at frontal, parietal, central, temporal, and occipital brain regions were utilized for electrocorticogram (ECoG) recordings. For signal analysis a recording from the left parietooccipital cortex (POC) was used. Electrodes introduced stereotactically into the rostral part of the reticular thalamic nucleus (RTN) and the dorsolateral thalamic nucleus (LD) of the left side were used for the electrothalamogram (EThG) recordings (Fig. 1.4 (a)). Unipolar signals were amplified and filtered (12-channel DC, 0.5–1,000 Hz bandpass filter, 50 Hz notch filter; Fa. Schwind, Erlangen) before sampled continuously (125 Hz) with a digital data acquisition system (GJB Datentechnik GmbH, Langewiesen). Four linked screw electrodes inserted into the nasal bone served as reference. ECoG and EThG recordings were checked visually to exclude artifacts.

1.4.2 Analysis of Time-Variant and Multivariate Causal Influences Within Distinct Thalamo-Cortical Networks

In order to quantify time-variant and multivariate causal influences in a distinct functional state of general brain activity, a representative example of deep sedation is chosen, characterized by existence of burst suppression patterns. Registrations from both thalamic leads (LD, RTN) and from the parietooccipital cortex (POC) have been utilized, which is known to respond early with patterns typical for gradual sedation including BSP [37]. In the present application, results for the Granger-causality index and the partial directed coherence are discussed, since a time-resolved extension of partial coherence is not considered and the directed transfer function approach leads to results similar to partial directed coherence.

For partial directed coherence analysis continuous registrations of 384 s duration were utilized to provide an overview of the entire recording (Fig. 1.4 (b)). For a closer investigation of the burst patterns, the analysis using the Granger-causality index was applied to triggered registrations of 3 s duration each, i.e. 1 s before and 2 s after burst onset (Fig. 1.4 (c)). In a total of 66 trials, trigger points were identified by visual inspection and were set at the burst onset. The deep sedation state was characterized by a distinct BSP in the POC lead as well as continuous high amplitude and low frequency activity in both thalamic leads.

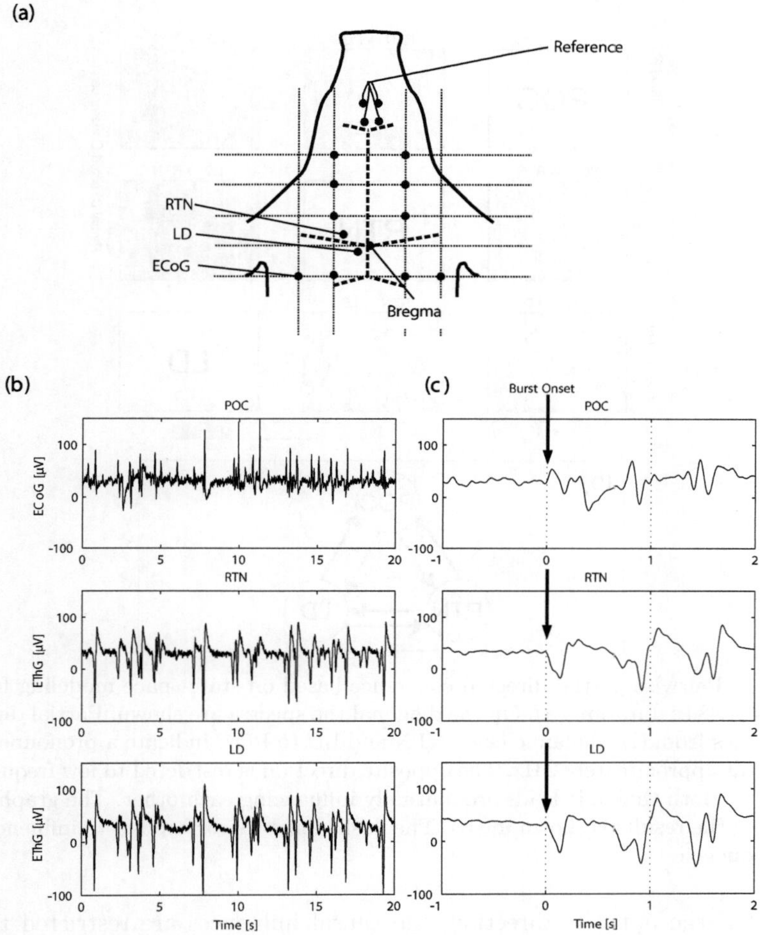

Fig. 1.4 (**a**) Schematic representation of skull electrode localizations. Dots indicate ECoG (electrocorticogram) recordings. POC indicates parietooccipital cortex recording and is used in the present investigation. Additionally, the RTN recording and LD recording were utilized recorded using EThG. (**b**) 20 s section of continuous original trace and (**c**) one representative trial of triggered original traces of brain electrical activity simultaneously recorded from cortical and thalamic structures of a juvenile pig under propofol-induced deep sedation

For the entire time series of 384 s duration, pairwise partial directed coherence analysis was performed to investigate time-varying changes in directed influences between both thalamic structures RTN and LD and the parietooccipital cortex (POC). The results are shown in Fig. 1.5 (a). The graph summarizing the influences is given in Fig. 1.5 (b). A strong and continuous influence is observed from both thalamic leads RTN and LD to POC at approximately

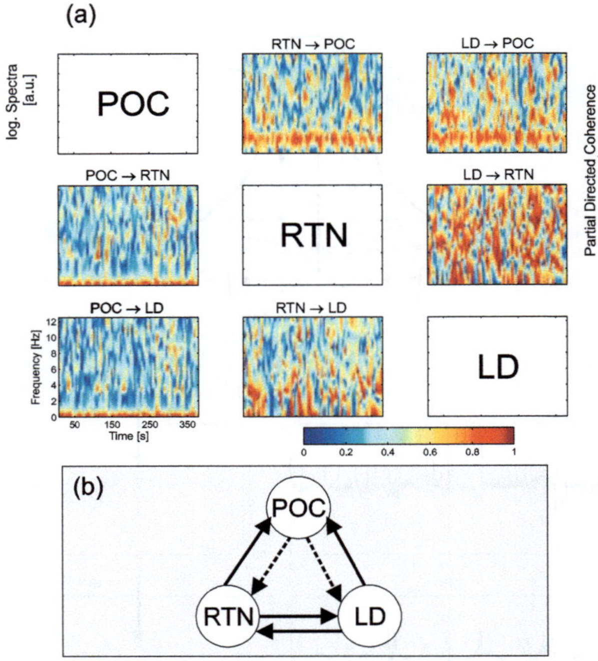

Fig. 1.5 Pairwise partial directed coherence based on state-space modeling for the signals of 384 s duration (**a**). On the diagonal the spectra are shown. Partial directed coherences from the thalamic leads RTN and LD to POC indicate a pronounced influence at approximately 2 Hz. The opposite direction is restricted to low frequencies (< 1 Hz). Both thalamic leads are mutually influencing each other. The graph summarizing the results is shown in (**b**). The dashed arrows correspond to influences for low frequencies

2 Hz. For the opposite direction, the causal influences are restricted to the low frequency range (<1 Hz) indicated by the dashed arrows in the graph. Furthermore, a directed influence is strongly indicated between the thalamic leads from LD to RTN, while the opposite direction shows a tendency to lower frequencies. The time-dependency is more pronounced in the interaction between both thalamic leads.

A clearer depiction of the interrelation structures occurring during the single burst patterns is presented in Fig. 1.6 by applying the Granger-causality index to segments of 3 s duration. For pairwise analysis between the three signals (Fig. 1.6 (a) and (b)), directed influences from both thalamic leads to the parietooccipital cortex are observed for broad time periods. At several, well-defined time points, causal influences are detected for the opposite direction and between both thalamic leads (dashed arrows). The interrelation between the thalamic leads remains significant for the multivariate analysis given in Fig. 1.6 (c) and (d). The directed influence from POC to LD and RTN

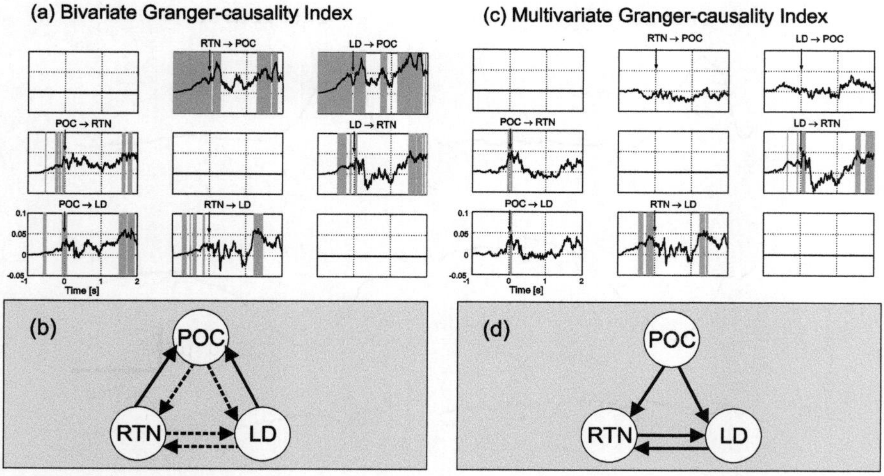

Fig. 1.6 Investigation of directed interrelations during the occurrence of burst patterns using the Granger-causality index in the time domain. Gray-colored regions indicate significant influences ($\alpha = 5\%$, one-sided). When applying pairwise analysis, directed influences from both thalamic leads LD and RTN to the parietooccipital cortex POC are detected (**a**). The results are summarized in the graph in (**b**). The dashed arrows corresponds to interactions lasting for short time intervals. The interrelation between the thalamic leads remains significant for the multivariate analysis (**c**). The directed influence from the parietooccipital cortex POC to the investigated thalamic structures is exclusively sustained at the burst onsets. The graph summarizing the results is given in (**d**)

is reduced to the burst onsets. From RTN and LD to the POC, no significant interrelation is traceable.

Results from the multivariate Granger-causality index cannot be directly correlated to the results obtained by the bivariate analysis. In particular, the missing interrelation from RTN and LD to POC is difficult to interpret with the knowledge of the bivariate results. One possible explanation might be an additional but unobserved process commonly influencing the three processes. This assumption is suggested by the results obtained from somatosensory evoked potential (SEP) analysis (Fig. 1.7). In contrast to previous opinions of a proposed functional disconnection of afferent sensory inputs to thalamo-cortical networks during interburst periods leading to a functional state of unconsciousness [1], SEP analysis indicates that even during this particular functional state a signal transduction appears from peripheral skin sensors via thalamo-cortical networks up to cortical structures leading to signal processing. Hence in principle, a subthalamically generated continuous input could be responsible for the pronounced influence in the low frequency band, as shown by partial directed coherence analysis. Such a low frequency component might not be observable by the Granger-causality index due to the missing selectivity for specific frequency bands.

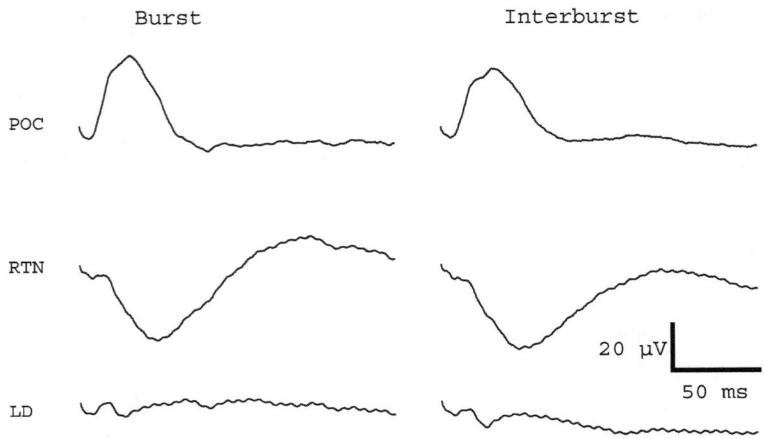

Fig. 1.7 Evoked activity derived from the parietooccipital cortex (POC, upper panel), rostral part of reticular thalamic nucleus thalamus (RTN, middle panel) and dorsolateral thalamic nucleus (LD, lower panel) owing to bipolar stimulation of the trigeminal nerve by a pair of hypodermic needles inserted on left side of the outer disc ridge of the porcine snout (rectangular pulses with constant current, duration of 70 µs, 1 Hz repetition frequency, 100 sweeps were averaged) in order to obtain somatosensory evoked potentials (SEP) during burst as well as interburst periods. Note a similar signal pattern during burst and interburst periods

Problems in the estimation procedure caused by, for instance, highly correlated processes or missing of important processes could also explain this effect [53]. Furthermore, the discrepancies between the bivariate and multivariate analysis could be due to the nonlinear behavior of the system. However, this possibility is not very likely, because spectral properties obtained in partial directed coherence analysis do not indicate a highly nonlinear behavior.

1.5 Discussion of Applicability of Multivariate Linear Analysis Techniques to Neural Signal Transfer

In the application of methods to neural signal transfer, for example in the analysis of neural coordination in either the normal or pathological brain, one should be aware not only of the potentials but also the limitations of the methods. For this purpose, the features of the different analysis techniques were analyzed by means of synthetic data simulated by various model systems [70].

On the basis of simulations, the performance of the four investigated analysis techniques, i.e. partial coherence with its corresponding phase spectrum (PC), the Granger-causality index (GCI), the directed transfer function

Table 1.1 Summary of the results obtained by the comparison of the four multivariate time series analysis techniques. To evaluate the performance, five aspects are considered. The brackets denote some specific limitations

	PC	GCI	DTF	PDC
Direct versus indirect interactions	+	+	–	+
Direction of influences	(–)	+	+	+
Specificity in absence of coupling	+	+	(+)	(+)
Nonlinearity in data	+	–	(+)	+
Influences varying with time		+	+	+

(DTF), and the partial directed coherence (PDC) are summarized with respect to five aspects (Table 1.1) [70], which are important when analyzing data from unknown processes:

- *Direct versus indirect interactions:* A differentiation between direct and indirect information transfer in multivariate systems is not possible by means of the directed transfer function. Therefore, the directed transfer function is not sensitive in this sense (minus sign in Table 1.1). The remaining multivariate analysis techniques are in general able to distinguish between direct and indirect interactions. Thus, the GCI, PDC, and PC are sensitive in distinguishing direct from indirect influences. Despite the high sensitivity in general, there might be some situations in which this characteristic is restricted, for instance in nonlinear, non-stationary systems.
- *Direction of influences:* All multivariate methods are capable of detecting the direction of influences. Partial coherence in combination with its phase spectrum is limited to high coherence values and to unidirectional influences between the processes. This shortcoming of partial coherence and partial phase spectrum is indicated by the minus sign in Table 1.1.
- *Specificity in absence of the influences:* All four analysis techniques reject interrelations in the absence of any influence between the processes, reflecting the high specificity of the methods. For the parametric approaches directed transfer function and partial directed coherence, a renormalization of the covariance matrix of the noise in the estimated vector auto-regressive model is required. Otherwise spurious interactions are detected. A significance level for both techniques should account for this. For the significance level for partial directed coherence, this dependence on the noise variance is explicitly considered. However, the renormalization is necessary to achieve a balanced average height of values of PDC and DTF in the case of an absence of an interaction at the corresponding frequency.
- *Nonlinearity in the data:* For the nonlinear coupled stochastic systems with pronounced frequencies, analysis techniques in the frequency domain are preferable. High model orders are required to describe the nonlinear system sufficiently with a linear vector auto-regressive model. Interpretation

of the results obtained by the directed transfer function and the Granger-causality index is more complicated since there has been no obvious significance level. The PC, PDC, and the DTF are sensitive in detecting interactions in nonlinear multivariate systems. The Granger-causality index does not reveal the correct interrelation structure.

- *Influences varying with time:* The Granger-causality index, directed transfer function, and the time-varying partial directed coherence detect various types of time-varying influences. Therefore they are sensitive for time-resolved investigations of non-stationary data.

This summary provides an overview of which analysis techniques are appropriate for specific applications or problems. However, the particular capabilities and limitations of a specific analysis technique do not simply point to drawbacks of the method in general. If for instance the major task is to detect directions of influences, the directed transfer function is applicable even if the differentiation, for example of direct or indirect interactions, is not possible.

Partial coherence as a non-parametric method is robust in detecting relationships in multivariate systems. Direct and indirect influences can be distinguished in linear systems and certain nonlinear stochastic system like the Rössler system. Since partial coherence is a non-parametric approach, it is possible to capture these influences without knowledge of the underlying dynamics. Furthermore, the statistical properties are well-known and critical values for a given significance level can be calculated in order to decide on significant influences. This is an important fact especially in applications to noisy neural signal transfer as measured by e.g. electroencephalography recordings. A drawback is that the direction of relationships can only be determined by means of phase spectral analysis. If spectral coherence is weak or restricted to a small frequency range, directions of influences are difficult to infer by means of partial phase spectral analysis. Additionally, mutual interactions between two processes are also hardly detectable utilizing partial phase spectra.

Defined in the time-domain, the Granger-causality index is favorable in systems where neither specific frequencies nor frequency-bands are exposed in advance. The Granger-causality index utilizes information from the covariance matrix. Weak interactions or narrow-band interactions are difficult to detect, since they can lead to only small changes in the covariance matrix. The Granger-causality index, estimated by means of the recursive least square algorithm, renders a methodology to trace interdependence structures in non-stationary data possible. This might become important in applications to brain neural networks, when the time course of transitions in neural coordination is of particular interest.

By means of the directed transfer function, directions of influences in multivariate dynamical systems are detectable. Nevertheless, in contrast to the remaining three analysis techniques, a differentiation between indirect and direct influences is in general not possible using the directed transfer function.

Analyzing brain networks, at least weakly nonlinear processes might be expected to generate the neural signals. In the application to the nonlinear stochastic systems, the directions of the couplings could be observed at the oscillation frequencies. The directed transfer function benefits from its property as an analysis technique in the frequency domain. Increasing the order of the fitted model system is sufficient to capture the main features of the system and thus to detect the interdependence structure correctly. Nevertheless, a matrix inversion is required for estimating the directed transfer function, which might lead to computational challenges especially if high model orders are necessary. In order to detect transitions in the coordination between neural signals, the directed transfer function is useful when applying a time-resolved parameter estimation procedure.

In the frequency domain, partial directed coherence is the most powerful analysis technique. By means of partial directed coherence, direct and indirect influences as well as their directions are detectable. The investigation of the paradigmatic model system of coupled stochastic Rössler oscillators has shown [70], that at least for nonlinearities, coupling directions can be inferred by means of partial directed coherence. Increasing the order of the fitted model is required to describe the nonlinear system by a linear vector auto-regressive model sufficiently. However, as the statistical properties of partial directed coherence and significance levels for the decision of significant influences are known, high model orders of the estimated vector auto-regressive model are less problematic. Using additionally time-resolved parameter estimation techniques, partial directed coherence is applicable to non-stationary signals. Using this procedure, influences in dependence on time and frequency are simultaneously detectable. Since in applications to neural networks it is usually unknown whether there are changes in neural coordination or whether such changes are of particular interest, respectively, time-resolved analysis techniques avoid possible false interpretations.

The promising results showing that most parametric, linear analysis techniques have revealed correct interaction structures in multivariate systems, indicate beneficial applicability to empirical data. Electrophysiological signals from thalamic and cortical brain structures representative for key interrelations within a network responsible for control and modulation of consciousness have been analyzed. Data obtained from experimental recordings of deep sedation with burst suppression patterns were used, which allows usage of data from a well-defined functional state including a triggered analysis approach. Partial directed coherence based on state space modeling allows for inference of the time- and frequency-dependence of the interrelation structure. The mechanisms generating burst patterns were investigated in more detail by applying the Granger-causality index. Besides a clear depiction of the system generating such burst patterns, the application presented here suggests that time dependence is not negligible.

1.6 Nonlinear Dynamics

So far, the linear methodology has been addressed. However, the field of nonlinear dynamics has brought to the forefront novel concepts, ideas, and techniques to analyze and characterize time series of complex dynamic systems. Especially synchronization analysis to detect interactions between nonlinear self-sustained oscillators has made its way into the daily routine in many investigations [43].

Following the observations and pioneering work of Huygens, the process of synchronization has been observed in many different systems such as systems exhibiting a limit cycle or a chaotic attractor. Several types of synchronization have been observed for these systems ranging from phase synchronization as the weakest form of synchronization via lag synchronization to generalized or complete synchrony [30, 39, 41, 46, 47].

Thereby, phase synchronization analysis has gained particular interest since it relies only on a weak coupling between the oscillators. It has been shown that some chaotic oscillators are able to synchronize their phases for considerably weak coupling between them [46]. To quantify the process of synchronization, different measures have been proposed [36, 45, 65]. Two frequently used measures are a measure based on entropy and a measure based on circular statistics, which is the so called mean phase coherence [36]. Both measures quantify the sharpness of peaks in distributions of the phase differences. In the following the mean phase coherence is introduced.

1.6.1 Self-Sustained Oscillators

While in the framework of linear systems as vector auto-regressive or moving-average processes are of particular interest, in nonlinear dynamics self-sustained oscillators play an important role. In general these oscillators can be formulated as

$$\dot{\mathbf{X}}(t) = f(\mathbf{X}(t), \alpha(t), U(t)) \;, \tag{1.28}$$

whereby $\mathbf{X}(t)$ has to be a more dimensional variable to ensure an oscillatory behavior. The external influence as well as the parameters can either be vector-valued or not.

Since especially the interaction between processes is considered here, the following description of a system of coupled oscillators

$$\dot{\mathbf{X}}_1(t) = \mathbf{f}_1(\mathbf{X}_1(t), \alpha_1) + \varepsilon_{1,2}\mathbf{h}_1\left(\mathbf{X}_1(t), \mathbf{X}_2(t)\right) \tag{1.29}$$

$$\dot{\mathbf{X}}_2(t) = \mathbf{f}_2(\mathbf{X}_2(t), \alpha_2) + \varepsilon_{2,1}\mathbf{h}_2\left(\mathbf{X}_2(t), \mathbf{X}_1(t)\right) \tag{1.30}$$

is more appropriate [43]. External driving is neglected in the following and the parameters are assumed to be constant with time. The coupling is present from oscillator j onto oscillator i if $\varepsilon_{i,j} \neq 0$. The functional relationship $\mathbf{h}_1(\cdot)$

and $\mathbf{h}_2(\cdot)$ of the coupling can thereby be arbitrary. In general it is even not necessary that it is a function. It can as well be a relation. However, here only well behaved functions are considered. Usually, diffusive coupling is used, i.e. $\mathbf{h}_1(\mathbf{X}_1(t), \mathbf{X}_2(t)) = (\mathbf{X}_2(t) - \mathbf{X}_1(t))$ and $\mathbf{h}_2$ accordingly. For $\varepsilon_{i,j} = 0$ the solution of the above system is expected to be a limit cycle for each oscillator in the sequel. This is to ensure a much simpler mathematical motivation of phase synchronization.

1.7 Phase Synchronization

To describe the interaction between coupled self-sustained oscillators, the notion of phase synchrony has gained particular interest. The phase $\Phi(t)$ of a limit cycle (periodic) oscillator is a monotonically increasing function with

$$\Phi(t)|_{t=pT} = p2\pi = p\omega T\,,$$

where p denotes the number of completed cycles, T is the time needed for one complete cycle, and ω the frequency of the oscillator. To define the phase also for values of the time $t \neq pT$, the following expression is used

$$\dot{\Phi}_i(t) = \omega_i\,,$$

whereby ω_i are the frequencies of the uncoupled oscillators with i denoting the i-th oscillator.

Few calculations show that a differential equation for the phase evolution

$$\Phi_j(\dot{\mathbf{X}}_j(t)) = \omega_j + \varepsilon_{j,i} \sum_k \frac{\partial \Phi_j}{\partial X_j^k} h_j^k(\mathbf{X}_1, \mathbf{X}_2) \tag{1.31}$$

can be obtained in the case of coupled oscillators as introduced above [43]. The superscript k denotes the k-th component of the corresponding vector.

For small $\varepsilon_{i,j}$ the above sum can be approximated by 2π periodic functions

$$\dot{\Phi}_1(t) = \omega_1 + \varepsilon_{1,2} H_1(\Phi_1, \Phi_2) \tag{1.32}$$

$$\dot{\Phi}_2(t) = \omega_2 + \varepsilon_{2,1} H_2(\Phi_2, \Phi_1) \tag{1.33}$$

which leads to

$$n\dot{\Phi}_1(t) - m\dot{\Phi}_2(t) = n\omega_1 - m\omega_2 + \varepsilon_{1,2}\tilde{H}_1(\Phi_1, \Phi_2) - \varepsilon_{2,1}\tilde{H}_2(\Phi_2, \Phi_1)$$

for some integers n and m [43]. The difference $n\dot{\Phi}_1(t) - m\dot{\Phi}_2(t)$ can be considered as a generalized phase difference starting from the simplest expression $\dot{\Phi}_1(t) - \dot{\Phi}_2(t)$ with $n, m = 1$.

In the case of $\varepsilon_{1,2} = \varepsilon_{2,1}$ and with the notion $\Phi_{1,2}^{n,m} = n\Phi_1 - m\Phi_2$ and $\Delta\omega = n\omega_1 - m\omega_2$ the above differential equation can be written as

$$\dot{\Phi}_{1,2}^{n,m}(t) = \Delta\omega + \varepsilon_{1,2} H(\Phi_{1,2}^{n,m}) \tag{1.34}$$

with some new 2π periodic function $H(\cdot)$.

This differential equation has only one fix point that is characterized by

$$\Delta\omega + \varepsilon_{1,2} H(\Phi_{1,2}^{n,m}) = 0 \,. \tag{1.35}$$

In this case the phase difference

$$\Phi_{1,2}^{n,m} = \text{const} \tag{1.36}$$

is constant over time. Thus, both phases evolve in exactly the same way after appropriate multiplication with some integers n and m. The system is then in the regime of $n : m$ phase synchronization.

To capture more realistic cases faced in several applications, the potential [43]

$$V(\Phi_{1,2}^{n,m}) = \Delta\omega \Phi_{1,2}^{n,m} + \varepsilon_{1,2} \int^{\Phi_{1,2}^{n,m}} H(x) \, \mathrm{d}x \tag{1.37}$$

of

$$\Delta\omega + \varepsilon_{1,2} H(\Phi_{1,2}^{n,m}) \tag{1.38}$$

is utilized. Depending on the parameters this potential is a monotonically increasing or decreasing function or it exhibits some minima caused by the 2π periodic function $H(\Phi_{1,2}^{n,m})$. Caused by the shape of the potential, it is referred to as *washboard* potential [43, 72]. An example of two potentials with the same frequency mismatch $\Delta\omega$ but different coupling strengths are presented in Fig. 1.8. While the coupling in (a) is sufficiently high to guarantee

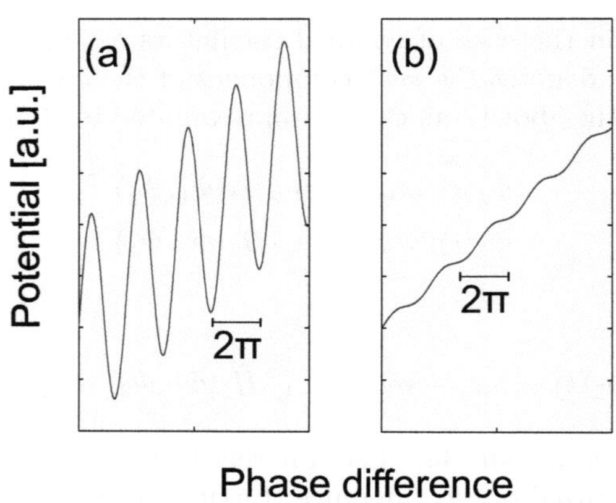

Fig. 1.8 Washboard potentials for the same frequency mismatch but higher coupling between the processes in (**a**) than in (**b**)

the existence of minima, the coupling in (b) leads to a potential that does not show extrema. Each of the minima in example (a) refers to a setting where phase synchronization is achieved.

If now some stochastic influence is added to the oscillators which thereby influences the phase difference dynamics, too, certain fluctuations within the minima are possible. These fluctuations can eventually lead to a "jump" from one minimum to another one. Caused by the fluctuations and the jumps from one minimum to another one

$$|\Phi_{1,2}^{n,m} \mod 2\pi|$$

is not constant any more but restricted by an appropriately chosen constant, i.e.

$$|\Phi_{1,2}^{n,m} \mod 2\pi| < \text{const} . \tag{1.39}$$

The notion of phase synchronization is still preserved in these cases but is regarded in a statistical sense.

In the case of chaotic oscillators, the notion of phase synchronization becomes even more interesting, since the above condition can be fulfilled although the amplitudes of the original signal stay uncorrelated [46]. In chaotic systems phase synchronization analysis, thus, yields novel insides in measurable interactions between these oscillators for very weak coupling strengths. Since the amplitudes stay uncorrelated for true phase synchrony, several other measures that are suggested to detect interactions fail.

The issue of estimating the amount of interaction between weakly coupled oscillators is addressed in the following section.

1.7.1 The Mean Phase Coherence – An Intuitive Introduction

If the investigated processes are known to be weakly coupled self-sustained oscillators, the above introduced condition for phase synchronization can be reformulated to yield a single number that quantifies phase synchrony. To motivate this quantity, the distribution of the phase differences $\Phi_{1,2}^{n,m} \mod 2\pi$, is investigated to ensure phase differences between $[-\pi, \pi]$.

If there is a sharp peak in the phase difference distribution of $\Phi_{1,2}^{n,m} \mod 2\pi$ the two phases perform a coherent motion, while the distribution is rather flat for independently evolving phases. Based on circular statistics the quantity [1, 34, 36, 65]

$$|R_{1,2}^{n,m}| = \left| \frac{1}{T} \sum_{t=1}^{T} \mathrm{e}^{i\Phi_{1,2}^{n,m}(t)} \right|$$

has been suggested. This quantity is normalized to $[0, 1]$ and becomes one for perfectly synchronized phases. The operation "mod 2π" can be skipped here as the complex exponential function is 2π periodic. The above quantity is sometimes referred to as *mean phase coherence* [36].

In applications, not the phases of the oscillators are observed but the time series of the amplitudes of some quantity. A possible methodology but not the only to derive phases from real-valued signals is the Hilbert transform leading to

$$V(t) = X(t) + iX_h(t) = A(t)\mathrm{e}^{i\Phi(t)} = A(t)Q(t) \tag{1.40}$$

and thereby $\Phi(t)$. Alternative approaches to obtain the phases are for instance based on wavelet transformations [3, 44].

1.7.2 The Mean Phase Coherence – A Mathematical Derivation

For linear systems, spectral coherence was introduced [8] to infer interactions between processes. Here, the concept of cross-spectral analysis is carried over to nonlinear synchronizing systems. To this aim, considering a two-dimensional dynamic process X_1, X_2, then the cross-spectrum $S_{X_1X_2}$ between X_1 and X_2 and the auto-spectra $S_{X_1X_1}$ of X_1 and $S_{X_2X_2}$ of X_2, respectively, can be estimated e.g. by smoothing the corresponding periodograms

$$\mathrm{Per}_{X_1X_2}(\omega) \propto \sum_t X_1(t)\mathrm{e}^{-\mathrm{i}\omega t} \sum_t X_2(t)\mathrm{e}^{\mathrm{i}\omega t} \tag{1.41}$$

and

$$\mathrm{Per}_{X_kX_k}(\omega) \propto \left|\sum_t X_k(t)\mathrm{e}^{-\mathrm{i}\omega t}\right|^2 , \quad k = 1,2 \tag{1.42}$$

after tapering the time series to avoid misalignment [5].

To carry the concept of cross-spectral analysis over to phase synchronization analysis, phase synchronization analysis is approached on the basis of the time series $Q(t)$ by identification of the function $A(t)$ as a taper window for $Q(t)$ (see (1.40)).

Plugging $Q_k(t) = \exp(\mathrm{i}\Phi_k(t))$ into (1.41) and (1.42) of the periodograms leads to

$$\mathrm{Per}_{Q_kQ_l}(\omega) \propto \sum_t Q_k(t)\mathrm{e}^{-\mathrm{i}\omega t} \sum_t Q_l(t)^*\mathrm{e}^{\mathrm{i}\omega t} \tag{1.43}$$

$$= \sum_{t,t'} \mathrm{e}^{\mathrm{i}\left(\Phi_k(t)-\Phi_l(t-t')\right)}\mathrm{e}^{-\mathrm{i}\omega t'} , \quad k,l = 1,2,\ k \neq l \tag{1.44}$$

and

$$\mathrm{Per}_{Q_kQ_k}(\omega) \propto \sum_{t,t'} \mathrm{e}^{\mathrm{i}\left(\Phi_k(t)-\Phi_k(t-t')\right)}\mathrm{e}^{-\mathrm{i}\omega t'} , \quad k = 1,2 , \tag{1.45}$$

respectively. The asterisk denotes complex conjugation.

To introduce a single number quantifying phase synchronization for coupled nonlinear oscillators 1 and 2, the periodogram values of all frequencies are summed up leading to

$$R_{1,2} = d \sum_{\omega} \mathrm{Per}_{Q_1 Q_2}(\omega) = \frac{1}{T} \sum_{t=1}^{T} \mathrm{e}^{\mathrm{i}(\Phi_1(t) - \Phi_2(t))} \tag{1.46}$$

with some appropriately chosen constant d ensuring $R_{1,1} = 1$. Only the phase differences $\Phi_{1,2}^{1,1}(t) = (\Phi_1(t) - \Phi_2(t))$ between the oscillators are contained in the expression for $R_{1,2}$.

The expression (1.46) is identical to the bivariate phase synchronization index

$$\left|R_{1,2}^{n,m}\right| = \left|\frac{1}{T} \sum_{t=1}^{T} \mathrm{e}^{\mathrm{i}\Phi_{1,2}^{n,m}(t)}\right| = \sqrt{\left\langle \sin \Phi_{1,2}^{n,m}(t) \right\rangle^2 + \left\langle \cos \Phi_{1,2}^{n,m}(t) \right\rangle^2} \tag{1.47}$$

introduced without foundation on cross-spectral analysis but on circular statistics in the previous section for $n = m = 1$ [34, 36, 46].

In the more general case, where the sampling rate is unequal to one, i.e. $t = t_i = i\Delta t$, for 1:1 synchronization, i.e. $m = n = 1$, it follows that

$$\left|R_{1,2}^{1,1}\right| = \sqrt{\langle \sin \phi_i \rangle^2 + \langle \cos \phi_i \rangle^2} = \frac{1}{N^2} \sum_{i,j=1}^{N} \cos(\phi_i - \phi_j) \tag{1.48}$$

with $\phi_i = \Phi_{1,2}^{1,1}(t_i)$.

Another approach to this problem is based on recurrence properties of the underlying dynamics and is given in Chap. 5.

1.8 Partial Phase Synchronization

Here, the concept of graphical models and partialization analysis applied to nonlinear synchronizing systems is introduced. To this aim, considering an N-dimensional dynamic process $X_1, \ldots, X_N$, the partial cross-spectral analysis can be achieved by inversion and renormalization of the spectral matrix $\mathbf{S}(\omega)$, as already discussed in detail in the first part of this chapter. The information about the linear interrelation between the processes X_k and X_l conditioned on the remaining examined processes $\mathbf{Y}$ is contained in the partial coherence (1.5)

$$\mathrm{Coh}_{X_k X_l|\mathbf{Y}}(\omega) = \frac{\left|S_{X_k X_l|\mathbf{Y}}(\omega)\right|}{\sqrt{S_{X_k X_k|\mathbf{Y}}(\omega)\, S_{X_l X_l|\mathbf{Y}}(\omega)}} \,. \tag{1.49}$$

Thus, for linear systems the auto- and cross-spectra enter the spectral matrix to estimate the partial auto- and cross-spectra leading to partial coherence

(1.49). For nonlinear synchronizing systems the synchronization indices are the natural choices to quantify interdependencies between processes as shown in Sect. 1.7.2. Caused by the similarity of the auto- and cross-spectra and the synchronization indices substituting the first by the latter in the spectral matrix leads to the multivariate extension of synchronization analysis.

This spectral matrix then becomes the *synchronization matrix*

$$\mathbf{R} = \begin{pmatrix} 1 & R_{1,2} & \dots & R_{1,N} \\ R^*_{1,2} & 1 & \dots & R_{2,N} \\ \vdots & \vdots & \ddots & \vdots \\ R^*_{1,N} & R^*_{2,N} & \dots & 1 \end{pmatrix} \tag{1.50}$$

with entries $R_{k,l} := R^{n,m}_{k,l}$ (1.47), which are the pairwise synchronization indices. Due to the proved analogy between spectral and synchronization theory, the inverse $\mathbf{PR} = \mathbf{R}^{-1}$ of the synchronization matrix $\mathbf{R}$ immediately leads to the definition of the $n:m$ *partial phase synchronization index*

$$R_{k,l|\mathbf{Y}} = \frac{|\mathbf{PR}_{kl}|}{\sqrt{\mathbf{PR}_{kk}\ \mathbf{PR}_{ll}}} \tag{1.51}$$

between X_k and X_l conditioned on the remaining processes which are described by $\{\mathbf{Y}|\mathbf{Y} = (X_y)_y,\ \ y = 1, \dots, N\,,\ y \neq k, l\}$. It replaces the partial coherence (1.49) for synchronizing systems. As for partial coherence, where the indirect interactions are characterized by an absent partial coherence accompanied by a bivariate significant coherence [10], the following holds: If the bivariate phase synchronization index $R_{k,l}$ is considerably different from zero, while the corresponding multivariate partial phase synchronization index $R_{k,l|\mathbf{Y}} \approx 0$, there is strong evidence for an indirect coupling between the processes X_k and X_l. Graphical models applying partial phase synchronization analysis are defined by:

An edge E between the oscillators k and l in a partial phase synchronization graph is missing, if and only if $R_{k,l|\mathbf{Y}}$ is small compared to $R_{k,l}$.

1.9 Application of Partial Phase Synchronization to a Model System

Three coupled stochastic Rössler oscillators

$$\dot{\xi}_j = \begin{pmatrix} \dot{X}_j \\ \dot{Y}_j \\ \dot{Z}_j \end{pmatrix} = \begin{pmatrix} -\omega_j\, Y_j - Z_j + \left[\sum_{i,i\neq j} \varepsilon_{j,i}\, (X_i - X_j)\right] + \sigma_j\, \eta_j \\ \omega_j\, X_j + a\, Y_j \\ b + (X_j - c)\, Z_j \end{pmatrix}, i, j = 1, 2, 3$$

are a genuine example of a system consisting of weakly coupled self-sustained stochastic oscillators. The parameters are set to $a = 0.15$, $b = 0.2$, $c = 10$,

$\sigma_j = 1.5$, $\omega_1 = 1.03$, $\omega_2 = 1.01$, and $\omega_3 = 0.99$ yielding a chaotic behavior in the deterministic case. For the noise term $\sigma_j \eta_j$ a standard deviation of $\sigma_j = 1.5$ is chosen and η_j is standard Gaussian distributed. Both the bidirectional coupling $\varepsilon_{1,3} = \varepsilon_{3,1}$ between oscillator ξ_1 and oscillator ξ_3 and the bidirectional coupling $\varepsilon_{1,2} = \varepsilon_{2,1}$ between oscillator ξ_1 and oscillator ξ_2 are varied between 0 and 0.3. Both synchronization phenomena, phase and lag synchronization, are contained in this range of coupling strengths. The oscillators ξ_2 and ξ_3 are not directly coupled since $\varepsilon_{2,3} = \varepsilon_{3,2} = 0$. The coupling scheme is summarized in Fig. 1.9 (a). However, caused by the indirect interaction between the oscillators ξ_2 and ξ_3 through the common oscillator ξ_1, the coupling scheme in Fig. 1.9 (b) is expected as a result of bivariate analysis.

In the following an example of 1 : 1 synchronization of the X-components is investigated. The bivariate synchronization index $R_{1,2}$ as well as $R_{1,3}$ increases when the corresponding coupling strength is increased, indicating phase synchronization (Fig. 1.10, upper triangular). Once a sufficient amount of coupling exists between oscillators ξ_1 and ξ_2 as well as between ξ_1 and ξ_3, a non-vanishing bivariate synchronization index $R_{2,3}$ between the not directly coupled oscillators ξ_2 and ξ_3 is observed (Fig. 1.10, upper triangular). This high but spurious phase synchronization is caused by the common influence from oscillator ξ_1 onto ξ_2 and ξ_3. The bivariate synchronization analysis suggests the coupling scheme between the three Rössler oscillators summarized in Fig. 1.10 (b), containing the additional but spurious edge between oscillator ξ_2 and ξ_3 denoted by the dashed line.

In Fig. 1.10 (a) (below the diagonal) the results of partial phase synchronization analysis are shown. While $R_{1,2|3}$ as well as $R_{1,3|2}$ are essentially unchanged compared to the bivariate synchronization indices, $R_{2,3|1}$ stays almost always below 0.1 and is therefore considerably smaller than $R_{2,3}$ in the area

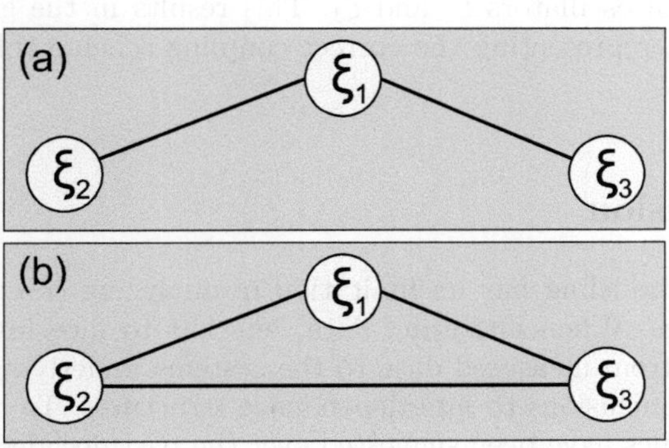

Fig. 1.9 (**a**) Graph for the simulated coupling scheme in the Rössler system. The direct coupling between oscillators ξ_2 and ξ_3 is absent. (**b**) Graph based on bivariate synchronization analysis. An additional but spurious edge between oscillator ξ_2 and ξ_3 is present

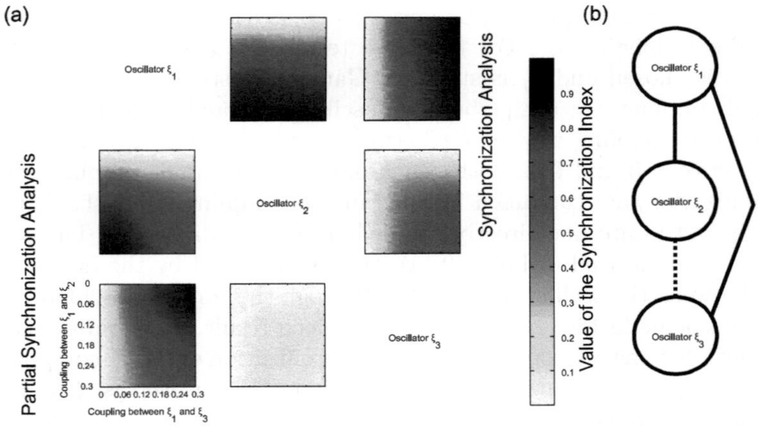

Fig. 1.10 (**a**) Phase synchronization and partial phase synchronization index. Coupling strengths between oscillators ξ_1 and ξ_2 and between oscillators ξ_1 and ξ_3 are varied between 0 and 0.3, for an absent coupling between ξ_2 and ξ_3. Values of the bivariate phase synchronization index (above the diagonal) and partial phase synchronization index (below the diagonal) are shown. When comparing the bivariate phase synchronization index $R_{2,3}$ with the partial phase synchronization index $R_{2,3|1}$ it becomes clear that the interaction between oscillator ξ_2 and ξ_3 is mediated by ξ_1 since $R_{2,3} \gg R_{2,3|1}$. (**b**) Graph for the simulated coupling scheme in the Rössler system. The direct coupling between oscillators ξ_2 and ξ_3 is absent. The additional but spurious edge between oscillator ξ_2 and ξ_3 correctly revealed by partial phase synchronization analysis is denoted by the dotted line

of spurious synchronization. This strongly indicates the absence of a direct coupling between oscillators ξ_2 and ξ_3. This results in the graph presented in Fig. 1.10 (b), representing the correct coupling scheme (black continuous edges).

1.10 Conclusion

First principle modeling has its limitation in analyzing complex systems in the Life Sciences. When analyzing data, one has to face inverse problems, i.e. conclusions from measured data to the systems underlying the measured time series or conclusions to interdependence structures. To this aim several analysis techniques have been suggested over the past decades. Ranging from linear to nonlinear techniques, from stochastic to deterministic, from one-dimensional to multidimensional systems, such techniques are widely applied to real world data.

Here, particular emphasis was laid on the multivariate linear and nonlinear approaches. The variety of analysis techniques that have been suggested to analyze multivariate linear systems leaves the practitioners alone with the decision which technique is to be used. Thus, several linear multivariate analysis techniques have been compared to investigate their conceptual properties. At the end a suggestion which technique should be used when was provided. This study is accomplished by an application to cortical and thalamic signals origination from juvenile pigs under deep sedation. Future work should be devoted to such a comparison including a wider class of time series analysis techniques, like for instance the multivariate phase synchronization approach.

Linear techniques seem to be not sufficient for some applications even though they do allow inference in for instance several nonlinear systems. Especially, a phenomenon referred to as phase synchronization has been observed for very weak coupling between self-sustained oscillators. A tailored analysis technique has been suggested in the literature to investigate phase synchrony. As demonstrated here the so called mean phase coherence to analyze synchronization is closely related to linear techniques, i.e. cross-spectral analysis. Thus, a motivation for the often used mean phase coherence has been provided.

Nonlinear signal analysis techniques should also work when investigating multivariate systems. Partial phase synchronization has been suggested in this Chapter to disentangle direct and indirect interactions in multivariate coupled synchronizing oscillatory systems. Application to more model systems and real-world data should be performed in the near future. Moreover, the actual desired goal is not only to differentiate direct and indirect interactions but also to infer the directionality of the coupling as in the linear case. To this aim, it might be promising to apply partial directed coherence or similar ideas to the phases of the oscillatory signals.

Multivariate analysis will become more and more prominent in future since more and more channels can be recorded simultaneously. The curse of dimensionality however limits the number of processes that can be analyzed simultaneously so far. The truism, the more advanced the techniques become the higher are the requirements of the analysis techniques, has to be beard in mind and dealt with. To this aim, complementing the existing techniques by novel techniques or adjustments to particular problems will be among the highest challenges in the future to analyze the vast amount of data that are recorded in the Life Sciences. Additionally, the assumptions that systems are stationary is not necessarily true. Thus, novel techniques or extensions are required that allow dealing with non-stationary data but still should be able to differentiate direct and indirect interactions and the directionality of interactions in both the linear and the nonlinear case.

Acknowledgement

We like to thank Dr. Reinhard Bauer and his group, Institute of Pathophysiology and Pathochemistry, University Hospital Jena, Germany, for providing us not only with the data from the animal model but also with excellent knowledge about underlying pathophysiological processes. Moreover, we like to thank Michael Eichler, University of Masstricnt, Netherlands, with whom we developed the concept of renormalized partial directed coherence.

This work was supported by the German Science Foundation (DFG SPP 1114).

References

[1] C. Allefeld and J. Kurths. An approach to multivariate phase synchronization analysis and its application to event-related potentials. *Int. J. Bif. Chaos*, 14:417–426, 2004.

[2] L. A. Baccala and K. Sameshima. Partial directed coherence: a new concept in neural structure determination. *Biol. Cybern.*, 84:463–474, 2001.

[3] A. Bandrivskyy, A. Bernjak, P. V. E. McClintock, and A. Stefanovska. Wavelet phase coherence analysis: application to skin temperature and blood flow. *Cardiovasc. Eng.*, 4:89–93, 2004.

[4] M. Bennett, M. F. Schatz, H. Rockwood, and K. Wiesenfeld. Huygens's clocks. *Proc. R. Soc. Lond. A*, 458:563–579, 2002.

[5] P. Bloomfield. *Fourier Analysis of Time Series: An Introduction.* Wiley, New York, 1976.

[6] S. Boccaletti, J. Kurths, G. Osipov, D. L. Valladares, and C. S. Zhou. The synchronization of chaotic systems. *Phys. Rep.*, 366:1–101, 2002.

[7] S. Boccaletti and D. L. Valladares. Characterization of intermittent lag synchronization. *Phys. Rev. E*, 62:7497–7500, 2000.

[8] D. R. Brillinger. *Time Series: Data Analysis and Theory.* Holden-Day, Inc., San Francisco, 1981.

[9] D. R. Brillinger. Remarks concerning graphical models for time series and point processes. *Rev. Econometrica*, 16:1–23, 1996.

[10] R. Dahlhaus. Graphical interaction models for multivariate time series. *Metrika*, 51:157–172, 2000.

[11] R. Dahlhaus and M. Eichler. Causality and graphical models for time series. In P. Green, N. Hjort, and S. Richardson, editors, *Highly Structured Stochastic Systems.* Oxford University Press, Oxford, 2003.

[12] R. Dahlhaus, M. Eichler, and J. Sandkühler. Identification of synaptic connections in neural ensembles by graphical models. *J. Neurosci. Methods*, 77:93–107, 1997.

[13] M. Eichler. A graphical approach for evaluating effective connectivity in neural systems. *Phil. Trans. Royal Soc. B*, 360:953–967, 2005.

[14] M. Eichler. Granger-causality and path diagrams for multivariate time series. *J. Econom.*, 137:334–353, 2007.

[15] M. Eichler. Graphical modeling of dynamic relationships in multivariate time series. In B. Schelter, M. Winterhalder, and J. Timmer, editors, *Handbook of Time Series Analysis.* Wiley, New York, 2006.

[16] M. Eichler, R. Dahlhaus, and J. Sandkühler. Partial correlation analysis for the identification of synaptic connections. *Biol. Cybern.*, 89:289–302, 2003.

[17] J. Granger. Investigating causal relations by econometric models and cross-spectral methods. *Econometrica 37*, 37:424–438, 1969.

[18] P. Grosse, M. J. Cassidy, and P. Brown. EEG–EMG, MEG–EMG and EMG–EMG frequency analysis: physiological principals and clinical applications. *Clin. Neurophysiol.*, 113:1523–1531, 2002.

[19] A. C. Harvey. *Forecasting Structural Time Series Models and the Kalman Filter.* Cambridge University Press, Cambridge, 1994.

[20] B. Hellwig, S. Häußler, M. Lauk, B. Köster, B. Guschlbauer, R. Kristeva-Feige, J. Timmer, and C. H. Lücking. Tremor-correlated cortical activity detected by electroencephalography. *Electroencephalogr. Clin. Neurophysiol.*, 111:806–809, 2000.

[21] B. Hellwig, S. Häußler, B. Schelter, M. Lauk, B. Guschlbauer, J. Timmer, and C. H. Lücking. Tremor correlated cortical activity in essential tremor. *Lancet*, 357:519–523, 2001.

[22] B. Hellwig, B. Schelter, B. Guschlbauer, J. Timmer, and C. H. Lücking. Dynamic synchronisation of central oscillators in essential tremor. *Clin. Neurophysiol.*, 114:1462–1467, 2003.

[23] D. Hemmelmann, W. Hesse, L. Leistritz, T. Wüstenberg, J. Reichenbach, and H. Witte. Example of a 4-dimensional image analysis: Time-variant model-related analysis of fast-fmri sequences. In *The 9th Korea-Germany Joint Workshop on Advanced Medical Images Processing*, pages 22–28, Seoul, 2006.

[24] W. Hesse, E. Möller, M. Arnold, and B. Schack. The use of time-variant EEG Granger causality for inspecting directed interdependencies of neural assemblies. *J. Neurosci. Methods*, 124:27–44, 2003.

[25] J. A. Hobson and E. F. Pace-Schott. The cognitive neuroscience of sleep: neuronal systems, consciousness and learning. *Nat. Rev. Neurosci.*, 3:679–693, 2002.

[26] N. L. Johnson, S. Kotz, and N. Balakrishnan. *Continuous Univariate Distributions.* Wiley, New York, 1995.

[27] E. G. Jones. Thalamic circuitry and thalamocortical synchrony. *Philos. Trans. R. Soc. Lond. B. Biol. Sci.*, 357:1659–1673, 2002.

[28] M. J. Kamiński and K. J. Blinowska. A new method of the description of the information flow in the brain structures. *Biol. Cybern.*, 65:203–210, 1991.

[29] J. Kent and T. Hainsworth. Confidence intervals for the noncentral chi-squared distribution. *J. Stat. Plan. Infer.*, 46:147–159, 1995.

[30] L. Kocarev and U. Parlitz. Generalized synchronization, predictability, and equivalence of unidirectionally coupled dynamical systems. *Phys. Rev. Lett.*, 76:1816–1819, 1996.

[31] R. Kus, M. Kamiński, and K. J. Blinowska. Determination of EEG activity propagation: pair-wise versus multichannel estimate. *IEEE Trans. Biomed. Eng.*, 51:1501–1510, 2004.

[32] L. Leistritz, W. Hesse, M. Arnold, and H. Witte. Development of interaction measures based on adaptive non-linear time series analysis of biomedical signals. *Biomed. Tech.*, 51(2):64–69, 2006. 0013-5585 (Print) Journal Article Research Support, Non-U.S. Gov't.

[33] L. Leistritz, W. Hesse, T. Wüstenberg, C. Fitzek, J. R. Reichenbach, and H. Witte. Time-variant analysis of fast-fmri and dynamic contrast agent mri sequences as examples of 4-dimensional image analysis. *Meth. Inf. Med.*, 45(6):643–650, 2006.

[34] K. Mardia and P. Jupp. *Directional Statistics*. Wiley, West Sussex, 2000.

[35] E. Möller, B. Schack, M. Arnold, and H. Witte. Instantaneous multivariate EEG coherence analysis by means of adaptive high-dimensional autoregressive models. *J. Neurosci. Methods*, 105:143–158, 2001.

[36] F. Mormann, K. Lehnertz, P. David, and C. E. Elger. Mean phase coherence as a measure for phase synchronization and its application to the EEG of epilepsy patients. *Physica D*, 144:358–369, 2000.

[37] E. Niedermeyer, D. L. Sherman, R. J. Geocadin, H. C. Hansen, and D. F. Hanley. The burst-suppression electroencephalogram. *Clin. Electroencephalogr.*, 30:99–105, 1999.

[38] G. V. Osipov, A. S. Pikovsky, M. G. Rosenblum, and J. Kurths. Phase synchronization effects in a lattice of nonidentical Roessler oscillators. *Phys. Rev. E*, 55:2353–2361, 1997.

[39] G.V. Osipov, J. Kurths, and C. Zhou. *Synchronization in Oscillatory Networks*. Complexity. Springer, Berlin, 2007.

[40] E. F. Pace-Schott and J. A. Hobson. The neurobiology of sleep: genetics, cellular physiology and subcortical networks. *Nat. Rev. Neurosci.*, 3:591–605, 2002.

[41] L. M. Pecora and T. L. Carroll. Synchronization in chaotic systems. *Phys. Rev. Lett.*, 64:821–824, 1990.

[42] A. Pikovsky, M. Rosenblum, and J. Kurths. Phase synchronization in regular and chaotic systems. *Int. J. Bif. Chaos*, 10:2291–2305, 2000.

[43] A. Pikovsky, M. Rosenblum, and J. Kurths. *Synchronization – A Universal Concept in Nonlinear Sciences.* Cambridge University Press, Cambridge, 2001.

[44] M. Le Van Quyen, J. Foucher, J. Lachaux, E. Rodriguez, A. Lutz, J. Martinerie, and F. J. Varela. Comparison of Hilbert transform and wavelet methods for the analysis of neuronal synchrony. *J. Neurosci. Methods*, 111:83–98, 2001.

[45] M. G. Rosenblum, A. Pikovsky, J. Kurths, C. Schäfer, and P. A. Tass. Phase synchronization: from theory to data analysis. In F. Moss and S. Gielen, editors, *Handbook of Biological Physics*, volume 4 of *Neuroinformatics*, pages 279–321, Elsevier, Amsterdam, 2001.

[46] M. G. Rosenblum, A. S. Pikovsky, and J. Kurths. Phase synchronization of chaotic oscillators. *Phys. Rev. Lett.*, 76:1804–1807, 1996.

[47] M. G. Rosenblum, A. S. Pikovsky, and J. Kurths. From phase to lag synchronization in coupled chaotic oscillators. *Phys. Rev. Lett.*, 78: 4193–4196, 1997.

[48] N. F. Rulkov, M. M. Sushchik, L. S. Tsimring, and H. D. I. Abarbanel. Generalized synchronization of chaos in directionally coupled chaotic systems. *Phys. Rev. E*, 51:980–994, 1995.

[49] K. Sameshima and L. A. Baccala. Using partial directed coherence to describe neuronal ensemble interactions. *J. Neurosci. Methods*, 94:93–103, 1999.

[50] B. Schack, G. Grießbach, M. Arnold, and J. Bolten. Dynamic cross spectral analysis of biological signals by means of bivariate ARMA processes with time-dependent coefficients. *Med. Biol. Eng. Comput.*, 33:605–610, 1995.

[51] B. Schack, P. Rappelsberger, S. Weiss, and E. Möller. Adaptive phase estimation and its application in EEG analysis of word processing. *J. Neurosci. Methods*, 93:49–59, 1999.

[52] B. Schack, S. Weiss, and P. Rappelsberger. Cerebral information transfer during word processing: where and when does it occur and how fast is it? *Hum. Brain Mapp.*, 19:18–36, 2003.

[53] B. Schelter, B. Hellwig, B. Guschlbauer, C. H. Lücking, and J. Timmer. Application of graphical models in bilateral essential tremor. *Proc. IFMBE (EMBEC)*, 2:1442–1443, 2002.

[54] B. Schelter, J. Timmer, and M. Eichler. Assessing the strength of directed influences among neural signals using renormalized partial directed coherence, submitted 2007.

[55] B. Schelter, M. Winterhalder, R. Dahlhaus, J. Kurths, and J. Timmer. Partial phase synchronization for multivariate synchronizing system. *Phys. Rev. Lett.*, 96:208103, 2006.

[56] B. Schelter, M. Winterhalder, M. Eichler, M. Peifer, B. Hellwig, B. Guschlbauer, C. H. Lücking, R. Dahlhaus, and J. Timmer. Testing for directed influences among neural signals using partial directed coherence. *J. Neurosci. Methods*, 152:210–219, 2006.

[57] B. Schelter, M. Winterhalder, B. Hellwig, B. Guschlbauer, C. H. Lücking, and J. Timmer. Direct or indirect? Graphical models for neural oscillators. *J. Physiol. Paris*, 99:37–46, 2006.

[58] B. Schelter, M. Winterhalder, J. Kurths, and J. Timmer. Phase synchronization and coherence analysis: sensitivity and specificity. *Int. J. Bif. Chaos*, in press, 2006.

[59] B. Schelter, M. Winterhalder, and J. Timmer, editors. *Handbook of Time Series Analysis.* Wiley-VCH, Berlin, 2006.

[60] S. J. Schiff, P. So, T. Chang, R. E. Burke, and T. Sauer. Detecting dynamical interdependence and generalized synchrony through mutual prediction in a neural ensemble. *Phys. Rev. E*, 54:6708–6724, 1996.

[61] S. M. Schnider, R. H. Kwong, F. A. Lenz, and H. C. Kwan. Detection of feedback in the central nervous system using system identification techniques. *Biol. Cybern.*, 60:203–212, 1989.

[62] R. H. Shumway and D. S. Stoffer. *Time Series Analysis and Its Application.* Springer, New York, 2000.

[63] M. Steriade. Impact of network activities on neuronal properties in corticothalamic systems. *J. Neurophysiol.*, 86:1–39, 2001.

[64] M. Steriade and I. Timofeev. Neuronal plasticity in thalamocortical networks during sleep and waking oscillations. *Neuron*, 37:563–576, 2003.

[65] P. A. Tass, M. G. Rosenblum, J. Weule, J. Kurths, A. Pikovsky, J. Volkmann, A. Schnitzler, and H. J. Freund. Detection of $n : m$ phase locking from noisy data: Application to magnetoencephalography. *Phys. Rev. Lett.*, 81:3291–3295, 1998.

[66] J. Timmer, M. Lauk, S. Häußler, V. Radt, B. Köster, B. Hellwig, B. Guschlbauer, C.H. Lücking, M. Eichler, and G. Deuschl. Cross-spectral analysis of tremor time series. *Int. J. Bif. Chaos*, 10:2595–2610, 2000.

[67] J. Timmer, M. Lauk, W. Pfleger, and G. Deuschl. Cross-spectral analysis of physiological tremor and muscle activity. I. Theory and application to unsynchronized EMG. *Biol. Cybern.*, 78:349–357, 1998.

[68] J. Volkmann, M. Joliot, A. Mogilner, A. A. Ioannides, F. Lado, E. Fazzini, U. Ribary, and R. Llinás. Central motor loop oscillations in Parkinsonian resting tremor revealed by magnetoencephalography. *Neurology*, 46:1359–1370, 1996.

[69] H. Voss, J. Timmer, and J. Kurths. Nonlinear dynamical system indentification from uncertain and indirect measurements. *Int. J. Bif. Chaos*, 14:1905–1933, 2004.

[70] M. Winterhalder, B. Schelter, W. Hesse, K. Schwab, L. Leistritz, D. Klan, R. Bauer, J. Timmer, and H. Witte. Comparison of linear signal processing techniques to infer directed interactions in multivariate neural systems. *Sig. Proc.*, 85:2137–2160, 2005.

[71] M. Winterhalder, B. Schelter, W. Hesse, K. Schwab, L. Leistritz, D. Klan, J. Timmer, and H. Witte. Detection of directed information flow in multidimensional biosignals. *Special Issue on Biosignal Processing, Biomed. Tech.*, 51:281–287, 2006.

[72] M. Winterhalder, B. Schelter, J. Kurths, A. Schulze-Bonhage, and J. Timmer. Sensitivity and specificity of coherence and phase synchronization analysis. *Phys. Lett. A*, 356:26–34, 2006.

[73] H. Witte, C. Schelenz, M. Specht, H. Jäger, P. Putsche, M. Arnold, L. Leistritz, and K. Reinhart. Interrelations between EEG frequency components in sedated intensive care patients during burst-suppression period. *Neurosci. Lett.*, 260:53–56, 1999.

2

Surrogate Data – A Qualitative and Quantitative Analysis

Thomas Maiwald[1], Enno Mammen[2], Swagata Nandi[3], and Jens Timmer[1]

[1] FDM, Freiburg Center for Data Analysis and Modeling, University of Freiburg, Eckerstr. 1, 79104 Freiburg, Germany
maiwald@fdm.uni-freiburg.de, jeti@fdm.uni-freiburg.de

[2] Department of Economics, University of Mannheim, L7, 3-5, 68131 Mannheim, Germany
emammen@rumms.uni-mannheim.de

[3] Theoretical Statistics and Mathematics Unit, Indian Statistical Institute, 7, S.J.S. Sansanwal Marg, New Delhi 110016, India
nandi@isid.ac.in

2.1 Motivation

The surrogates approach was suggested as a means to distinguish linear from nonlinear stochastic or deterministic processes. The numerical implementation is straightforward, but the statistical interpretation depends strongly on the stochastic process under consideration and the used test statistic. In the first part, we present quantitative investigations of level accuracy under the null hypothesis, power analysis for several violations, properties of phase randomization, and examine the assumption of uniformly distributed phases. In the second part we focus on level accuracy and power characteristics of Amplitude Adjusted Fourier–Transformed (AAFT) and improved AAFT (IAAFT) algorithms. In our study AAFT outperforms IAAFT. The latter method has a similar performance in many setups but it is not stable in general. We will see some examples where it breaks down.

2.2 Introduction

In a statistical test, for a given data set a hypothesis is formulated, whose validity has to be examined. This *null hypothesis* cannot be verified or falsified with 100% accuracy. Instead, only an upper bound for the probability is given, that the null hypothesis is rejected although it holds true. For this a real valued *test statistic* T is derived from the data, with known distributions under the possible specifications of the null hypothesis. If the null hypothesis contains only one specification (*simple hypothesis*) and the value of T is for example larger than the $1-\alpha$ quantile of the null distribution, the null hypothesis will be rejected and a false rejection will occur only with probability α. Often, the null distribution of T is not simple and the distribution of the underlying specification has to be estimated and/or approximated. One possibility is to use *surrogate* methods. They have been suggested for the null hypothesis of a linear Gaussian process transformed by an invertible nonlinear function, see [20].

This chapter examines the different approaches to generate surrogate data and their properties. The basic idea of all surrogate methods is to randomize the Fourier phases of the underlying process. In Sect. 2.3 we illustrate how the nature of a process changes if the phases are randomized. For this purpose we show plots where the amount of randomization is continuously increased. Section 2.4 summarizes all suggested surrogate approaches and illustrates their qualitative behavior. Section 2.5 presents a simulation study where the level accuracy and power performance of surrogate data tests is compared for different processes, test statistics and surrogate methods. Section 2.6 takes a closer look on two methods for generating surrogate data. It compares the AAFT and the IAAFT approach. The chapter ends with short conclusions in Sect. 2.7.

2.3 Phase Randomization – Effects and Assumptions

The key procedure of all surrogate methods is to randomize the Fourier phases. It is argued that linear Gaussian processes do not possess asymptotically any information in the Fourier phases, since they are comprehensively determined by their mean and autocovariance function, which corresponds one-to-one via the Fourier transformation to the power spectrum. Hence, realizations of a linear Gaussian process should differ essentially only in their Fourier phases which is utilized by the surrogates approach: New realization of a linear Gaussian process based on one realized time series can be obtained by drawing new values for the Fourier phases. Since no information is expected in the Fourier phases, the underlying distribution for the new phase values is the uniform distribution on $[0, 2\pi]$.

This section investigates the assumption of uniformly distributed Fourier phases and takes a qualitative look on the effects of phase randomization on linear and nonlinear time series.

2.3.1 Effects of Phase Randomization

Let $\mathbf{x} = (x_1, ..., x_N)$ be a given time series of length N with mean 0. Its Fourier transformation is

$$f(\omega) = \frac{1}{\sqrt{2\pi N}} \sum_{t=1}^{N} \mathrm{e}^{-i\omega t} x_t, \quad -\pi \le \omega \le \pi \,. \tag{2.1}$$

If the transformation is calculated for discrete frequencies $\omega_j = 2\pi/N$, $j = 1, 2, ..., N$, the original time series can be regained by the back transformation:

$$x_t = \sqrt{\frac{2\pi}{N}} \sum_{j=1}^{N} \mathrm{e}^{i\omega_j t} f(\omega_j), \quad t = 1, 2, ..., N. \tag{2.2}$$

The FT-surrogates method constructs a new time series y_t with the same periodogram and otherwise statistically independent from x_t [19]. The Fourier amplitudes $|f(\omega_j)|$ are fixed and the Fourier phases $\varphi(\omega_j) = \arg(f(\omega_j))$ are replaced by uniform distributed random numbers $\varphi_{\text{rand}}(\omega_j) \in [0,\ 2\pi]$. The new realization is given by

$$y_t = \sqrt{\frac{2\pi}{N}} \sum_{j=1}^{N} \mathrm{e}^{i\omega_j t} \, |f(\omega_j)| \, \mathrm{e}^{i\varphi_{\text{rand}}(\omega_j)}. \tag{2.3}$$

To track the effect of phase randomization, we increase randomization strength continuously from 0% to 100%. For this random numbers

$$\varphi_{\text{rand}}(\omega_j) = \varphi(\omega_j) + a\, u(\omega_j)$$

with $u(\omega_j) \in \mathcal{U}[-\pi,\ \pi]$ are drawn. Parameter $a \in [0,\ 1]$ changes the phase randomization strength. Figure 2.1(a) shows, that three test statistics of Sect. 2.5, time invariance, prediction error and kurtosis do not change for the autoregressive process with increasing phase randomization. For the nonlinear stochastic Van-der-Pol oscillator already small changes of the Fourier phases lead to significant different values of the test statistics (b). Figure 2.1(c–f) exemplifies the influence of phase randomization on time series directly. The autoregressive process as a linear process does not change visually with increasing phase randomization (c). The nonlinear Lorenz system looses already for 20% phase randomization its characteristical oscillations (d). With 100% phase randomization, even the ear switch is not visible any more. The time series equals a realization of an autoregressive process with small coherence length. The deterministic Rössler system is shown in total length (e). It possesses a sharp harmonic component leading to higher harmonics in the power spectrum. If the phase relation of these harmonics is lost due to phase randomization, a beat occurs resulting in strong correlations of the time series up to the whole time series length. This holds even stronger for the zig-zag curve (f). The time series of a stationary, nonlinear, deterministic process becomes a time series which is indistinguishable from a realization of an non stationary, cyclic, linear process.

The correlation dimension in Fig. 2.2 differs for the nonlinear processes logistic map, Lorenz system and stochastic Van-der-Pol oscillator. Whereas for the logistic map already a 1% phase randomization leads to an unbounded correlation dimension on small scales, this happens for the Lorenz systems not until a randomization of 10%. The stochastic Van-der-Pol does not show any changes with increasing phase randomization – as a stochastic process it has an unbounded correlation dimension even for the original process.

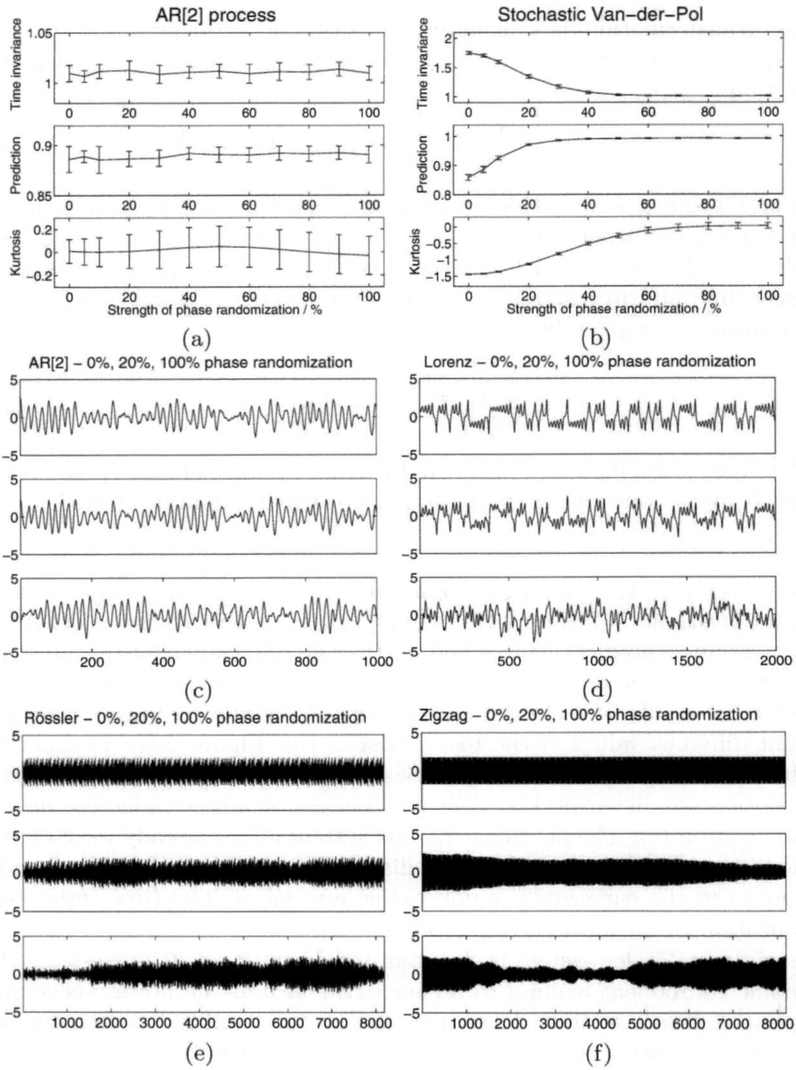

Fig. 2.1 Influence of increasing phase randomization

2.3.2 Distribution of Fourier Phases

Figure 2.3 shows the cumulative Fourier phase distribution for realizations of an AR[2] process and the Lorenz system. For the AR[2], the cumulative distribution is very close to the straight line corresponding to perfect uniform distribution. For the Lorenz system, most phases are larger than 2 – the distribution is unbalanced. The deviation from a uniform distribution can

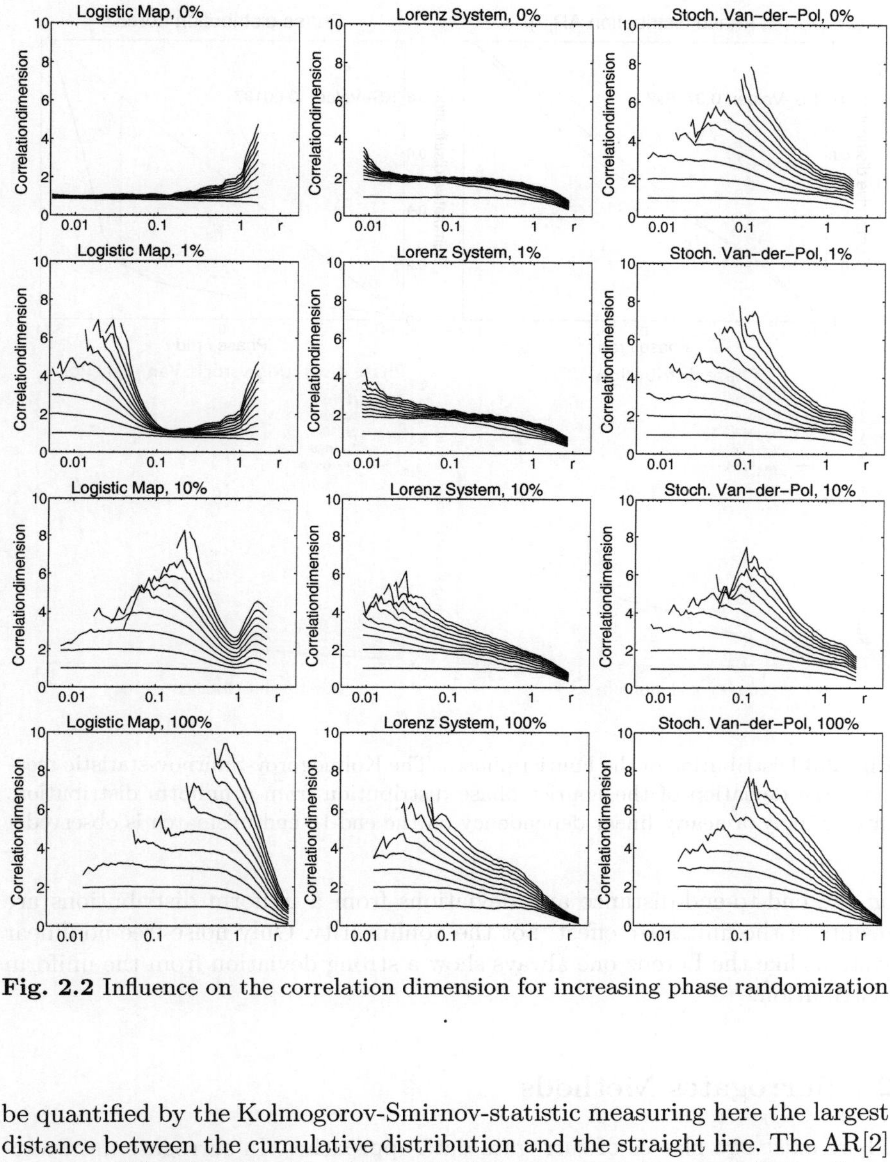

Fig. 2.2 Influence on the correlation dimension for increasing phase randomization

be quantified by the Kolmogorov-Smirnov-statistic measuring here the largest distance between the cumulative distribution and the straight line. The AR[2] leads to a KS-value of 0.075, whereas the Lorenz system has a nearly ten times larger value of 0.6. To examine whether this result is representative, a simulation with 1000 AR[2] of length between 1024 and 16384 has been done and the KS-value calculated. Figure 2.3 exhibits a nearly linear relationship between the mean KS-value and the end-to-end-distance $|x_N - x_1|$. Since a similar simulation for the nonlinear stochastic Van-der-Pol oscillator shows the same effect, the phase distribution of linear and nonlinear systems depends strongly

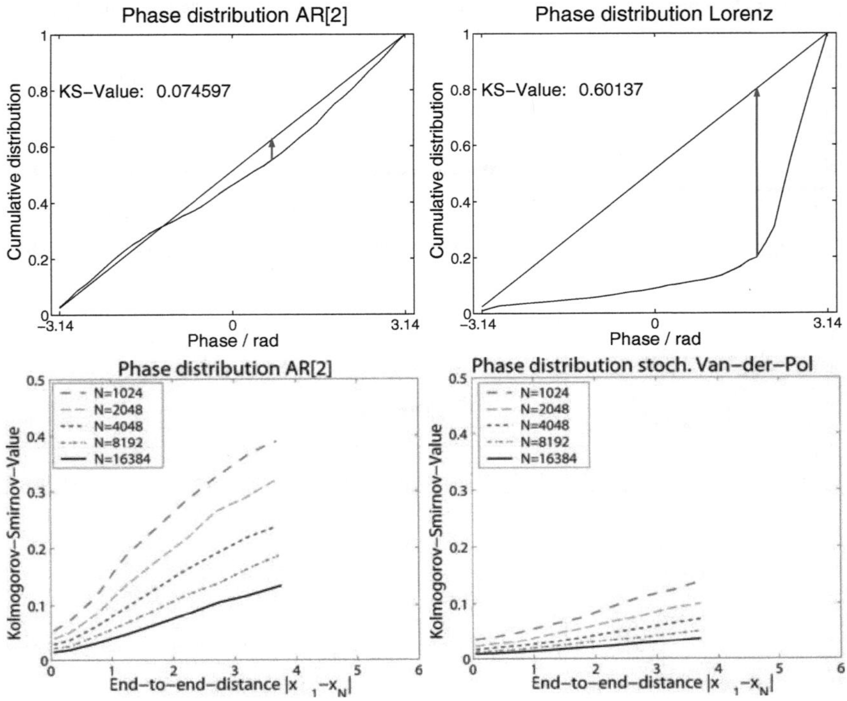

Fig. 2.3 Distribution of the Fourier phases. The Kolmogorov-Smirnov-statistic measures the deviation of the Fourier phase distribution from a uniform distribution. Interestingly, a nearly linear dependency on the end-to-end-mismatch is observed

on the end-to-end-distance and deviations from a uniform distributions are results of the finite size effect, not the nonlinearity. Only noise-free nonlinear systems like the Lorenz one always show a strong deviation from the uniform distribution.

2.4 Surrogates Methods

There exist two classes of surrogate data approaches. In surrogate data testing one reproduces all the linear properties of the process by preserving the second order properties of the observed data. This can be done by preserving the observed sample mean and sample auto-covariances (or by preserving the observed sample mean and periodogram values at Fourier frequencies $\omega_j = 2\pi j/n$). The approach can be implemented by first Fourier transforming the data set, then randomizing the phases and finally inverting the transformed data. Then resamples will have the same linear properties as the data set, see Sect. 2.3 II. The term "surrogate data" was first introduced by [19] and the method became popular after that. But the basic idea and related approaches were discussed in a number of earlier publications, see [6, 8].

Several other ways exist to generate surrogate data with asymptotically the same autocovariance function as a given time series. Essentially they differ in conserving the original periodogram by construction or generating a new one derived from the estimated power spectrum.

Fourier transformed surrogate data method generates resamples for the null hypothesis that the process is a stationary Gaussian linear process. This method is not suitable for the hypothesis that the data are not Gaussian. This can be encountered in several applications. In practice, the null hypothesis of linear Gaussian processes is rather restrictive as only very few data pass the test that they are normally distributed. Fourier-transformed surrogates are, by construction, asymptotically jointly normally distributed, thus surrogates will have a different distribution than the observed data, when the data deviate from normality. For such cases the more general null hypothesis has been proposed that the process is a linear stationary process transformed by a static (invertible) nonlinear function $h(.)$. More explicitly, the observed process $\{x_t\}$ is generated by a transformation:

$$x_t = h(z_t)\,,$$

where z_t is a Gaussian stationary process. For this extended null hypothesis methods have been proposed that transform the original data to a Gaussian amplitude distribution before the surrogate method is applied. A back transformation to the original amplitudes realizes the nonlinear measurement function.

Classical Monte-Carlo-simulation is the counterpart of the parameter-free surrogate approach. Here, the problem arises to select the right model, e.g., the right order of an autoregressive process. If the right model is known, the Monte-Carlo-approach should be the used.

The basic method to generate Fourier transformation surrogates (FT) has been described in Sect. 2.3. For an overview on resampling methods that have been used in the statistical literature on time series analysis see also [12].

2.4.1 Amplitude Adjusted FT-Surrogates (AAFT)

If the data derive under an extended null hypothesis from a linear, Gaussian process measured via an invertible nonlinear function h, the Gaussian amplitude distribution is lost in general. Since only Gaussian distributed linear processes are uniquely given by their autocovariance function and mean value, the measurement function h has to be estimated in order to calculate surrogates for $\hat{h}^{-1}(\mathbf{x})$. The surrogate data will be *measured* with $\hat{h}$ to be comparable to the original data. In detail:

1. Ranking of the original data.
2. Generation of ranked, Gaussian distributed random numbers

3. The k-th value of the sorted original series is replaced with the k-th value of the sorted Gaussian data and the order of the original data is re-established.
4. The data are now Gaussian distributed and FT-Surrogates can be calculated.
5. The k-th largest value of the FT-Surrogates is replaced with the k-th largest value of the original time series. Note, that the original amplitude distribution is exactly maintained.

Asymptotically for $N \to \infty$ the first three steps of the transformation are equivalent to the application of the true inverse function h^{-1} itself. The procedure was suggested as *Amplitude Adjusted Fourier Transform*–Surrogates [19] and its effect is illustrated in Fig. 2.4. Test statistics like the skewness or kurtosis based on the amplitude distribution have by construction exactly the same value for the original and the surrogate time-series. A more detailed description of the AAFT algorithm can be found in Sect. 2.5 where also the convergence of the fitted transformations is discussed.

For finite data sets a flattened power spectrum is observed for AAFT-Surrogates compared to original data – the spectrum "whitens". The reason is, that the estimator of the inverse function $\hat{h}^{-1}(\cdot)$ does not match exactly $h^{-1}(\cdot)$ for finite N [17]. The differences $\delta(x_t) = \hat{h}^{-1}(x_t) - h^{-1}(x_t)$ are essentially independent of t and possess as an uncorrelated process a white spectrum. Therefore, application of the estimated inverse function adds uncorrelated random numbers to the original data,

$$\hat{h}^{-1}(x_t) = h^{-1}(x_t) + \delta(x_t) .$$

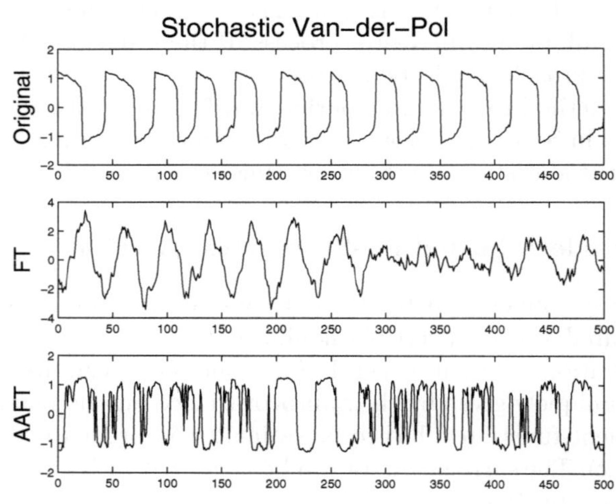

Fig. 2.4 FT- and AAFT-Surrogates. The bimodal distribution of the stochastic Van-der-Pol oscillator is not kept by the FT-Surrogates method, but by the AAFT-method. Both methods yield time inverse invariant time series

2.4.2 Iterated Amplitude Adjusted FT–Surrogates (IAAFT)

To reduce the whitening effect of AAFT–surrogates, an iterative approach has been suggested, which asymptotically yields the same periodogram and amplitude distribution as the original data [16]. First, AAFT-Surrogates are generated. Then, in the Fourier domain, the periodogram values are replaced by the periodogram of the original process, the phases are kept. In time domain, the amplitudes are adjusted to the original process. These two steps are iterated until periodogram and amplitude distributions of original and generated time series are equal except for a given tolerance. A description of the IAAFT algorithm can be found in Sect. 2.5 where this approach is compared with the AAFT algorithm.

2.4.3 Digitally Filtered Shuffled Surrogates (DFS)

Some authors criticize that the FT-, AAFT- and IAAFT-surrogates methods conserve not only mean and autocovariance function of the underlying process, but also properties of the given realization itself, since only Fourier phases and not the Fourier amplitudes are changed. In this way, the new data possess less variability than new realizations of the original process [5]. Alternatively *Digitally Filtered Shuffled Surrogates* have been suggested.

The DFS–Surrogates approach is based on the following steps [5]:

1. Estimation of the power spectrum by averaging periodograms of several data segments.
2. Estimation of the autocovariance function by an inverse Fourier transformation of the estimated spectrum. This yields the "response"-function.
3. Convolution of a random permutation of the original time series with the response-function. This corresponds to a multiplication in the Fourier domain, where the transformed of the randomized original time series is multiplied to the estimated spectrum.
4. Adapting the amplitude distribution to the original one.

Similar to the AAFT-Surrogates, the amplitude distribution is adjusted in the last step leading to a whitened spectrum. Again, an iterative procedure, the *iDFS-method*, can be applied to reduce this effect.

(Fourier based) surrogate data tests, AAFT, IAAFT, DFS and iDFS belong to the class of constrained realization approaches. In constrained realization approaches one avoids fitting of model parameters and one does not assume any model equation. Instead of fitting a model (e.g., a finite order AR process), one generates resamples that have a certain set of properties in common with the observed data set. An alternative is based on fitting the model of the data and generating data from the fitted model.

2.4.4 New Periodogram from Estimated Spectrum (NPS)

We suggest a new method to generate surrogate data, which is based on the statistical properties of the periodogram

$$I(\omega) \sim \frac{1}{2} S(\omega) \chi_2^2 ,$$

i.e., the periodogram is for given frequency ω distributed like a χ^2-distribution with two degrees of freedom around the power spectrum $S(\omega)$. New realizations of the process can be obtained by

1. estimation of the power spectrum, e.g., via a smoothed periodogram,
2. generation of a periodogram realization by multiplication of χ_2^2–distributed random numbers to the spectrum,
3. drawing new phases as uniform random numbers on $[0,\ 2\pi]$,
4. calculation of an inverse Fourier transformation.

This approach differs slightly from frequency domain bootstrap that does not apply Step 3, see [4]. Again, an iterative approach is imaginable for the extended null hypothesis of a linear, Gaussian process with an invertible, nonlinear measurement function.

2.4.5 Fixed Fourier Phases (FPH)

In a sense *orthogonal* to the FT-surrogates, one could keep the Fourier phases and draw only a new periodograms, like for the NPS–surrogates. This is actually no surrogates approach, but it illustrates the influence of Fourier phases and amplitudes.

2.4.6 Classical Monte-Carlo-Simulation

If the model of the underling stochastic process under the null hypothesis is known, new realizations of the process can be generated after estimating all parameters θ by means of the given data set. The estimation minimizes the square error

$$\sum_{i=1}^{N} (x_i - y_i(\theta))^2$$

for the given realization $\mathbf{x} = (x_1, ..., x_N)$ and a parameterized time series $\mathbf{y}(\theta) = (y_1(\theta), ..., y_N(\theta))$. The difficulty is to select the right model. Linear processes can be formulated as: A parameter $\theta = (p, q, a_1, ..., a_p, b_1, ..., b_q)$ exist, for which

$$X_t = \sum_{i=1}^{p} a_i X_{t-i} + \sum_{i=0}^{q} b_i \varepsilon_{t-i} , \quad \varepsilon_t \sim \mathcal{N}(0,1) . \tag{2.4}$$

The model order p, q has to be determined. Since finite, invertible ARMA$[p, q]$ processes can be written as infinite ARMA$[\infty, 0]$ or ARMA$[0, \infty]$ processes, selection of the wrong model class can lead to infinite parameters which have to be determined. If the right model class is known, calculation of the partial

ACF or ACF for AR or MA processes respectively, can be used to determine the process order.

In this study we fit an AR[80] process to the original time series and generate AR-Fit–surrogates by

$$y_t = \sum_{i=1}^{p} a_i y_{t-i} + \varepsilon_t \tag{2.5}$$

with random start values and after the transient behavior is over.

2.5 Statistical Properties of the Surrogate Methods – A Comparative Simulation Study

This section contains simulation results on surrogates based hypothesis tests. The simulations have been carried out for a representative class of processes under the null hypothesis and under the alternative and for a variety of test statistics.

2.5.1 Processes Under the Null Hypothesis

Linear Gaussian processes can be written as [2, 14]

$$X_t = \sum_{i=1}^{p} a_i X_{t-i} + \sum_{i=0}^{q} b_i \varepsilon_{t-i}\,, \quad \varepsilon_t \sim \mathcal{N}(0,1)\,. \tag{2.6}$$

These processes are linear, but not cyclic, a property which is implicitly assumed by the Fourier transformation on a finite data set. We suggest a cyclic process with the periodogram of an autoregressive process:

$$X_t = A + \sqrt{\frac{2\pi}{N}} \sum_{j=1}^{M} \sqrt{B_j^2 + C_j^2}\, \cos(\omega_t j + \theta_j)\,, \quad t = 1, ..., N$$

with $A \sim \mathcal{N}(0,1)$, $B_j, C_j \sim \mathcal{N}(0, \sigma_j^2)$, $\theta_j \sim \mathcal{U}[0, 2\pi]$ and $\sigma_j^2 = I_Y(\omega_j)$. I_Y is the periodogram of an AR[2] process. A, B_j, C_j and θ_j are independent for $j = 1, ..., M$. Besides, $\omega_t = 2\pi t/N$, $M = (N-1)/2$ for odd N and $M = (N-2)/2$ for even N. Note, that

$$B_j^2 + C_j^2 \sim \sigma_j^2 \chi_2^2$$

with mean $2\sigma_j^2$. The periodogram of the process reads $I_X(\omega_k) = (B_k^2 + C_k^2)/4$ for $k = 1, ...M$. The same approach with an exponentially decreasing periodogram with $\sigma_j^2 = \exp(-j/M)$ leads to a cyclic Gaussian process (see Fig. 2.5).

The extension of the null hypothesis with invertible, nonlinear measurement function h is investigated in this study with $h(x) = x^3$ and an AR[2].

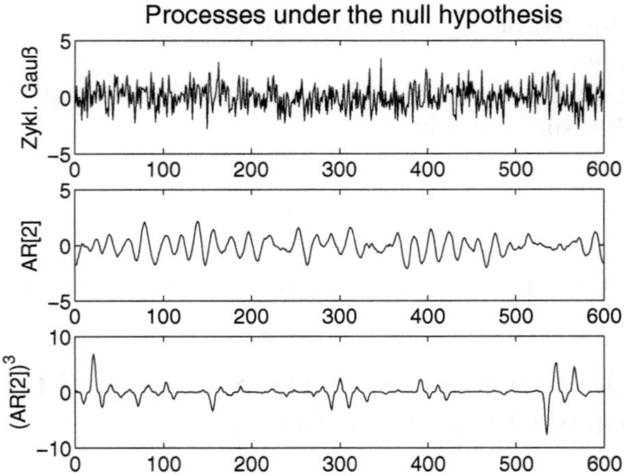

Fig. 2.5 Processes under the null hypothesis. Cyclic Gaussian process with exponentially decreasing spectrum, AR[2] process and AR[2] process measured via $h(x) = x^3$

2.5.2 Processes Under the Alternative: Nonlinear Deterministic and Stochastic Systems

Deterministic systems can be written as differential systems oder difference equations. All consequent time points derive exactly from the initial values. If the distance between neighbored trajectories increases exponentially, the system is chaotic. Here we use the logistic map as example for a chaotic difference equation [7]

$$x_i = r x_{i-1}(1 - x_{i-1}), \quad 3.6... < r \leq 4 .$$

For $r = 4$ chaotic behavior occurs and the spectrum is not distinguishable from white noise. All information about the nonlinearities is saved in the phases, making this toy system interesting for our study (see Fig. 2.6).

The most famous differential systems with chaotic behavior are the Lorenz and Rössler systems [11, 15]:

$$\begin{aligned}\dot{x} &= \sigma(y - x)\\ \dot{y} &= -y + x(r - z)\\ \dot{z} &= xy - bz\end{aligned}$$

here with $\sigma = 10$, $b = 8/3$ and $r = 40$, and

$$\begin{aligned}\dot{x} &= -y - z\\ \dot{y} &= x + ay\\ \dot{z} &= b + (x - c)z\end{aligned}$$

here with $a = 0.1$, $b = 0.1$ and $c = 18$.

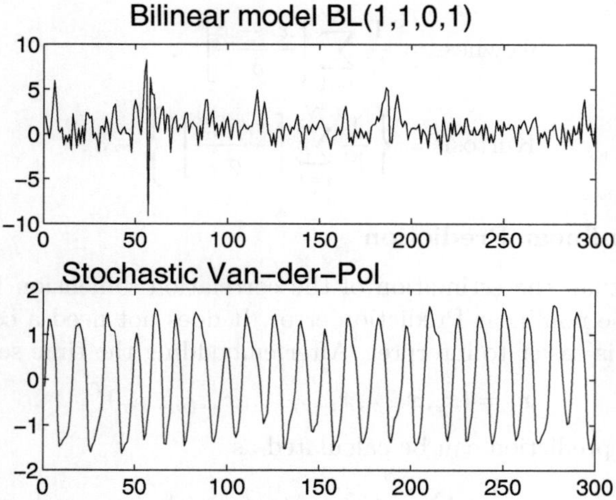

Fig. 2.6 Bilinear model and stochastic Van-der-Pol oscillator

A stable limit cycle is given by the Van-der-Pol oscillator [22]

$$\ddot{x} = \mu(1 - x^2)\dot{x} - x, \quad \mu > 0 .$$

Bilinear models are the extension of ARMA models. We use the bilinear model BL(1,0,1,1) [18]

$$X_i = aX_{i-1} + bX_{i-1}\varepsilon_{i-1} + \varepsilon_i ,$$

with $a = b = 0.4$. The deterministic Lorenz, Rössler and Van-der-Pol systems can be disturbed with additive noise leading, e.g., to

$$\ddot{x} = \mu(1 - x^2)\dot{x} - x + \varepsilon, \quad \mu > 0 .$$

2.5.3 Test Statistics

This subsection presents published and new developed test statistics, which are used to examine the qualitative and quantitative properties of the surrogates methods in the next main section. Since linear and nonlinear processes have to be distinguished from each other, emphasis is laid on test statistics which are sensitive for this difference.

Skewness and Kurtosis

Deviations from Gaussian amplitude distributions can be measured with the centered third and fourth moment, skewness and kurtosis:

$$\text{Skewness} = \frac{1}{N}\sum_{i=1}^{N}\left[\frac{x_i - \bar{x}}{\sigma}\right]^3 ,$$

$$\text{Kurtosis} = \left\{\frac{1}{N}\sum_{i=1}^{N}\left[\frac{x_i - \bar{x}}{\sigma}\right]^4\right\} - 3 .$$

Error of Nonlinear Prediction

More robust than the estimation of the correlation dimension is the the estimation of the nonlinear Prediction error. It does not need a certain scaling behavior and is easier to interpret. After embedding the time series to

$$\boldsymbol{x}_i = (x_i, x_{i-\tau}, x_{i-2\tau}, ..., x_{i-(m-1)\tau})^T \tag{2.7}$$

the nonlinear prediction can be calculated as

$$F_{\varepsilon,\tau}(\boldsymbol{x}_i) = \frac{1}{N_\varepsilon(\boldsymbol{x}_i)}\sum_j \boldsymbol{x}_{j+\tau} \qquad \forall j \neq i \text{ with } ||\boldsymbol{x}_i - \boldsymbol{x}_j|| < \varepsilon .$$

For every $\boldsymbol{x}_i$, the point $\hat{\boldsymbol{x}}_{i+s} = F_{\varepsilon,s}(\boldsymbol{x}_i)$ is predicted. $N_\varepsilon(\boldsymbol{x}_i)$ is the number of all neighbors of $\boldsymbol{x}_i$ with a distance less than ε. As test statistic the well defined Prediction error is used:

$$\gamma(m,\tau,\varepsilon) = \left(\frac{1}{N}\sum ||\boldsymbol{x}(t+\tau) - F_{\varepsilon,\tau}(\boldsymbol{x}(t))||^2\right)^{1/2} .$$

Time Invariance

We suggest a very powerful test statistic to measure the time invariance:

$$\max\left\{\frac{\#\{|x_{i+1}| > |x_i|\}}{\#\{|x_{i+1}| < |x_i|\}}, \frac{\#\{|x_{i+1}| < |x_i|\}}{\#\{|x_{i+1}| > |x_i|\}}\right\}$$

after demeaning $\mathbf{x}$. Application to the stochastic Van-der-Pol oscillator illustrates the power of this test statistic (Fig. 2.7). Even with $\mu = 0.01$ time invariance test statistic detects the nonlinearity whereas it is not visible for $\mu = 0.1$. Besides, this test statistic does not depend on the end-to-end-distance $|x_1 - x_N|$.

Normalized End-to-End-Distance

The Fourier transform considers a finite time series as part of an infinite, periodical time series. A jump between the first and last data point $|x_N - x_1|$ bigger than the mean point-to-point-distance leads to a sawtooth behavior of the infinite long time series. This needs an exact adjustment of the Fourier phases and also changes the Fourier amplitudes.

$$\text{Normalized end-to-end-distance} = \frac{|x_1 - x_N|}{\frac{1}{N-1}\sum_{i=2}^{N}|x_i - x_{i-1}|} .$$

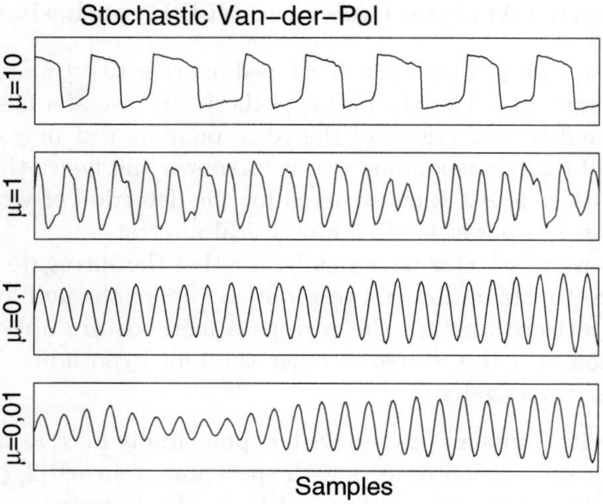

Fig. 2.7 Sensitivity of time invariance. Displayed are realizations of the stochastic Van-der-Pol oscillator with different values for μ, corresponding to a different strength of the nonlinearity. The test statistic time invariance is able to detect the nonlinearity even for $\mu = 0.01$

Biggest Jump

The maximum distance between two consecutive points is given by

$$\max_{j=2,\ldots,N} |x_{j+1} - x_j| \, .$$

Smoothness

The smoothness of a time series can be quantified by:

$$\text{Smoothness} = \frac{\frac{1}{N-1}\sum_{i=2}^{N} |x_i - x_{i-1}|}{\frac{1}{N}\sum_{i=1}^{N} |x_i - \bar{x}|} \, .$$

Normalized Distribution of Fourier Phases

The test statistic *distribution of Fourier phases* measured via the Kolmogorov-Smirnov-statistic has already been introduced in Sect. 2.3. To reduce the linear dependency on the end-to-end-distance, the KS-value is divided by the end-to-end-distance for the normalized version.

2.5.4 Qualitative Analysis Under the Null Hypothesis

The surrogates approach has been developed in order to estimate the distribution of a test statistic under the null hypothesis. In case of a false approximation the size and critical region of the corresponding test may change. Then the behavior of the corresponding test is unknown and hence the test useless. Furthermore, since a test is constructed for the detection of violations of the null hypothesis, the power is of high practical interest.

This section investigates numerically, whether the surrogate methods generate the correct distribution of a test statistic under the null hypothesis. We analyze qualitatively whether the surrogate methods are able to reproduce the distribution of a test statistic under the null hypothesis. The following hypotheses are investigated:

1. Linear cyclic Gaussian process with exponentially decreasing spectrum
2. Linear cyclic Gaussian process with spectrum of an AR[2] process
3. Linear Gaussian process, realized with an AR[2] process
4. Linear Gaussian process measured via a nonlinear invertible measurement function, here realized with an AR[2] process and $h(x) = x^3$.

For ten realizations of every process under null hypothesis 200 surrogate time series are generated and several test statistics are calculated yielding values $t_1, ..., t_N$. Their cumulative distribution

$$F(t) = \frac{\#\{t_i,\, i = 1, ..., N \,|\, t_i \leq t\}}{N} \tag{2.8}$$

is qualitatively compared to the one of the test statistic based on 200 realizations of the underlying process itself. Figure 2.8 shows the result: For some combinations of null hypothesis, surrogate method and test statistic, the original distribution displayed in the bottom panel can be regained in the upper ten panels. Here, this occurs for the combination linear Gaussian process with FT-Surrogates and the test statistic *time invariance*. For the combination linear Gaussian process with AAFT-Surrogates and the test statistic *prediction error* the original distribution is not maintained but, even worse, the distribution depends on the value of the test statistic for the original time series marked in each panel by the vertical line.

The qualitative characterization has been done for six surrogate methods, eleven test statistic and four null hypothesis. The results are summarized in Tables 2.1–2.4. The ✓ symbols a correct reproduction of the test statics distribution, a false reproduction is marked by $\times$, and $\circ$ has been used for ambiguous cases.

The first null hypothesis leads to a correct distribution for most combinations. Except for a few cases the phase fixing FPH-method does not yield a correct distribution, as expected. The methods AAFT and DFS scale the amplitude distributions to the distribution of the original time series resulting in always the same value for the test statistics skewness and kurtosis.

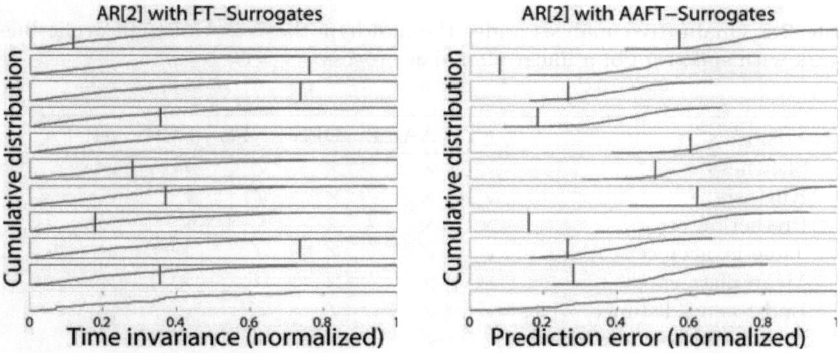

Fig. 2.8 Cumulative distributions of test statistic values under the null hypothesis. Surrogate methods should asymptotically be able to reproduce the complete cumulative distribution of any test statistic under the null hypothesis. However it turns out, that this depends strongly on the chosen null hypothesis, test statistic and surrogate method. The new developed time invariance measure on the left is able to reproduce the original cumulative distribution (panel 11 at the bottom) for each of ten AR[2] realizations (panels 1–10). Using the prediction error as test statistic and AAFT surrogates show a strong dependency of the test statistic value of the original data realization, marked by vertical lines

Since a constant value does not reflect the statistic fluctuations, the original distribution is not regained.

Similar results yield the simulations for the second null hypothesis, the cyclic Gaussian process with periodogram of a linear Gaussian process. Only the AR-Fit method has problems, which is based on the vanishing

Table 2.1 Qualitative analysis under the null hypothesis 1: Gaussian cyclic linear process with exponentially decreasing spectrum

	FT	AAFT	DFS	NPS	FPH	AR-Fit
Skewness	✓	×	×	✓	×	✓
Kurtosis	✓	×	×	✓	×	✓
Prediction error	✓	✓	✓	✓	✓	✓
Time invariance	✓	✓	✓	✓	✓	✓
Mean distance	×	×	×	○	×	×
End-to-end-distance	✓	✓	✓	✓	×	✓
Norm. end-to-end-distance	✓	✓	✓	✓	×	✓
Biggest jump	✓	✓	✓	✓	✓	✓
Smoothness	×	×	×	○	×	×
Phase distribution	✓	✓	✓	×	×	✓
Norm. phase distribution	✓	✓	○	○	×	✓

Table 2.2 Qualitative analysis under the null hypothesis 2: Gaussian cyclic linear process with spectrum of a linear Gaussian process

	FT	AAFT	DFS	NPS	FPH	AR-Fit
Skewness	✓	×	×	✓	✓	✓
Kurtosis	✓	×	×	✓	✓	✓
Prediction error	×	×	×	×	×	×
Time invariance	✓	✓	✓	✓	✓	✓
Mean distance	×	×	×	×	×	×
End-to-end-distance	✓	✓	✓	✓	×	×
Norm. end-to-end-distance	✓	✓	✓	✓	×	×
Biggest jump	✓	✓	✓	✓	∘	✓
Smoothness	×	×	×	×	×	×
Phase distribution	✓	✓	✓	×	×	×
Norm. phase distribution	✓	∘	✓	∘	∘	×

end-to-end-distance for the cyclic process. This behavior is not reproduced by the generated AR[80].

Very good results yield the AR-Fit method for the third null hypothesis, the linear Gaussian process realized via an autoregressive process, since the right model class was used for the Monte-Carlo-approach. The *actual* surrogates methods drop behind indicating their demand for cyclicity.

The fourth null hypothesis, the linear Gaussian process measured via an invertible, nonlinear function is for most combinations already part of the "alternative", since a Gaussian distribution is assumed. Even the AAFT method, which was constructed for this setting, does not succeed in combination with most test statistics.

Table 2.3 Qualitative analysis under the null hypothesis 3: Linear Gaussian process

	FT	AAFT	DFS	NPS	FPH	AR-Fit
Skewness	✓	×	×	✓	✓	✓
Kurtosis	✓	×	×	✓	×	✓
Prediction error	×	×	×	×	×	✓
Time invariance	✓	✓	✓	✓	✓	✓
Mean distance	∘	×	×	×	×	×
End-to-end-distance	×	×	×	×	×	✓
Norm. end-to-end-distance	×	×	✓	×	×	✓
Biggest jump	✓	✓	×	✓	∘	✓
Smoothness	×	×	×	∘	×	×
Phase distribution	×	×	×	×	×	✓
Norm. phase distribution	∘	∘	∘	×	∘	✓

Table 2.4 Qualitative analysis under the null hypothesis 4: Linear Gaussian process measured with nonlinear function

	FT	AAFT	DFS	NPS	FPH	AR-Fit
Skewness	×	×	×	×	×	×
Kurtosis	×	×	×	×	×	×
Prediction error	×	✓	×	×	×	×
Time invariance	✓	✓	✓	✓	✓	✓
Mean distance	×	∘	×	×	×	×
End-to-end-distance	×	×	×	×	×	✓
Norm. end-to-end-distance	×	∘	×	×	×	∘
Biggest jump	×	×	×	×	×	×
Smoothness	×	×	×	×	×	×
Phase distribution	×	×	×	×	×	∘
Norm. phase distribution	∘	∘	∘	×	×	∘

The qualitative analysis can be summarized as follows: The accuracy of the reproduction of the test statistics distribution under the null hypothesis depends *strongly* on the chosen combination of null hypothesis, surrogates method and test statistic. Only for one test statistic, the new developed time invariance test statistic, all surrogate methods and null hypotheses lead to a correct distribution. For the normalized end-to-end-distance *no* combination was successful.

2.5.5 Quantitative Analysis of Asymptotic, Size and Power

This section investigates numerically, which power the surrogate methods have for different alternatives. The power is analyzed depending on data length and strength of violation, in order to examine the asymptotic behavior and to establish a ranking of the different surrogate methods. A two-sided test with a significance level of 6% is constructed for the following simulations. For every original time series 99 surrogates are generated and the value of a given test statistic calculated. Is the value for the original time series under the first three or last three of the ranked 100 test statistic values, the null hypothesis is rejected. This procedure is repeated for 300 original time series.

The quantitative analysis of the asymptotic determines the power depending on the length of the time series. For the time invariance test statistic and time series of the logistic map, every surrogate method FT, AAFT, DFS, NPS, IAAFT and iDFS show a similar behavior (Fig. 2.9). For 256 or more data points, the null hypothesis is rejected in all cases and for all methods. The same time series evaluated with the test statistic smoothness lead to a ranking of the surrogate methods: The FT-surrogates reach maximum power already for 256 data points, NPS-surrogates for 1024, AAFT-surrogates for 8192, DFS- and iDFS-surrogates need about 10000 data points to reject the

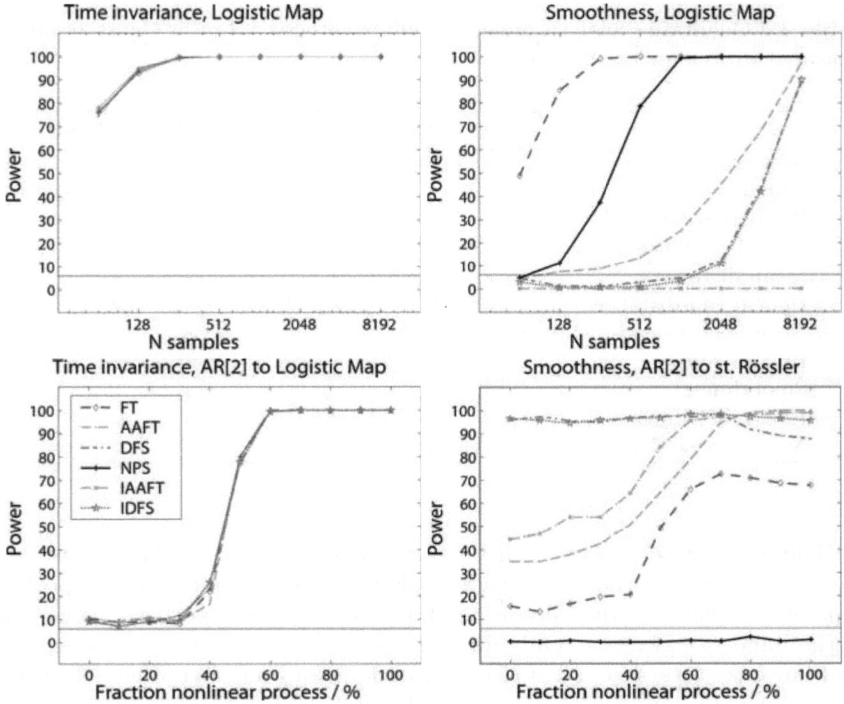

Fig. 2.9 Quantitative analysis of asymptotic, size and power

false null hypothesis with a power of 100%. The IAAFT-surrogates have no power at all. Unfortunately, this ranking is not reproduced in other, not shown combinations of alternatives and test statistics, indicating the irregular behavior of tests based on surrogate methods.

Necessary requirement for the usefulness of a test is on the one hand a correct size and on the other hand a good power. This section investigates the power of surrogates based tests for linearity with variable strength of the null hypothesis violation. First, two stochastic processes X_0 and X_A are combined to a new process

$$X = (1 - a)X_0 + aX_A,\ a \in [0;\ 1]\,. \tag{2.9}$$

The process X_A violates the null hypothesis, which is fulfilled by X_0. A representative selection of processes X_A is chosen, in order to investigate violations of all kind. The parameter a increases from 0 to 1 and consequently the influence of the nonlinear process increases from 0% to 100%.

Figure 2.9 shows two cases to discuss the behavior of surrogates based linearity tests. For the test statistic time invariance all surrogate methods lead to a similar behavior and reject the null hypothesis for 50%–60% of the nonlinear process. The size is a bit too high. For the test statistic smoothness, the size is not correct for all surrogate methods except for the NPS method, which has on the other hand no power for the violation with the stochastic Rössler system. The worst case of nearly 100% rejection of the true null hypothesis based on the DFS- and iDFS-surrogates indicates again the irregular behavior of surrogates based tests.

2.5.6 Summary of the Simulation

The nonlinearity of the stochastic Van-der-Pol oscillator can be changed via the parameter μ. The cases of Fig. 2.10 summarize the behavior of surrogates based linearity tests:

- For some test statistics like *time invariance*, every method works similarly and in a plausible way (a).
- Some test statistics have a different behavior which can be derived from the surrogates construction like for the *kurtosis* in combination with amplitude adjusting surrogates (b).

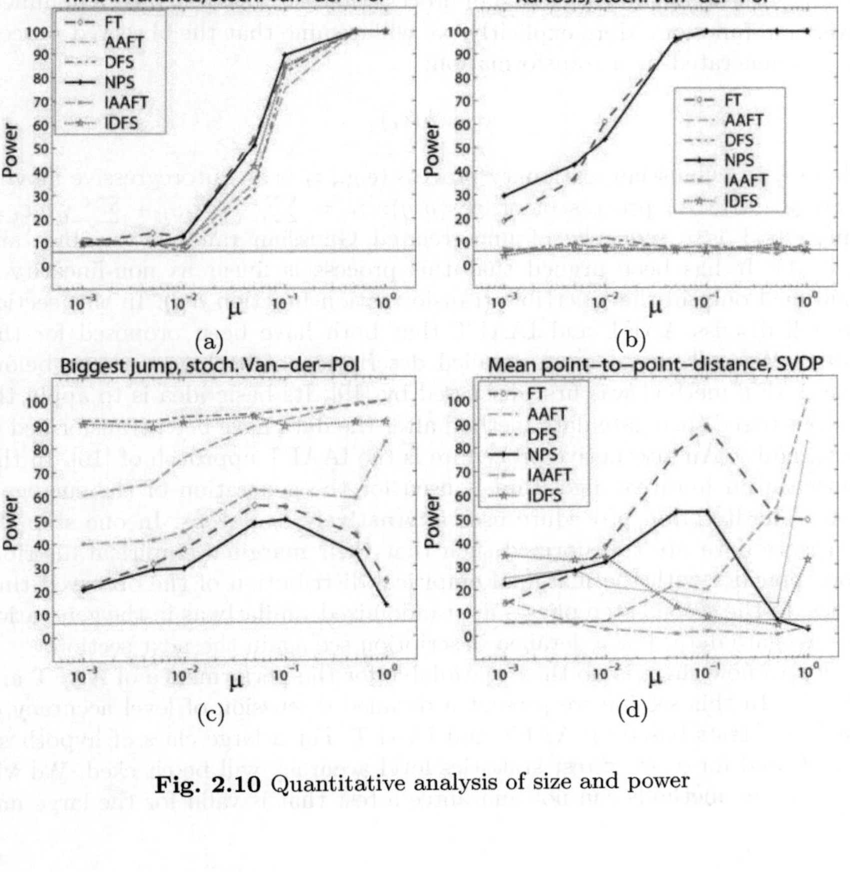

Fig. 2.10 Quantitative analysis of size and power

- For some test statistic, like for the *biggest jump*, a performance order of the methods can be established (c). Unfortunately, this order is not maintained but reversed in other combinations.
- And finally, for some test statistics like the *mean point to point distance*, no regular or reasonable behavior of the surrogates methods occurs (d). This would lead in a statistical test to spurious results and wrong consequences.

2.6 The AAFT and the IAAFT Approach – A Detailed Comparison

In this section, we study level accuracy and power properties of tests based on Amplitude Adjusted Fourier-Transformed (AAFT) surrogates and on Iterated AAFT (IAAFT) surrogates. We will see that both methods work rather reliable as long as the process do not have long coherence times. IAAFT is outperformed by AAFT: It has a similar performance in many setups but it is not stable in general. We will see some examples where it breaks down.

The AAFT method and the IAAFT approach are designed for the general null hypothesis of a linear Gaussian process that is transformed by a nonlinear invertible function. More explicitly, we will assume that the observed process $\{x_t\}$ is generated by a transformation:

$$x_t = h(z_t)\,,$$

where z_t is a Gaussian stationary process (e.g., z_t is an autoregressive moving average (ARMA) process of order (p, q): $z_t = \sum_{j=1}^{p} a_i z_{t-i} + \sum_{j=0}^{q} b_i \varepsilon_{t-i}$, where $\{\varepsilon_t\}$ is a sequence of uncorrelated Gaussian random variables and $b_0 = 1$). It has been argued that this process is linear as non-linearity is contained only in the invertible transformation function $h(.)$. In this section we will discuss AAFT and IAAFT that both have been proposed for this model. We will give a more detailed description of both procedures below. The AAFT method was first discussed by [19]. Its basic idea is to apply the Fourier based surrogate data method after the data have been transformed to Gaussianity. An alternative procedure is the IAAFT approach of [16]. In this approach an iterative algorithm is used for the generation of the surrogate data. The iteration procedure uses alternatively two steps. In one step the surrogate data are transformed such that their marginal empirical distribution coincides with the marginal empirical distribution of the observed time series. In the second step phases are randomized similarly as in the generation of surrogate data. For a detailed description see again the next section.

Up to now there is no theory available for the performance of AAFT and IAAFT. In this section we present a detailed discussion of level accuracy of statistical tests based on AAFT and IAAFT. For a large class of hypothesis models and for a set of test statistics level accuracy will be checked. We will see that the methods can not guarantee a test that is valid for the large null hypothesis of transformed Gaussian stationary processes. The methods do not perform well for time series with long coherence times. But we will also present examples of well behaved standard Gaussian linear models where IAAFT turn out as very unstable. We now give a detailed description of the AAFT and IAAFT algorithms.

2.6.1 The AAFT and the IAAFT Approach – Definition

The method of Amplitude Adjusted Fourier-Transformed (AAFT) proceeds as follows:

1. Let $\mathbf{r} = (r_1, r_2, \ldots, r_n)$ be the rank vector of an observational data vector $(x_1, \ldots, x_n)$ and $(x_{(1)}, \ldots, x_{(n)})$ be its ordered sample.
2. Generate a sample $w_1, ..., w_n$ of i.i.d. standard normal random variables and denote its ordered sample by $(w_{(1)}, ..., w_{(n)})$. Put $\widehat{z}_t = w_{(r_t)}$, $t = 1, \ldots, n$. Then $\widehat{z}_t$ has the same rank in $\widehat{z}_1, ..., \widehat{z}_n$ as x_t in $x_1, ..., x_n$.
3. Obtain phase-randomized surrogates of $\widehat{z}_t$, say z_t^*, $t = 1, \ldots, n$.
4. The surrogate data sample x_t^* of x_t is defined by $x_t^* = x_{(r_t^*)}$. Here r_t^* is the rank of z_t^* in the series $z_1^*, \ldots, z_n^*$.

In Steps 1 and 2, the observed data are transformed to normal variables. This is done by using the random transformation $\widehat{z}_t = \widehat{g}(x_t) = \widehat{\Phi}^{-1}\widehat{F}(x_t)$. Here $\widehat{\Phi}$ is the empirical distribution function of the normal random variables $w_1, ..., w_n$ and $\widehat{F}$ is the empirical distribution function of $x_1, ..., x_n$. In Step 3, the transformed data are phase randomized: Fourier-transformed surrogates are generated. In the final step, the surrogate data are transformed back to the original values. This is done by sorting the observed data according to the ranks of the Fourier transformed surrogates: The phase randomized data $z_1^*, ..., z_n^*$ are transformed by using the random transformation $x_t^* = \widehat{g}^-(z_t^*) = \widehat{F}^{-1}\widehat{G}(z_t^*)$, where now $\widehat{G}$ is the empirical distribution function of $z_1^*, ..., z_n^*$. The resulting data $x_1^*, ..., x_n^*$ are called AAFT surrogates. Note that in this procedure the transformation $\widehat{g}^-$ in Step 4 is typically not the inverse of the transformation $\widehat{g}$ used in Step 2. They differ because $\widehat{F}$ differs from $\widehat{G}$. In particular, the transformation $\widehat{g}$ does not depend on the surrogate sample whereas $\widehat{g}^-$ does.

The basic model assumptions imply that the transformed data $\widehat{z}_t$ and the FT-based surrogates z_t^* follow approximately a linear Gaussian process. Thus, it may be reasonable directly to base the statistical inference on these time series and to check if $\widehat{z}_t$ follows a linear Gaussian process. This would suggest to calculate a test statistic for the transformed series $\widehat{z}_t$ and to calculate critical levels for this test by using the surrogates z_t^*. We will consider both type of tests, tests based on the original time series x_t and their AAFT surrogates x_t^* and tests based on $\widehat{z}_t$ and z_t^*.

Iterative AAFT (IAAFT) is an iterative algorithm based on AAFT. It was proposed by [16] as an improved algorithm of AAFT. This algorithm is used to generate resamples for the same hypothesis as AAFT, i.e., the hypothesis of a transformed linear Gaussian process. IAAFT is a method to produce surrogate data which have the same power spectrum and the same empirical distribution as the observed data. Note that this aim is not achieved by AAFT. The IAAFT algorithm proceeds iteratively. It iteratively corrects deviations in the spectrum and deviations in the empirical distribution (between the surrogates and the original time series). The algorithm proceeds in the following steps:

1. Generate an ordered list of the sample ($x_{(1)} \leq \ldots \leq x_{(n)}$) and calculate the periodogram $I_k^2 = |\sum_{t=0}^{n-1} x_t e^{it\omega_k}|^2$, $\omega_k = 2\pi k/n$, $k = 1, \ldots, n$.
2. Initialization step: generate a random permutation (sample without replacement) $\{x_t^{a,(0)}\}$ of the data $\{x_t\}$.
3. Iteration steps:

   (i) At the j-th iteration, take the Fourier transform of $\{x_t^{a,(j)}\}$, replace the squared amplitudes by I_k^2 (without changing the phases) and transform back by application of the inverse Fourier transform:

$$x_t^{b,(j)} = 1/\sqrt{n} \sum_{s=0}^{n-1} \frac{\widehat{x}_s^{a,(j)}}{|\widehat{x}_s^{a,(j)}|} |I_s| \exp(-it\omega_s),$$

   where $\widehat{x}_t^{a,(j)} = 1/\sqrt{n} \sum_{s=0}^{n-1} x_s^{a,(j)} \exp(-it\omega_s)$ is the discrete Fourier transform of $\{x_s^{a,(j)}\}$.

   (ii) The resulting series in (i) is rescaled back to the original data:

$$x_t^{a,(j+1)} = x_{(r_t^j)},$$

   where r_t^j is the rank of $x_t^{b,(j)}$ in $x_1^{b,(j)}, \ldots, x_n^{b,(j)}$.
4. Repeat (i) and (ii) in Step 3 until the relative difference in the power spectrum is sufficiently small. The limiting values x_t^* of $x_t^{b,(j)}$ (or $x_t^{a,(j)}$) are called IAAFT surrogates.

2.6.2 Comparison of AAFT and IAAFT – Numerical Experiments

In this section, results based on numerical experiments are presented. We have conducted simulations for the following models.

$M_1 : x_t$ is an i.i.d. sequence with distribution χ_1^2.

$M_2 : x_t$ is an i.i.d. sequence with distribution $U[0, 1]$.

$M_3 : x_t = z_t^3, z_t$ is a circular process with $\sigma_j^2 = \exp(-j/m)$.

$M_4 : x_t = z_t, z_t$ is a circular process with $\sigma_j^2 = \exp(-j/8)$.

$M_5 : x_t = z_t^3, z_t$ is a circular process with $\sigma_j^2 = \exp(-j/8)$.

$M_6 : x_t = z_t^3, z_t = 1.4z_{t-1} - 0.48z_{t-2} + \varepsilon_t$.

$M_7 : x_t = z_t^3, z_t = 1.8z_{t-1} - 0.81z_{t-2} + \varepsilon_t$.

In all models the residuals ε_t are i.i.d. and have a standard normal distribution. The circular processes in M_4, M_5 and M_6 are generated as follows:

$$z_t = A + c \sum_{j=1}^{m} \sqrt{B_j^2 + C_j^2} \cos(\omega_j t + \theta_j), \qquad t = 1, \ldots, n.$$

Here $A \sim \mathcal{N}(0, \sigma^2)$, $B_j, C_j \sim \mathcal{N}(0, \sigma_j^2)$, $\theta_j \sim U[0, 2\pi]$ and A, B_j, C_j, and θ_j are independent. Furthermore, we used the notation $\omega_j = 2\pi j/n$, $c = \sqrt{2\pi/n}$, $m = (n-2)/2$. Note that after application of a transformation the circular Gaussian processes are still circular but in general not Gaussian distributed.

All models are transformed stationary Gaussian processes. Models M_1 and M_2 are i.i.d. processes. Trivially, both can be written as $x_t = h(z_t)$ with z_t i.i.d. standard gaussian. M_3 is a transformed Gaussian circular process. For this circular process $\sigma_j^2 = \exp(-j/m), m = n - 1/2$ and it exhibits a performance near to an i.i.d. sequence. Models M_4 and M_5 are circular processes with $\sigma_j^2 = \exp(-j/8)$. Model M_4 is a Gaussian circular process and M_5 is a nonlinear transformation of M_4. The fast decay of σ_j^2 leads to long coherence times. Model M_6 is a transformed AR(2) process transformed by $h(x) = x^3$. The underlying AR(2) process has all roots inside the unit circle (0.6 and 0.8). M_7 represents a transformed Gaussian AR process of order two. The underlying AR(2) process has roots 0.9 close to unity.

In the implementation of the AAFT algorithm the transformation function $h(.)$ and its inverse $h^{-1}(.)$ are estimated. One may expect that AAFT works as well as a classical surrogate data test if these estimates are accurate. For this reason, we checked the accuracy of these estimates for models M_1 – M_7. We consider a single realization $(x_1, \ldots, x_n)$ of each model with $n = 256$. In Figs. 2.11–2.14 the series are plotted against $(\widehat{z}_1, \ldots, \widehat{z}_n)$ (horizontal axis) as solid lines. In the same plots, the true function $h(.)$ as a function of z_t is plotted as dotted line. For models M_3, M_5–M_7, $h(x) = x^3$ and for M_4, $h(x) = x$. M_1 and M_2 are i.i.d. processes. So for these i.i.d. processes, the true transformation function is given as $h(x) = F^{-1}\Phi(x)$, where $\Phi(.)$ denote the distribution function of the standard normal distribution and F denotes

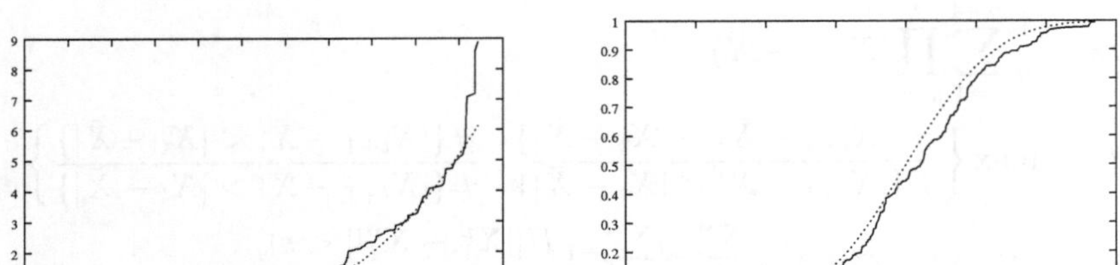

Fig. 2.11 The estimated (solid line) and true (dotted line) transformation functions for models M_1 and M_2

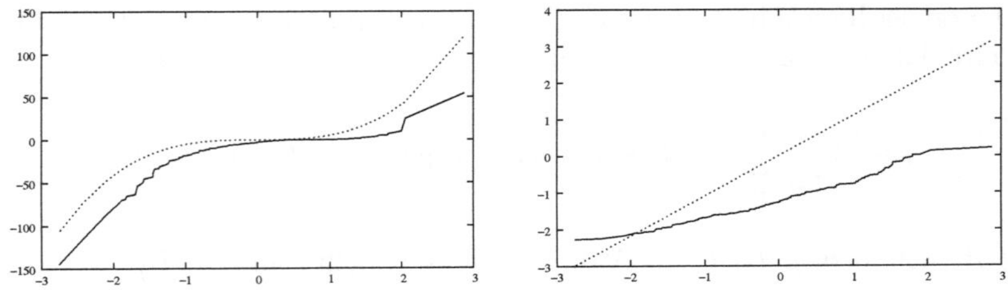

Fig. 2.12 The estimated (solid line) and true (dotted line) transformation functions for models M_3 and M_4

the distribution function of χ^2_1 or $U[0,1]$, respectively. We observe that for the two i.i.d. processes M_1 and M_2, the transformation function is estimated very accurately. Approximately, this also holds for M_3 and M_6 (note that the z-axes differ and always they include extreme points of a standard normal law). For the other models the estimate behaves poorly, in particular in the tails.

We have performed numerical experiments for the following test statistics:

$$T_1 = \frac{1}{n}\sum_{t=1}^{n-1}(X_t X_{t+1}^2 - X_t^2 X_{t+1}),$$

$$S_1 = \frac{\#\{X_t > X_{t+1}\}}{n}, \qquad S_2 = n - 1 - S_1,$$

$$T_2 = S_1,$$

$$T_3 = \frac{|S_2 - S_1|}{S_1 + S_2},$$

$$T_4 = \max_\tau Q(\tau), \qquad Q(\tau) = \frac{\sum_{t=\tau+1}^n (X_{t-\tau} - X_t)^3}{[\sum_{t=\tau+1}^n (X_{t-\tau} - X_t)^2]^{3/2}},$$

$$T_5 = \frac{1}{n}\sum_{t=1}^{n-2}\prod_{k=0}^{2}(X_{t+k} - \bar{X}),$$

$$T_6 = \frac{1}{n}\sum_{t=1}^{n-4}\prod_{k=0}^{4}(X_{t+k} - \bar{X}),$$

$$T_7 = \max\left\{\frac{\#\{|X_{t+1} - \bar{X}| > |X_t - \bar{X}|\}}{\#\{|X_{t+1} - \bar{X}| < |X_t - \bar{X}|\}}, \frac{\#\{|X_{t+1} - \bar{X}| < |X_t - \bar{X}|\}}{\#\{|X_{t+1} - \bar{X}| > |X_t - \bar{X}|\}}\right\}$$

$$T_8 = C_n(r), \qquad C_n(r) = \frac{\sum_{i=2}^n \sum_{j=1}^i I(||\mathbf{X}_i^\nu - \mathbf{X}_j^\nu|| < r)}{n(n-1)/2}.$$

Here, I is the indicator function and $||\mathbf{X}|| = \max_k |X_k|$. The vector $\mathbf{X}_i^\nu = (X_{i-(\nu-1)d}, X_{i-(\nu-2)d}, \ldots, X_i)^T$ is an element of the phase space with embedding dimension ν and delay time d. We have used delay time $d = 2$.

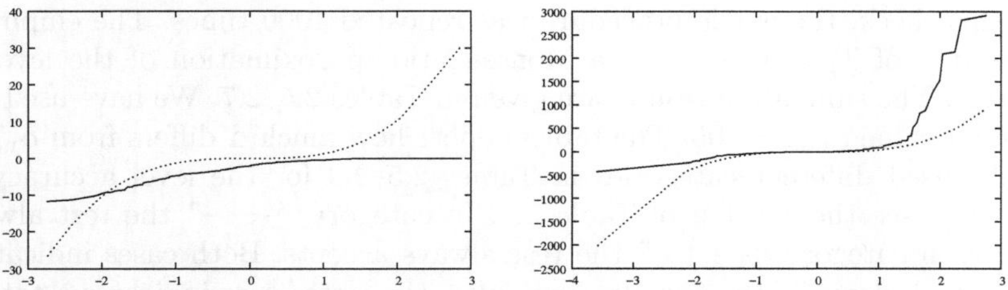

Fig. 2.13 The estimated (solid line) and true (dotted line) transformation functions for models M_5 and M_6.

The results are reported for embedding dimension $\nu = 4$. We have used different values of r for different models. The test statistics T_1, ..., T_4 and T_7 are statistics for checking time asymmetry. The tests T_5 and T_6 are based on joint higher order central moments and they test the nonlinearity of the dynamics. T_8 is the correlation sum. It is the sample analog of the correlation integral. Other physically meaningful measures (for example, the correlation dimension, the maximum Lyapunov exponent, etc.) have been proposed to check the nonlinear chaotic behavior of a data generating process. But there is no automatic implementation of these test statistics. Thus, it is difficult to incorporate these statistics in a simulation study. This is the reason why we have considered correlation sums which can be computed by an automatic scheme for different values of the threshold parameter r.

In the simulations we generated data $\mathbf{x_n} = (x_1 \ldots, x_n)$ from Models M_1–M_7 and we calculated the test statistics $T_j(\mathbf{x_n})$ for $j = 1, \ldots, 6$. We used sample size $n = 512$. For each simulated $\mathbf{x_n}$, 1000 surrogate resamples $\mathbf{x_n^*}$ were generated. For each of these resamples, we calculated test statistics $T_j(\mathbf{x_n^*})$, $j = 1, \ldots, 6$. The surrogate data test rejects the hypothesis of a linear stationary Gaussian process, if $T_j(\mathbf{x_n}) > k^*_{j\alpha}$, where $k^*_{j\alpha}$ denotes the $(1-\alpha)$-th quantile of $T_j(\mathbf{x_n^*})$. The first aim of our simulations is to check the level accuracy of this test, i.e., to check if the rejection probability on the hypothesis is approximately equal to the nominal level α_{nom}: $P[T_j(\mathbf{x_n}) > k_{j\alpha}] \approx \alpha_{\text{nom}}$.

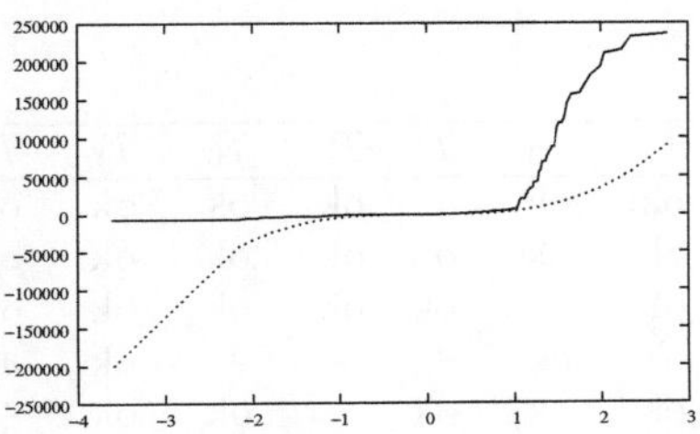

Fig. 2.14 The estimated (solid line) and true (dotted line) transformation function for M_7.

For this check, the whole procedure was repeated 1000 times. The empirical fraction $\hat{\alpha}$ of $T_j(\mathbf{x_n}) > k_{j\alpha}$ is a Monte-Carlo-approximation of the level of the test. The simulation results are given in Tables 2.5–2.7. We have used the nominal value $\alpha_{\text{nom}} = .05$. The tables report how much $\hat{\alpha}$ differs from α_{nom}.

We used different categories in Tables 2.5–2.9 for the level accuracy of the tests, see the caption of Table 2.5. In category "– – –" the test always rejects, in category "++++" the test always accepts. Both cases indicate a total break down of the test. In particular, the latter case indicates a totally erroneous performance of the test. Category "–" indicates that the test is too conservative. In particular, the test will have poor power for neighbored alternatives. We call level accuracies "–", "ok" and "+" reasonable. Categories "++" and "+++" again indicate a break down of the test. The test behaves like a blind test: it has a performance comparable to a random number that is produced without looking at the data.

Tables 2.5 and 2.6 summarize the results for the AAFT surrogate data tests. Table 2.5 gives the results when the test statistics have been applied to the untransformed data x_t (AAFT I) and Table 2.6 shows the performance when the test statistics have been calculated for the transformed data $\widehat{z}_t$ (AAFT II). Both tests work quite well for Models M_1–M_6. There is no big difference in the level accuracy of AAFT I and AAFT II. In Models M_1–M_6, AAFT I is too liberal ("++", "+++") in two cases and too conservative in one case ("–"). AAFT II has a slightly poorer performance: it is too liberal in one case and too conservative in six cases. Both procedures outperform IAAFT, see Table 2.7. In Models M_1–M_6, IAAFT is in five cases too liberal and in 9 cases too conservative. Furthermore, in 8 of these cases it totally breaks down: it always rejects or it always accepts. On the other hand, the AAFT I procedure never totally breaks down and the AAFT II procedure only in one case. This suggests that IAAFT is not stable. This may be caused by the iterative nature of the algorithm. The algorithm sometimes runs into

Table 2.5 *Level accuracy for AAFT surrogate data tests based on (x_t, x_t^*). The level accuracy is marked by "–" if the Monte-Carlo estimate $\hat{\alpha}$ of the level is 0.0, "–" if $0 < \hat{\alpha} \leq .015$, "–" if $.015 < \hat{\alpha} \leq .03$, "ok" if $.03 < \hat{\alpha} \leq .075$, "+" if $.075 < \hat{\alpha} \leq .125$, "++" if $.125 < \hat{\alpha} \leq .250$, "+++" if $.250 < \hat{\alpha} < 1$, and "++++" if $\hat{\alpha} = 1$. The nominal level is .05*

Model	T_1	T_2	T_3	T_4	T_5	T_6	T_7	T_8	r
M_1	ok	ok	ok	ok	ok	ok	ok	ok	.1
M_2	ok	ok	ok	ok	ok	ok	ok	ok	.2
M_3	ok	ok	ok	ok	ok	ok	ok	ok	.2
M_4	–	ok	ok	+	+	+	ok	+++	.1
M_5	–	ok	ok	ok	+	ok	ok	+	.1
M_6	–	ok	ok	+	ok	ok	ok	++	2.
M_7	ok	++	++	+	++	++	++	ok	.5

Table 2.6 Level accuracy for AAFT surrogate data tests based on $(\widehat{z}_t, z_t^*)$ with nominal level 0.05. The level accuracy is marked as in Table 2.5

Model	T_1	T_2	T_3	T_4	T_5	T_6	T_7	T_8	r
M_1	ok	ok	ok	ok	ok	ok	ok	ok	.1
M_2	ok	ok	ok	ok	ok	ok	ok	ok	.2
M_3	ok	ok	ok	ok	ok	ok	ok	ok	.2
M_4	ok	ok	ok	ok	−	−	ok	+	.1
M_5	ok	ok	ok	ok	−	−	ok	+	.2
M_6	++	ok	ok	ok	−	−	ok	+	.5
M_7	+++	++	++	+	−	−	++	+++	.2

a totally misleading generation of surrogate data. Even for i.i.d. data (Models M_1 and M_2) IAAFT breaks totally down, here in five out of 16 cases. This means that here IAAFT does not capture the transformation of the data: in case of large deviations from normality the basic idea of IAAFT to have the same power spectrum as the observed data forces the method to biased estimates of the transformation.

Both, AAFT I and AAFT II, work perfectly well for i.i.d. data and for Model M_3. This performance can be easily theoretically verified. If the underlying process $x_1, ..., x_n$ is an i.i.d. sequence then their ranks $(r_1, ..., r_n)$ have a uniform distribution on the set of all permutations of $(1, ..., n)$. Thus $\widehat{z}_1, ..., \widehat{z}_n$ is a random permutation of an i.i.d. sample of standard normal variables. Thus it is also an i.i.d. sample of standard normal variables, and in particular it is a circular stationary linear Gaussian process. It has been shown in [3] that the method of phase randomized surrogates has exact finite sample level for circular stationary processes, see also [13]. This immediately implies that AAFT I has exact finite sample level if the underlying process is i.i.d.. Furthermore, it also implies that AAFT II has exact finite sample level. This can be seen as follows. Consider first a deterministic sequence $u_1, ..., u_n$ and a test

Table 2.7 Level accuracy for IAAFT surrogate data tests with nominal level 0.05. The level accuracy is marked as in Table 2.5

Model	T_1	T_2	T_3	T_4	T_5	T_6	T_7	T_8
M_1	+	ok	ok	ok	ok	ok	−	−
M_2	−	−	−	−	ok	ok	+++	−
M_3	+	ok	ok	ok	ok	ok	−	ok
M_4	ok	ok	ok	ok	ok	ok	++++	ok
M_5	ok	ok	ok	−	ok	ok	++++	+++
M_6	ok	ok	ok	ok	−	−	−	++
M_7	+	++	++	ok	−	−	−	−

statistic T. Order $u_1, ..., u_n$ in the same order as $\hat{z}_1, ..., \hat{z}_n$ and as its surrogates $z_1^*, ..., z_n^*$. This gives two samples. Calculate now T for the two samples. This can interpreted as AAFT I applied to a modified test statistic. Thus it achieves exact finite sample level. The procedure would be exactly equal to AAFT II if $u_1, ..., u_n$ would be chosen as $x_1, ..., x_n$. This choice is nonrandom but the argument carries over because the order statistic $x_{(1)} \leq ... \leq x_{(n)}$ is independent of $\hat{z}_1, ..., \hat{z}_n$.

Model M_3 is a circular stationary model with short coherence time. Both AAFT procedures work well for this model. This does not extend to circular stationary model with long coherence times. This can be seen in the results for Models M_4 and M_5. In some cases, the AAFT procedures have very poor level accuracies for these models. In particular this shows that in this respect AAFT behaves differently as (phase randomized) surrogate data testing that has exact level for all circular stationary (Gaussian) processes. The additional estimation of the transformation h that is incorporated in AAFT can lead to level inaccuracies in case of long coherence times. The poor performance of this estimation was illustrated in the plots for Models M_4 and M_5 in Figs. 2.12 and 2.13.

All procedures, AAFT and IAAFT performed poorly in Model M_7. AAFT I and IAAFT achieved reasonable results in 3 out of 8 cases, AAFT II only in one case. M_7 is a near to unit root process. Reference [13] used simulations for the same process without transformation to illustrate a poor performance of phase randomized surrogate data tests for near to unit root processes. They considered two modifications of surrogate data testing. One was based on taking subsamples of the series with small gap between the first and last observation. The end to end mismatch correction leads to drastic improvements. Motivated by this result, we tried the same modification for both AAFT procedures for Model M_7. The results are reported in Tables 2.8 and 2.9. In Table 2.8 the mismatch correction procedure is applied to the observed series x_t, whereas in Table 2.9 it is applied to the transformed series $\hat{z}_t$. For these experiments, we used the sample size $N = 512$ and $K_1 = K_2 = 40$. (The starting point of the subseries is chosen among the first K_1 observations, the last point among the last K_2 observations, for details see [13]). There was no big difference if the method was applied two x_t or to $\hat{z}_t$. There was also a gain after

Table 2.8 Level accuracy for AAFT surrogate data tests with end to end correction. The end to end correction was applied to the untransformed data x_t. The nominal level is 0.05. The level accuracy is marked as in Table 2.5

	T_1	T_2	T_3	T_4	T_5	T_6	T_7	T_8	r
(x_t, x^*)	−−	+	+	ok	++	++	+	ok	.5
$(\hat{z}_t, z^*)$	+	+	+	+	−−−	−−−	ok	+++	.2

Table 2.9 Level accuracy for AAFT surrogate data tests with end to end correction. The end to end correction was applied to the transformed data $\widehat{z}_t$. The nominal level is 0.05. The level accuracy is marked as in Table 2.5

	T_1	T_2	T_3	T_4	T_5	T_6	T_7	T_8	r
(x_t, x^*)	−−	ok	+	ok	++	++	+	ok	.5
$(\widehat{z}_t, z^*)$	+	ok	+	+	−−−	−−	+	+++	.2

application of the end to end correction, but not quite so impressive as for surrogate data tests. Now both AAFT methods worked for 5 out of 8 cases.

We also checked the performance of AAFT for the test statistics T_1,...,T_8 on the following alternatives:

- A_1: Logistic map: $x_{t+1} = 4x_t(1 - x_t)$, $x_0 \in (0, 1)$.
- A_2: Reversal logistic map: $x_t = 0.5[1 + \varepsilon_t(1 - x_{t-1})^{\frac{1}{2}}]$ where ε_t is equal to 1 or -1 with probability $\frac{1}{2}$ and $x_0 \sim Beta(.5, .5)$.
- A_3: Tent map: $x_{t+1} = \begin{cases} 2x_t & \text{if} \quad 0 \le x_t \le .5 \\ 2(1 - x_t) & \text{if} \quad .5 \le x_t \le 1 \end{cases}$.
- A_4: Reversal tent map: $x_t = .5(1 - \varepsilon_t + \varepsilon_t x_{t-1})$ where ε_t is equal to 1 or -1 with probability $\frac{1}{2}$ and $x_0 \sim U[0, 1]$.

Here we have considered two pairs of models (A_1, A_2) and (A_3, A_4). A_1 and A_3 are two purely deterministic models and A_2 and A_4 are their stochastic counterparts. After time reversion the stochastic models A_2 and A_4 are identical to A_1 or A_3, respectively. For a discussion of stochastic processes with time reversal deterministic processes see [1, 9, 10, 21].

We performed similar simulations for A_i, $i = 1, ..., 4$ as for Models M_1,...,M_8. The results of the simulations are summarized in Tables 2.10 and 2.11. We see that both AAFT tests, the test based on (x_t, x_t^*) and the test based on $(\widehat{z}_t, z_t^*)$ have a quite similar performance. For the test statistics T_1–T_4, T_7 and T_8 they have nearly the same power. As above, they show a different performance for T_5 and T_6. The tests show quite different behavior. Three tests (T_3, T_7 and T_8) always have a power near to one. The tests based

Table 2.10 Estimated power of AAFT tests based on (x_t, x_t^*) with nominal level 0.05

Model	T_1	T_2	T_3	T_4	T_5	T_6	T_7	T_8	r
A_1	1.00	.000	1.00	.999	.093	.043	.983	1.00	.1
A_2	.000	1.00	1.00	.059	.082	.027	.998	1.00	.1
A_3	1.00	.000	1.00	1.00	.235	.073	1.00	1.00	.1
A_4	.000	1.00	1.00	.067	.145	.175	1.00	1.00	.1

Table 2.11 Estimated power of AAFT tests based on $(\hat{z}_t, z_t^*)$ with nominal level 0.05

Model	T_1	T_2	T_3	T_4	T_5	T_6	T_7	T_8	r
A_1	.994	.000	1.00	.999	.321	.693	.980	1.00	.1
A_2	.000	1.00	1.00	.054	.333	.718	1.00	1.00	.1
A_3	1.00	.000	1.00	1.00	.479	.664	1.00	1.00	.1
A_4	.000	1.00	1.00	.068	.368	.739	1.00	1.00	.1

on T_1 and T_4 have power near to one for the deterministic models A_1 and A_3 and have rejection probability less than the nominal level for the stochastic models A_2 and A_4. The test statistic T_2 behaves the other way around. It rejects always for A_2 and A_4 and it never rejects for A_1 and A_3. The test statistics T_5 and T_6 are not stable in their performances.

2.6.3 Comparison of AAFT and IAAFT – A Summary

In the previous subsection we investigated level accuracy and power characteristics of Amplitude Adjusted Fourier–Transformed (AAFT) and improved AAFT (IAAFT) algorithms. Both approaches are methods to get a statistical test for the hypothesis of a transformed stationary linear Gaussian processes. In both methods surrogate data are generated and tests are based on the comparison of test statistics evaluated with original observations and with surrogate data. In our study AAFT outperforms IAAFT in keeping the level on the hypothesis. AAFT works quite well as long as the coherence time is not too long. IAAFT is not stable in general. In many cases it totally breaks down: it always rejects or it always accepts on the hypothesis. In case of long coherence times the performance of AAFT can be improved by using end to end corrections. But the improvements are not so impressive as for phase randomized surrogate data.

2.7 Conclusions

This chapter contains a detailed description of the performance for a range of surrogate methods in a variety of settings. In a simulation study with a more general focus we showed that the performance strongly depends on the chosen combination of test statistic, resampling method and null hypothesis. For one test statistic, the new introduced time invariance test statistic, all surrogate methods lead to accurate levels. In a more detailed comparison of AAFT and IAAFT, both methods perform well as long as the coherence time of the process is not too large. In this case, AAFT has a more reliable and more accurate level.

References

[1] M.S. Bartlett. Chances or chaos. *Journal of Royal Statistical Society, Series A*, 153:321–347, 1990.

[2] P.J. Brockwell and R.A. Davis. *Time Series: Theory and Methods.* Springer, New York, 1987.

[3] K.S. Chan. On the validity of the method of surrogate data. *Fields Institute Communications*, 11:77–97, 1997.

[4] R. Dahlhaus and D. Janas. A frequency domain bootstrap for ratio statistics in time series. *Annals of Statistics*, 24:1934–1963, 1996.

[5] K. Dolan and M.L. Spano. Surrogates for nonlinear time series analysis. *Physical Review E*, 64, 2001.

[6] S. Elgar, R.T. Guza, and R.J. Seymour. Groups of waves in shallow water. *Journal of Geophysical Research*, 89:3623–3634, 1984.

[7] M.J. Feigenbaum. Quantitative universality in a class of nonlinear transitions. *Journal of Statistical Physics*, 19:25–52, 1978.

[8] P. Grassberger. Do climatic attractors exist? *Nature*, 323:609, 1986.

[9] A.J. Lawrance. Chaos: but not in both directions! *Statistics and Computing*, 11:213–216, 2001.

[10] A.J. Lawrance and N.M. Spencer. Curved chaotic map time series models and their stochastic reversals. *Scandinavian Journal of Statistics*, 25: 371–382, 1998.

[11] E.N. Lorenz. Deterministic aperiodic flow. *Journal of the Atmospheric Sciences*, 20:130, 1963.

[12] E. Mammen and S. Nandi. Bootstrap and resampling. In Y. Mori J. E. Gentle, W. Härdle, editor, *Handbook of Computational Statistics*, pages 467–496. Springer, New York, 2004.

[13] E. Mammen and S. Nandi. Change of the nature of a test when surrogate data are applied. *Physical Review E*, 70:16121–32, 2004.

[14] M.B. Priestley. *Spectral analysis and time series.* Academic Press, London, 1989.

[15] O.E. Roessler. An equation for continuous chaos. *Physical Letters A*, 57:397–381, 1976.

[16] T. Schreiber and A. Schmitz. Improved surrogate data for nonlinearity tests. *Physical Review Letters*, 77:635–638, 1996.

[17] T. Schreiber and A. Schmitz. Surrogate time series. *Physica D*, 142: 346–382, 2000.

[18] T. Subba Rao and M.M. Gabr. *An Introduction to Bispectral Analysis and Bilinear Time Series Models.* Springer, New York, 1984.

[19] J. Theiler, S. Eubank, A. Longtin, B. Galdrikian, and J.D. Farmer. Testing for nonlinearity in time series: the method of surrogate data. *Physica D*, 58:77–94, 1992.

[20] J. Timmer. What can be inferred from surrogate data testing? *Physical Review Letters*, 85:2647, 2000.

[21] H. Tong and B. Cheng. A note on the one-dimensional chaotic maps under time reversal. *Advances in Applied Probability*, 24:219–220, 1992.
[22] B. van der Pol. On oscillation-hysteresis in a simple triode generator. *Philosophical Magazine*, 43:700–719, 1922.

3

Multiscale Approximation

Stephan Dahlke[1], Peter Maass[2], Gerd Teschke[3], Karsten Koch[1], Dirk Lorenz[2], Stephan Müller[5], Stefan Schiffler[2], Andreas Stämpfli[5], Herbert Thiele[4], and Manuel Werner[1]

[1] FB 12, Mathematics and Computer Sciences, Philipps-University of Marburg, Hans-Meerwein-Strasse, Lahnberge, D-35032 Marburg, Germany
{dahlke,koch,werner}@mathematik.uni-marburg.de
[2] Center for Industrial Mathematics (ZeTeM), FB 3, University of Bremen, Postfach 330440, D-28334 Bremen, Germany
{pmaass,dlorenz,schiffi}@math.uni-bremen.de
[3] Zuse Institute Berlin (ZIB), Takustrasse 7, D-14195 Berlin-Dahlem, Germany
teschke@zib.de
[4] Bruker Daltonics GmbH, Fahrenheitstrasse 4, D-28359 Bremen, Germany
Herbert.Thiele@bdal.de
[5] F. Hoffmann-La Roche AG, Grenzacherstrasse 124, CH-4070 Basel, Switzerland
{stephan.mueller,andreas.staempfli}@roche.com

3.1 Introduction

During the last two decades wavelet methods have developed into powerful tools for a wide range of applications in signal and image processing. The success of wavelet methods is based on their potential for resolving local properties and to analyze non-stationary structures. This is achieved by multiscale decompositions, e.g., a signal or image is mapped to a phase space parametrized by a time/space- and a scale/size/resolution parameter. In this respect, wavelet methods offer an alternative to classical Fourier- or Gabor-transforms which create a phase space consisting of a time/space- frequency parametrization. Hence, wavelet methods are advantageous whenever local, non-stationary structures on different scales have to be analyzed.

The diversity of wavelet methods, however, requires a detailed mathematical analysis of the underlying physical or technical problem in order to take full advantage of wavelet methods. This first of all requires to choose an appropriate wavelet. The construction of wavelets with special properties is still a central problem in the field of wavelet and multiscale analysis. We will review a recently developed construction principle for multivariate multiwavelets in the next section. As a second step one needs to develop tools for analyzing the result of the wavelet decomposition. Recently, nonlinear wavelet methods (shrinkage techniques) have been developed for solving ill-posed operator equations and inverse problems in imaging. The success of wavelet methods in this field is a consequence of the following facts:

- Weighted sequence norms of wavelet expansion coefficients are equivalent in a certain range (depending on the regularity of the wavelets) to Sobolev and Besov norms.
- For a wide class of operators their representation in the wavelet basis is nearly diagonal.
- The vanishing moments of wavelets remove smooth parts of a function and give rise to very efficient compression strategies.

This will be demonstrated in a section on applications in signal and image processing, where we highlight the potential of wavelet methods for nonlinear image decomposition and deconvolution tasks. The numerical results include evaluations of real life data from MALDI/TOF mass spectroscopy from Bruker Daltonics GmbH, Bremen, and Hoffmann-La Roche AG, Basel.

3.2 Multiwavelets

In this section we present a construction principle for multivariate multiwavelets, which remedies some fundamental drawbacks of classical wavelet constructions.

The general setting can be described as follows. Let M be an integer $d \times d$ scaling matrix which is *expanding*, i.e., all its eigenvalues have modulus larger than one. If for a finite set $\mathcal{I}$ the system

$$\psi_{i,j,\beta}(x) := m^{j/2}\psi_i(M^j x - \beta), \quad i \in \mathcal{I},\ j \in \mathbb{Z},\ \beta \in \mathbb{Z}^d, \tag{3.1}$$

where $m = |\det M|$, is a basis of $L_2(\mathbb{R}^d)$, then $\{\psi_{i,j,\beta}\}_{i\in\mathcal{I},j\in\mathbb{Z},\beta\in\mathbb{Z}^d}$ is called a *wavelet basis.* Within this classical setting there are still some serious bottlenecks. It has turned out that some desirable properties cannot be achieved at the same time. For instance, it would be optimal to construct an *orthonormal* basis that is also *interpolating,* since orthonormality gives rise to very efficient decomposition and reconstruction algorithms, and the interpolation property yields a Shannon-like sampling theorem. However, it can be checked that for sufficiently smooth wavelets such a construction is impossible [42].

To overcome this problem, a more general approach that provides more flexibility is needed.

One way is to consider *multiwavelets,* i.e., a collection of function *vectors* $\Psi^{(n)} := \left(\psi_0^{(n)}, \ldots, \psi_{r-1}^{(n)}\right)^\top \in L_2(\mathbb{R}^d)^r$, $0 < n < m$, for which

$$\left\{\psi_0^{(n)}\left(M^j \cdot -\beta\right), \ldots, \psi_{r-1}^{(n)}\left(M^j \cdot -\beta\right) \,\middle|\, j \in \mathbb{Z}, \beta \in \mathbb{Z}^d, 0 < n < m\right\}$$

forms a (Riesz) basis of $L_2(\mathbb{R}^d)$. Compared to the classical scalar setting, this notion of wavelets is much more general.

This section is organized as follows. In Subsect. 3.2.1, we briefly recall the basic properties of scaling functions and wavelets as far as they are needed for our purposes. Then, in Subsect. 3.2.2, we describe a new construction method that enables us to construct orthogonal families of multiwavelets with sufficiently high smoothness which are in addition also interpolating. This is a very surprising result which clearly demonstrates the usefulness of the multiwavelet approach, since nothing similar can be done in the classical wavelet setting. Finally, in Subsect. 3.2.3, our approach is further extended by presenting construction principles for biorthogonal pairs of symmetric compactly supported interpolating scaling vectors with nice approximation and smoothness properties. The results in this subsection have been published in a series of papers [27, 28, 29, 30], we refer to these papers for further details.

3.2.1 General Setting

Refinable Function Vectors

Let $\Phi := (\phi_0, \ldots, \phi_{r-1})^\top$, $r > 0$, be a vector of $L_2(\mathbb{R}^d)$-functions which satisfies a *matrix refinement equation*

$$\Phi(x) = \sum_{\beta \in \mathbb{Z}^d} A_\beta \Phi(Mx - \beta), \quad A_\beta \in \mathbb{R}^{r \times r}, \tag{3.2}$$

with the *mask* $A := (A_\beta)_{\beta \in \mathbb{Z}^d}$, then Φ is called (A, M)*-refinable.*

We shall always assume that the mask has only a finite number of non-vanishing entries, $A \in \ell_0(\mathbb{Z}^d)^{r \times r}$, and these entries are denoted by

$$A_\beta = \begin{pmatrix} a_\beta^{(0,0)} & \cdots & a_\beta^{(0,r-1)} \\ \vdots & \ddots & \vdots \\ a_\beta^{(r-1,0)} & \cdots & a_\beta^{(r-1,r-1)} \end{pmatrix}. \tag{3.3}$$

Applying the Fourier transform component-wise to (3.2) yields

$$\widehat{\Phi}(\omega) = \frac{1}{m} \mathbf{A}(\mathrm{e}^{-\mathrm{i}M^{-\top}\omega}) \widehat{\Phi}(M^{-\top}\omega), \quad \omega \in \mathbb{R}^d, \tag{3.4}$$

where $\mathrm{e}^{-i\omega}$ is a shorthand notation for $(\mathrm{e}^{-i\omega_1}, \ldots, \mathrm{e}^{-i\omega_d})^\top$. The *symbol* $\mathbf{A}(z)$ is the matrix valued Laurent series with entries

$$a_{i,j}(z) := \sum_{\beta \in \mathbb{Z}^d} a_\beta^{(i,j)} z^\beta, \quad z \in {}^d,$$

and ${}^d := \{z \in \mathbb{C}^d : |z_i| = 1, i = 1, \ldots, d\}$ denotes the d-dimensional torus. All elements of ^{d} have the form $z = \mathrm{e}^{-i\omega}$, $\omega \in \mathbb{R}^d$, thus we have $z^\beta = \mathrm{e}^{-\mathrm{i}\langle \omega, \beta \rangle}$, and for $\xi \in \mathbb{R}^d$ we use the notation $z_\xi := \mathrm{e}^{-\mathrm{i}(\omega+2\pi\xi)}$. In addition, we define $z^M := \mathrm{e}^{-\mathrm{i}M^\top\omega}$ such that $(z^M)^\beta = z^{M\beta}$ and $z_\xi^M := \mathrm{e}^{-\mathrm{i}M^\top(\omega+2\pi\xi)}$.

One central aim is the construction of families of *interpolating* m-scaling vectors Φ with compact support, i.e., all components of Φ are at least continuous and satisfy

$$\phi_n\left(M^{-1}\beta\right) = \delta_{\rho_n,\beta} \quad \text{for all} \quad \beta \in \mathbb{Z}^d,\, 0 \le n < m\,, \tag{3.5}$$

where $R := \{\rho_0, \ldots, \rho_{m-1}\}$ denotes a complete set of representatives of $\mathbb{Z}^d/M\mathbb{Z}^d$. Note that the interpolation condition (and the length of the scaling vector) is determined by the determinant of the scaling matrix. One advantage of interpolating scaling vectors is that they give rise to a Shannon-like sampling theorem as follows. For a compactly supported function vector $\Phi \in L_2(\mathbb{R}^d)^m$, let us define the shift-invariant space

$$S(\Phi) := \Big\{ \sum_{\beta\in\mathbb{Z}^d} u_\beta \Phi(\cdot - \beta) \,\big|\, u \in \ell(\mathbb{Z}^d)^{1\times m} \Big\}.$$

A direct computation shows that, if Φ is a compactly supported interpolating m-scaling vector, then for all $f \in S(\Phi)$ the representation

$$f(x) = \sum_{\beta\in\mathbb{Z}^d} \sum_{i=0}^{m-1} f\left(\beta + M^{-1}\rho_i\right) \phi_i(x-\beta) \tag{3.6}$$

holds. The interpolation requirement is quite strong and implies the following necessary condition on the mask.

Lemma 3.1 *Let $\rho_k \in M\mathbb{Z}^d$, then the mask of an interpolating m-scaling vector has to satisfy*

$$a^{(i,k)}_{M\alpha+\rho_j-M^{-1}\rho_k} = \delta_{0,\alpha}\delta_{i,j} \quad \textit{for all } \alpha \in \mathbb{Z}^d,\, 0 \le i,j < m\,.$$

For simplicity of notation, we shall assume $\rho_0 = 0 \in \mathbb{Z}^d$ without loss of generality. Then the above lemma implies that the symbol of an interpolating m-scaling vector has to have the form

$$\mathbf{A}(z) = \begin{pmatrix} z^{\rho_0} & a^{(0,1)}(z) & \cdots & a^{(0,m-1)}(z) \\ \vdots & \vdots & \ddots & \vdots \\ z^{\rho_{m-1}} & a^{(m-1,1)}(z) & \cdots & a^{(m-1,m-1)}(z) \end{pmatrix}. \tag{3.7}$$

For the case $m = 2$ we can choose $R = \{0, \rho\}$ and obtain

$$\mathbf{A}(z) = \begin{pmatrix} 1 & a^{(0)}(z) \\ z^\rho & a^{(1)}(z) \end{pmatrix}. \tag{3.8}$$

Multiwavelets

Next we want to briefly explain how a multiwavelet basis can be constructed, provided that a suitable (interpolating) refinable function vector is given. As in the classical setting, a multiwavelet basis can be constructed by means of a *multiresolution analysis* which is a sequence $(V_j)_{j\in\mathbb{Z}}$ of closed subspaces of $L_2(\mathbb{R}^d)$ which satisfies:

(MRA1) $V_j \subset V_{j+1}$ for each $j \in \mathbb{Z}$,

(MRA2) $g(x) \in V_j$ if and only if $g(Mx) \in V_{j+1}$ for each $j \in \mathbb{Z}$,

(MRA3) $\bigcap_{j\in\mathbb{Z}} V_j = \{0\}$,

(MRA4) $\bigcup_{j\in\mathbb{Z}} V_j$ is dense in $L_2(\mathbb{R}^d)$, and

(MRA5) there exists a vector $\Phi \in L_2(\mathbb{R}^d)^r$, called the *generator*, such that

$$V_0 = \overline{\operatorname{span}\{\phi_i(x-\beta) \mid \beta \in \mathbb{Z}^d, 0 \le i < r\}}.$$

Let W_0 denote an algebraic complement of V_0 in V_1 and define $W_j := \{g(M^j\cdot) \mid g \in W_0\}$. Then, one immediately obtains that $V_{j+1} = V_j \oplus W_j$ and consequently, due to (MRA3) and (MRA4), $L_2(\mathbb{R}^d) = \bigoplus_{j\in\mathbb{Z}^d} W_j$. If one finds function vectors $\Psi^{(n)} \in L_2(\mathbb{R}^d)^r$, $0 < n < m$, such that the integer translates of the components of all $\Psi^{(n)}$ are a basis of W_0, then, by dilation, one obtains a multiwavelet basis of $L_2(\mathbb{R}^d)$. Since $W_0 \subset V_1$, each $\Psi^{(n)}$ can be represented as

$$\Psi^{(n)}(x) = \sum_{\beta\in\mathbb{Z}^d} B_\beta^{(n)} \Phi(Mx-\beta) \tag{3.9}$$

for some $B^{(n)} \in \ell(\mathbb{Z}^d)^{r\times r}$. By applying the Fourier transform component-wise to (3.9), one obtains

$$\widehat{\Psi}^{(n)}(\omega) = \frac{1}{m}\mathbf{B}^{(n)}(\mathrm{e}^{-\mathrm{i}M^{-\top}\omega})\widehat{\Phi}(M^{-\top}\omega), \quad \omega \in \mathbb{R}^d,$$

with the symbol

$$\mathbf{B}^{(n)}(z) := \sum_{\beta\in\mathbb{Z}^d} B_\beta^{(n)} z^\beta, \quad z \in {}^d.$$

Therefore, the task of finding a stable multiwavelet basis can be reduced to constructing the symbols $\mathbf{B}^{(n)}(z)$.

Consequently, to obtain some multiwavelets, we first have to find a way to construct a suitable MRA. Under mild conditions, (MRA1) and (MRA2) imply that the function vector Φ in (MRA5) satisfies a refinement equation of the form (3.2). Therefore, refinable function vectors are the natural candidates for generators. Fortunately, it can be shown that any compactly supported interpolating scaling vector indeed generates an MRA, therefore, the whole construction problem is reduced to finding suitable interpolating scaling vectors. For further information, the reader is referred to [14, 33, 41].

Approximation Order

In order to obtain efficient numerical algorithms, the power of the MRA to approximate (sufficiently smooth) functions is essential. For a compactly supported function vector $\Phi \in L_2(\mathbb{R}^d)^r$ and $h > 0$, let

$$S_h(\Phi) := \left\{ f\left(\frac{\cdot}{h}\right) \mid f \in S(\Phi) \cap L_2(\mathbb{R}^d) \right\}$$

denote the space of all h-dilates of $S(\Phi) \cap L_2(\mathbb{R}^d)$. Φ (or $S(\Phi)$) is said to provide *approximation order* $k > 0$ if the Jackson-type inequality

$$\inf_{g \in S_h(\Phi)} \|f - g\|_{L_2} = \mathcal{O}(h^k), \quad \text{as } h \to 0,$$

holds for all f contained in the Sobolev space $H^k(\mathbb{R}^d)$. The approximation properties of a scaling vector are closely related to its ability to reproduce polynomials. A function vector $\Phi : \mathbb{R}^d \longrightarrow \mathbb{C}^r$ with compact support is said to provide *accuracy order* $k + 1$, if $\pi_k^d \subset S(\Phi)$, where π_k^d denotes the space of all polynomials of total degree less or equal than k in $\mathbb{R}^d$. It was shown by Jia, see [23], that if a compactly supported scaling vector Φ has linear independent integer translates, then the order of accuracy is equivalent to the approximation order provided by Φ.

A mask $A \in \ell_0(\mathbb{Z}^d)^{r \times r}$ of an r-scaling vector with respect to a scaling matrix M satisfies the *sum rules of order* k if there exists a set of vectors $\{y_\mu \in \mathbb{R}^r \mid \mu \in \mathbb{Z}_+^d, |\mu| < k\}$ with $y_0 \neq 0$ such that for some uniquely determined numbers $w(\mu, \nu)$

$$\sum_{0 \le \nu \le \mu} (-1)^{|\nu|} \left(\sum_{\beta \in \mathbb{Z}^d} \frac{(M^{-1}\rho + \beta)^\nu}{\nu!} A_{\rho + M\beta}^\top \right) y_{\mu - \nu} = \sum_{|\nu| = |\mu|} w(\mu, \nu) y_\nu \tag{3.10}$$

holds for all $\mu \in \mathbb{Z}_+^d$ with $|\mu| < k$ and all $\rho \in R$. It was proven in [8, 24] that if the mask of a compactly supported scaling vector Φ satisfies the sum rules of order k, then Φ provides accuracy of order k.

3.2.2 Multivariate Orthonormal Interpolating Scaling Vectors

In this subsection we want to derive a construction method to obtain a multiwavelet basis $\Psi^{(n)}$ which is orthogonal and interpolating. As already outlined above, the whole construction can be reduced to the task of finding a suitable refinable scaling vector. In particular, we will focus on scaling matrices with $|\det(M)| = 2$. Then, in the interpolating setting, we obtain $r = m = 2$, and since the number of multiwavelets is determined by m, cf. Subsect. 3.2.1, this approach enables us to construct a basis of $L_2(\mathbb{R}^d)$ generated by one single mother multiwavelet consisting of two functions only. Therefore, the final goal in this section is to construct a multiwavelet basis that satisfies

$$\langle \psi_i, m^{j/2}\psi_{i'}(M^j \cdot -\beta)\rangle = c\delta_{i,i'}\delta_{0,j}\delta_{0,\beta}, \quad i,i'=0,1,\ j\in\mathbb{Z},\ \beta\in\mathbb{Z}^d\,, \tag{3.11}$$

with a constant $c > 0$, as well as $\Psi\left(M^{-1}\beta\right) = \begin{pmatrix}\delta_{0,\beta}\\ \delta_{\rho,\beta}\end{pmatrix}$ for all $\beta \in \mathbb{Z}^d$ and $R = \{0,\rho\}$.

Main Ingredients

As already explained, the starting point is an interpolating scaling vector whose integer translates of all component functions are mutually orthogonal, i.e.,

$$\langle \phi_i, \phi_j(\cdot - \beta)\rangle = \hat{c}\delta_{i,j}\delta_{0,\beta}\,, \quad i,j = 0,1,\ \beta\in\mathbb{Z}^d\,, \tag{3.12}$$

with a constant $\hat{c} > 0$. By using Fourier transform, it can be checked that the symbol $\mathbf{A}(z)$ of an orthonormal scaling vector has to satisfy

$$\sum_{\tilde{\rho}\in\widetilde{R}} \mathbf{A}\left(z_{M^{-\top}\tilde{\rho}}\right)\overline{\mathbf{A}\left(z_{M^{-\top}\tilde{\rho}}\right)}^{\top} = m^2\,\mathbf{I}_r\,, \tag{3.13}$$

where $\tilde{R}$ denotes a complete set of representatives of $\mathbb{Z}^d/M^\top\mathbb{Z}^d$. For the special case of an interpolating 2-scaling vector with compact support we obtain the following simplified conditions.

Theorem 3.2 *Let* $\mathbf{A}(z)$ *be the symbol of an interpolating* 2*-scaling vector with mask* $A\in\ell_0(\mathbb{Z}^d)^{2\times 2}$. $\mathbf{A}(z)$ *satisfies* (3.13) *if and only if the symbol entries* $a^{(0)}(z)$ *and* $a^{(1)}(z)$ *in* (3.8) *satisfy*

$$\left|a^{(0)}(z)\right|^2 + \left|a^{(0)}\left(z_{M^{-\top}\tilde{\rho}}\right)\right|^2 = 2 \tag{3.14}$$

and

$$a^{(1)}(z) = \pm z^{\alpha}\overline{a^{(0)}(z_{M^{-\top}\tilde{\rho}})} \tag{3.15}$$

for some $\alpha\in[\rho]$, *where* $[\rho]$ *denotes the coset of* ρ, *and with* $\widetilde{R} = \{0,\tilde{\rho}\}$.

So any construction of orthonormal and interpolating scaling functions has to start with the necessary conditions stated in Theorem 3.2. In order to obtain a useful result, also the approximation order, i.e., the sum rules have to be taken into account. For an orthonormal interpolating scaling vector with $m = 2$ we obtain the following simplification.

Theorem 3.3 *If we choose*

$$a^{(1)}(z) = z^{\rho}\sum_{\beta\in\mathbb{Z}^d}(-1)^{[\rho](\beta)}a_\beta z^{-\beta}\,, \quad a^{(0)}(z) = \sum_{\beta\in\mathbb{Z}^d} a_\beta z^\beta$$

in (3.15)*, then for an orthonormal interpolating* 2*-scaling vector the sum rules of order* k *are reduced to*

$$(M^{-2}\rho)^{\mu} = \sum_{\beta\in\mathbb{Z}^d} a_\beta \left(-M^{-1}\beta\right)^{\mu} ,$$

$$(M^{-2}\rho)^{\mu} = \sum_{\beta\in\mathbb{Z}^d} a_\beta \left(M^{-1}\beta\right)^{\mu} (-1)^{[\rho](\beta)}$$

for all $\mu \in \mathbb{Z}_+^d$, $|\mu| < k$, *with* $R = \{0, \rho\}$.

Explicit Construction

Based on the results in the previous subsection, we suggest the following construction principle:

1. Choose a scaling matrix M with $|\det(M)| = 2$ and the nontrivial representative ρ of $\mathbb{Z}^d/M\mathbb{Z}^d$ such that $R = \{0, \rho\}$.
2. Start with the first symbol entry $a^{(0)}(z) = \sum_{\beta\in\Lambda} a_\beta z^\beta$ by choosing the support $\Lambda \subset \mathbb{Z}^d$ of $(a_\beta)_{\beta\in\Lambda}$. Centering the coefficients around a_0 provides the best results, therefore, we suggest the choice of $\Lambda = [-n, n]^d \cap \mathbb{Z}^d$.
3. According to Theorem 3.2 the second symbol entry $a^{(1)}(z)$ has to have the form $a^{(1)}(z) = \pm z^\alpha \overline{a^{(0)}(z_{M^{-\top}\tilde{\rho}})}$ with $\alpha \in [\rho]$. Based on our observations we suggest to choose $\alpha = \rho$ and a positive sign, since this seems to provide the highest regularity and the smallest support.
4. Apply the orthogonality condition (3.14) to the coefficient sequence $(a_\beta)_{\beta\in\Lambda}$. This will consume about one half of the degrees of freedom.
5. Finally, apply the sum rules of Theorem 3.3 up to the highest possible order to the coefficient sequence $(a_\beta)_{\beta\in\Lambda}$.

Starting with an index set $\Lambda = \{-n, \dots, n\}^2$, we obtain a sequence of scaling vectors denoted by Φ_n with increasing accuracy order and regularity. For the special case of the *quincunx matrix* M_q, defined by

$$M_q := \begin{pmatrix} 1 & -1 \\ 1 & 1 \end{pmatrix} ,$$

let us denote the resulting scaling vector by Φ_n^q. In Table 3.1 the properties of the constructed examples are shown. Note that for $n \geq 2$ all our solutions have critical Sobolev exponents strictly larger than one. Therefore, by the Sobolev embedding theorem, see [1], all these scaling vectors are at least continuous.

For $n = 5$ we obtain an example that is continuously differentiable. The corresponding functions are shown in Fig. 3.1. The reader should note that these scaling vectors are very well localized.

Table 3.1 Accuracy order and critical Sobolev exponent $\mathfrak{s}$ of the Φ_n^q

n	Accuracy order	$\mathfrak{s}(\Phi_n^q)$
0	1	0.238
1	1	0.743
2	2	1.355
3	3	1.699
4	3	1.819
5	4	2.002

Multiwavelets

Once a suitable scaling vector is found, the construction of an associated multiwavelet basis is easy.

Theorem 3.4 *Let* $\mathbf{A}(z)$ *be the symbol of a compactly supported orthonormal interpolating* 2*-scaling vector* Φ*. Furthermore, let the function vector* Ψ *be defined by* (3.9)*, where* $\mathbf{B}(z)$ *is given by*

$$\mathbf{B}(z) = \begin{pmatrix} 1 & -a^{(0)}(z) \\ z^{\rho} & -a^{(1)}(z) \end{pmatrix} \tag{3.16}$$

with $a^{(0)}(z)$ *and* $a^{(1)}(z)$ *as in* (3.8)*. Then,* $\sqrt{2}\Psi$ *gives rise to an orthonormal multiwavelet basis, and* Ψ *is also interpolating.*

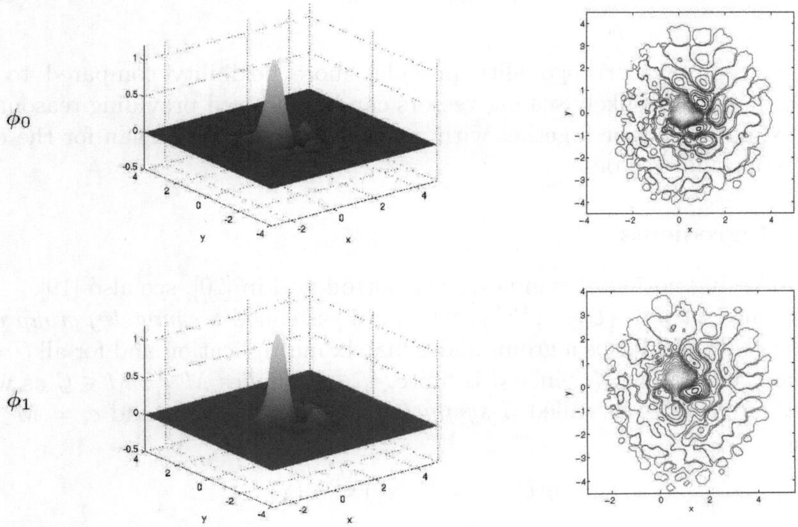

Fig. 3.1 Component functions of Φ_5^q

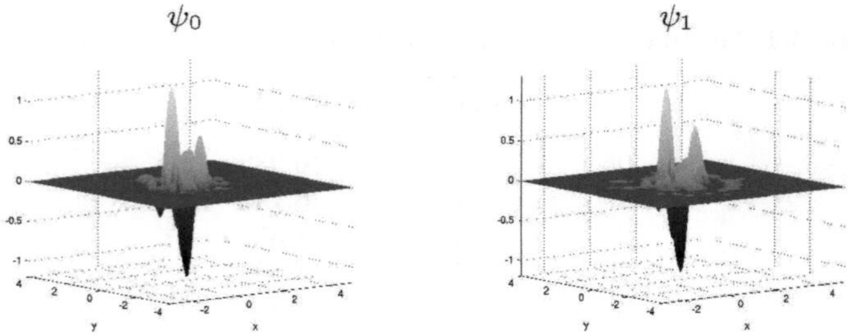

Fig. 3.2 Multiwavelets corresponding to Φ_5^q

One example of such an interpolating multiwavelet corresponding to our scaling vector Φ_5^q is shown in Fig. 3.2.

3.2.3 Multivariate Symmetric Interpolating Scaling Vectors

In this subsection our aim is to extend the approach obtained in the previous subsection by incorporating an additional property, namely symmetry. Since symmetry is hard to achieve in the orthonormal setting, we focus on the construction of a pair of *biorthogonal* scaling vectors $(\Phi, \widetilde{\Phi})$, i.e., $\left\langle \phi_i, \widetilde{\phi}_j(\cdot - \beta) \right\rangle = c \cdot \delta_{i,j}\delta_{0,\beta}, \quad 0 \le i, j < r$, holds for all $\beta \in \mathbb{Z}^d$ and a constant $c > 0$, the interpolation property being solely satisfied by the primal scaling vectors. A necessary condition for Φ and $\widetilde{\Phi}$ to be biorthogonal is that their symbols $\mathbf{A}(z)$ and $\widetilde{\mathbf{A}}(z)$ satisfy

$$\sum_{\tilde{\rho} \in \widetilde{R}} \mathbf{A}(z_{M^{-\top}\tilde{\rho}}) \overline{\widetilde{\mathbf{A}}(z_{M^{-\top}\tilde{\rho}})}^{\top} = m^2 \, \mathbf{I}_m \,. \tag{3.17}$$

The concept of biorthogonality provides more flexibility compared to orthonormality, and likely scaling vectors can be obtained providing reasonable approximation power together with much smaller supports than for the case of orthonormal vectors.

Main Ingredients

The following notion of symmetry was introduced in [20], see also [19].

A finite set $\mathcal{G} \subset \left\{ U \in \mathbb{Z}^{d \times d} \,\middle|\, |\det U| = 1 \right\}$ is called a *symmetry group with respect to* M if $\mathcal{G}$ forms a group under matrix multiplication and for all $U \in \mathcal{G}$ we have $MUM^{-1} \in \mathcal{G}$. Since $\mathcal{G}$ is finite, $U \in \mathcal{G}$ implies $M^{-1}UM \in \mathcal{G}$ as well. A scaling vector Φ is called $\mathcal{G}$*-symmetric*, if for $0 \le i < m$ and $c_i = M^{-1}\rho_i$ holds

$$\phi_i(U(x - c_i) + c_i) = \phi_i(x) \,.$$

Since all elements of $\mathcal{G}$ are integer matrices, the notion of symmetry can be used for sequences as well. In particular it holds

Proposition 3.5 *Let $\mathcal{G}$ be a symmetry group with respect to M and let Φ be a $\mathcal{G}$-symmetric interpolating m-scaling vector with mask $A \in \ell_0(\mathbb{Z}^d)^{r\times r}$. If $U[\rho] = [\rho]$ holds for all $\rho \in R$ and for all $U \in \mathcal{G}$, then the mask entries $(a_\beta^{(i,j)})_{\beta\in\mathbb{Z}^d}$ are $\mathcal{G}$-symmetric with centers $\rho_i - M^{-1}\rho_j =: c(i,j)$, meaning that*

$$a_\beta^{(i,j)} = a_{U(\beta-c(i,j))+c(i,j)}^{(i,j)},$$

for all $\beta \in \mathbb{Z}^d$ and all $U \in \mathcal{G}$, $0 \le i,j < m$.

Proposition 3.5 shows that a symmetric interpolating scaling vector is completely determined by a small part of its mask.

Furthermore, we are able to give the following decomposition of the supports of our masks.

Proposition 3.6 *Let $a^{(i,j)} \in \ell_0(\mathbb{Z}^d)$, $0 \le i,j < m$, be $\mathcal{G}$-symmetric with centers $c(i,j) := \rho_i - M^{-1}\rho_j$. Then there exist finite sets $\Omega_j \subset \mathbb{Z}^d$ such that $\Omega_j + M^{-1}\rho_j$ is $\mathcal{G}$-symmetric (i.e., $U(\Omega_j + M^{-1}\rho_j) \subset \Omega_j + M^{-1}\rho_j$ for all $U \in \mathcal{G}$), and $\operatorname{supp}(a^{(i,j)}) \subset \Omega_j + \rho_i$. Furthermore, there exist sets $\Lambda_j \subset \Omega_j$ such that we have the disjoint decomposition*

$$\Omega_j + \rho_i = \bigcup_{\beta\in\Lambda_j+\rho_i} \bigcup_{U\in G_{\beta-c(i,j)}} \{U(\beta - c(i,j)) + c(i,j)\}. \qquad (3.18)$$

Similar to Theorem 3.2, using that Φ is intended to be interpolating, the biorthogonality condition can be considerably simplified.

Proposition 3.7 *Let $(\Phi, \widetilde{\Phi})$ be a pair of dual m-scaling vectors with masks $(A_\beta), (\widetilde{A}_\beta) \in \ell_0(\mathbb{Z}^d)^{m\times m}$. If Φ is interpolating, then the biorthogonality condition* (3.17) *holds if and only if*

$$\widetilde{a}_{\rho_i - M\alpha}^{(j,0)} + \sum_{n=1}^{m-1} \sum_{\beta\in\mathbb{Z}^d} a_\beta^{(i,n)} \widetilde{a}_{\beta-M\alpha}^{(j,n)} = m \cdot \delta_{0,\alpha}\delta_{i,j}, \quad 0 \le i,j < m, \qquad (3.19)$$

holds for all $\alpha \in \mathbb{Z}^d$.

Thus, given the mask of a primal interpolating scaling vector, the biorthogonality condition leads to simple linear conditions on the dual mask.

Similar to the orthonormal case, before we can incorporate the sum rules (3.10) into our construction, the vectors y_μ have to be determined, see [30] for details on how this can be realized.

Explicit Construction

In this section we give an explicit construction method for the masks of symmetric interpolating scaling vectors on $\mathbb{R}^d$ with compact support, as well as for the masks of the dual scaling vectors which are also symmetric and compactly supported. We start with the primal side:

1. Choose the scaling matrix M and a complete set of representatives $R = \{0, \rho_1, \ldots, \rho_{m-1}\}$ of $\mathbb{Z}^d/M\mathbb{Z}^d$. Choose an appropriate symmetry group $\mathcal{G}$.
2. To determine the support of the mask $A \in \ell_0(\mathbb{Z}^d)^{m\times m}$, choose the sets Ω_j in Proposition 3.6 for $1 \leq j < m$ and compute some minimal generating sets $\Lambda_j \subset \Omega_j$. Thus, we start with $m \cdot \sum_{j=1}^{m-1} |\Lambda_j|$ degrees of freedom.
3. Apply a proper sum rule order k (i.e., as high as possible) taking into account the symmetry conditions in Proposition 3.5.
4. Find the best solution.

If the sets Λ_j are not too large, we have to deal with a moderate number of linear equations only. Hence, the system in step 3 can be solved analytically. In general, this system is under-determined and thus, as step 4 of our scheme, we can use these remaining degrees of freedom to maximize the regularity of the corresponding scaling vector Φ.

Given the mask of a symmetric interpolating scaling vector, the mask of a dual scaling vector can be obtained as follows:

1. For $0 \leq i < m$ choose the symmetry center c_i of $\widetilde{\phi}_i$. Due to the biorthogonality of Φ and $\widetilde{\Phi}$, the choice $c_i = M^{-1}\rho_i$ suggests itself.
2. Determine the support of $\widetilde{A} \in \ell_0(\mathbb{Z}^d)^{m\times m}$ by choosing $\widetilde{\Omega}_j$, $0 \leq j < m$, and compute some minimal generating sets $\widetilde{\Lambda}_j \subset \widetilde{\Omega}_j$ corresponding to Proposition 3.6. Thus, we have $m \cdot \sum_{j=0}^{m-1} |\widetilde{\Lambda}_j|$ degrees of freedom.
3. Apply the biorthogonality condition (3.19) to the coefficient sequence $(\widetilde{A}_\beta)_{\beta\in\mathbb{Z}^d}$ with respect to the symmetry conditions on the dual mask.
4. Choose a proper sum rule order $\widetilde{k}$ and compute the vectors $\widetilde{y}_\mu$, $|\mu| < \widetilde{k}$.
5. Apply the sum rules of order $\widetilde{k}$ to the coefficient sequence $\widetilde{A}$ with respect to the symmetry conditions on the dual mask.
6. Proceed analogously to step 4 for the primal vectors.

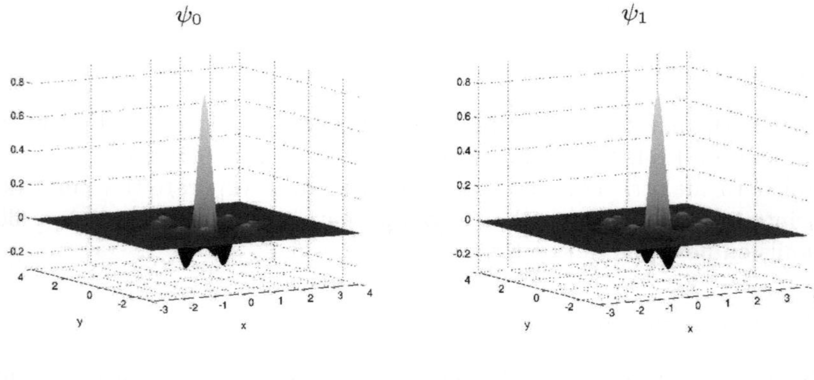

Fig. 3.3 Primal multiwavelets corresponding to M_q with six vanishing moments and critical Sobolev index $\mathfrak{s} = 3.664$

In [30] it has been shown that for a pair of compactly supported biorthogonal r-scaling vectors biorthogonal multiwavelets $\Psi^{(n)}$ and $\tilde{\Psi}^{(n)}$, $1 \leq n < m$, can be obtained by solving appropriate matrix extension problems. For instance, for the quincunx matrix M_q and the symmetry group

$$\mathcal{G} := \left\{ \pm\mathbf{I}, \pm\begin{pmatrix} 0 & -1 \\ 1 & 0 \end{pmatrix}, \pm\begin{pmatrix} 1 & 0 \\ 0 & -1 \end{pmatrix} \right\},$$

we obtain the primal multiwavelets shown in Fig. 3.3.

3.3 Multiscale Approximation in Image Processing

In this section, we discuss some aspects of multiscale approximation in the context of signal and image processing. Essentially, we are focusing on problems where it is reasonable to assume that the solution has a sparse expansion with respect to a wavelet basis. The starting point is always a wavelet-based variational formulation of the underlying signal or image restoration problem, which incorporates a Besov-penalty for ensuring a sparse approximation. The main ingredient to solve the regularized variational problem is therefore the norm equivalence between Besov norms and sequence norms for the orthonormal wavelet decomposition:

$$\|f\|_{H^s(\mathbb{R})} < \infty \Leftrightarrow \sum_k |\langle f, \varphi_{J,k}\rangle|^2 + \sum_{j\geq J}\sum_k 2^{2js}\, |\langle f, \psi_{j,k}\rangle|^2 < \infty \quad (3.20)$$

$$\begin{aligned}\|f\|_{B^s_p(L_p(\mathbb{R}))} < \infty \Leftrightarrow 2^{J(1/2-1/p)} &\sum_k |\langle f, \varphi_{J,k}\rangle|^p \\ &+ \sum_{j\geq J}\sum_k 2^{pjs} 2^{j(\frac{p}{2}-1)}\, |\langle f, \psi_{jk}\rangle|^p < \infty \end{aligned} \quad (3.21)$$

see [9]. By using tensor products, an analogous result can also be derived for the multivariate case, see the appendix in [13]. A first result of this type was presented in [17], where the definition of an appropriate surrogate functional led to an iterated soft shrinkage procedure. The shrinkage is due to the ℓ_p-penalization term in the variational formulation and leads to sparse signal representation. The importance of sparse representations for various tasks in image processing such as compression, denoising, deblurring and texture analysis has been highlighted in various papers, which also led to substantial generalizations for solving this type of variational problems in image processing, [2, 3, 38]. In this section emphasis is placed on the special case of simultaneously denoising, decomposing and deblurring as well as some particular deconvolution tasks for peak-like objects. We also discuss the potential of these methods for analyzing real life data from 1 D and 2 D applications in mass spectroscopy.

3.3.1 Simultaneous Decomposition, Deblurring and Denoising of Images by Means of Wavelets

We follow approaches presented by Vese–Osher and Osher–Solé–Vese, see [35, 37] and discuss a special class of variational functionals that induce a decomposition of images into oscillating and cartoon components and possibly an appropriate "noise" component; the cartoon part is, ideally, piecewise smooth with possibly abrupt edges and contours; the texture part on the other hand "fills" in the smooth regions in the cartoon with, typically, oscillating features. Osher, Solé and Vese propose to model the cartoon component by the space BV; this induces a penalty term that allows edges and contours in the reconstructed cartoon images, leading however to a numerically intensive PDE based scheme.

Our hope is to provide a computationally thriftier algorithm by using a wavelet-based scheme that solves not the same but a very similar variational problem, in which the BV-constraint is replaced by a $B^1_1(L_1)$-term. This relies on the fact that elementary methods based on wavelet shrinkage solve similar extremal problems where $BV(\Omega)$ is replaced by the Besov space $B^1_1(L_1(\Omega))$. Since $BV(\Omega)$ can not be simply described in terms of wavelet coefficients, it is not clear that $BV(\Omega)$ minimizers can be obtained in this way. Yet, it is shown in [10], exploiting $B^1_1(L_1(\Omega)) \subset BV(\Omega) \subset B^1_1(L_1(\Omega)) - \mathit{weak}$, that methods using Haar systems provide near $BV(\Omega)$ minimizers. So far there exists no similar result for general (in particular smoother) wavelet systems.

We shall nevertheless use wavelets that have more smoothness/vanishing moments than Haar wavelets, because we expect them to be better suited to the modeling of the smooth parts in the cartoon image. Though we may not obtain provable "near-best-BV-minimizers", we hope to nevertheless be "not far off". This approach allows us, moreover, to incorporate bounded linear blur operators into the problem so that the minimization leads to a simultaneous decomposition, deblurring and denoising.

Wavelet-Based Variational Formulation and Iterative Strategy for Image Decomposition

The basic idea of the variational formulation of the decomposition problem goes back to the famous total variation framework of Rudin et al. [39] and was improved in a series of papers, see e.g., [34, 35, 36, 37], finally amounting to the following minimization problem

$$\begin{gathered}\inf_{u,v} G_2(u,v)\,, \quad \text{where} \\ G_2(u,v) = \int_\Omega |\nabla u| + \lambda \|f-(u+v)\|^2_{L_2(\Omega)} + \mu \|v\|^2_{H^{-1}(\Omega)}\,,\end{gathered} \tag{3.22}$$

where u stands for the cartoon part and v for the oscillatory part of a given image f. In general, one drawback is that the minimization of (3.22) leads to numerically intensive schemes. Instead of solving problem (3.22) by means of finite difference schemes, we propose a wavelet-based treatment by replacing $BV(\Omega)$ by the Besov space $B_1^1(L_1(\Omega))$. Incorporating, moreover, a bounded linear operator K, we end up with the following variational problem:

$$\begin{aligned} &\inf_{u,v} \mathcal{F}_f(v,u)\,, \quad \text{where} \\ &\mathcal{F}_f(v,u) = \|f - K(u+v)\|^2_{L_2(\Omega)} + \gamma\|v\|^2_{H^{-1}(\Omega)} + 2\alpha|u|_{B_1^1(L_1(\Omega))}\,. \end{aligned} \tag{3.23}$$

At first, we may observe the following

Lemma 3.8 *If the null-space $\mathcal{N}(K)$ of the operator K is trivial, then the variational problem* (3.23) *has a unique minimizer.*

In order to solve problem (3.23) by means of wavelets we have to switch to the sequence space formulation. When K is the identity operator the problem simplifies to

$$\inf_{u,v} \left\{ \sum_{\lambda \in J} \left(|f_\lambda - (u_\lambda + v_\lambda)|^2 + \gamma 2^{-2|\lambda|}|v_\lambda|^2 + 2\alpha|u_\lambda| \right) \right\}, \tag{3.24}$$

where $J = \{\lambda = (i,j,k) : k \in J_j, j \in \mathbb{Z}, i = 1,2,3\}$ is the index set used in our separable setting. The minimization of (3.24) is straightforward, since it decouples into easy one-dimensional minimizations. This results in an explicit shrinkage scheme, presented also in [15, 16]:

Proposition 3.9 *Let f be a given function. The functional* (3.24) *is minimized by the parameterized class of functions $\tilde{v}_{\gamma,\alpha}$ and $\tilde{u}_{\gamma,\alpha}$, given by the following nonlinear filtering of the wavelet series of f:*

$$\tilde{v}_{\gamma,\alpha} = \sum_{\lambda \in J_{j_0}} (1 + \gamma 2^{-2|\lambda|})^{-1} \left[f_\lambda - S_{\alpha(2^{2|\lambda|}+\gamma)/\gamma}(f_\lambda) \right] \psi_\lambda$$

and

$$\tilde{u}_{\gamma,\alpha} = f_{\langle j_0 \rangle} + \sum_{\lambda \in J_{j_0}} S_{\alpha(2^{2|\lambda|}+\gamma)/\gamma}(f_\lambda)\psi_\lambda\,,$$

where S_t denotes the soft-shrinkage operator, J_{j_0} all indices λ for scales larger than j_0 and $f_{\langle j_0 \rangle}$ is the approximation at the coarsest scale j_0.

In the case where K is not the identity operator the minimization process results in a coupled system of nonlinear equations for the wavelet coefficients u_λ and v_λ, which is not as straightforward to solve. To overcome this problem, we proceed as follows. We first solve the quadratic problem for v, and then construct an iteration scheme for u. To this end, we introduce the differential operator $T := (-\Delta)^{1/2}$. Setting $v = Th$, problem (3.23) reads as

$$\inf_{(u,h)} \mathcal{F}_f(h,u)\,, \quad \text{with}$$
$$\mathcal{F}_f(h,u) = \|f - K(u+Th)\|^2_{L_2(\Omega)} + \gamma\|h\|^2_{L_2(\Omega)} + 2\alpha|u|_{B^1_1(L_1(\Omega))}\,. \tag{3.25}$$

Minimizing (3.25) with respect to h results in

$$\tilde{h}_\gamma(f,u) = (T^*K^*KT+\gamma)^{-1}T^*K^*(f-Ku)$$

or equivalently

$$\tilde{v}_\gamma(f,u) = T(T^*K^*KT+\gamma)^{-1}T^*K^*(f-Ku)\,.$$

Inserting this explicit expression for $\tilde{h}_\gamma(f,u)$ in (3.25) and defining

$$f_\gamma := T_\gamma f\,, \qquad T^2_\gamma := I - KT(T^*K^*KT+\gamma)^{-1}T^*K^*\,, \tag{3.26}$$

we obtain

$$\mathcal{F}_f(\tilde{h}_\gamma(f,u),u) = \|f_\gamma - T_\gamma Ku\|^2_{L_2(\Omega)} + 2\alpha|u|_{B^1_1(L_1(\Omega))}\,. \tag{3.27}$$

Thus, the remaining task is to solve

$$\inf_u \mathcal{F}_f(\tilde{h}_\gamma(f,u),u)\,, \quad \text{where}$$
$$\mathcal{F}_f(\tilde{h}_\gamma(f,u),u) = \|f_\gamma - T_\gamma Ku\|^2_{L_2(\Omega)} + 2\alpha|u|_{B^1_1(L_1(\Omega))}\,. \tag{3.28}$$

Proposition 3.10 *Suppose that K is a linear bounded operator modeling the blur, with K maps $L_2(\Omega)$ to $L_2(\Omega)$ and $\|K^*K\|<1$. Moreover, assume T_γ is defined as in (3.26), and the functional $\mathcal{F}^{\mathrm{sur}}_f(\tilde{h},u;a)$ is defined by*

$$\mathcal{F}^{\mathrm{sur}}_f(\tilde{h}_\gamma(f,u),u;a) = \mathcal{F}_f(\tilde{h}_\gamma(f,u),u) + \|u-a\|^2_{L_2(\Omega)} - \|T_\gamma K(u-a)\|^2_{L_2(\Omega)}\,.$$

Then, for arbitrarily chosen $a \in L_2(\Omega)$, the functional $\mathcal{F}^{\mathrm{sur}}_f(\tilde{h}_\gamma(f,u),u;a)$ has a unique minimizer in $L_2(\Omega)$. The minimizing element is given by

$$\tilde{u}_{\gamma,\alpha} = \mathbf{S}_\alpha(a + K^*T^2_\gamma(f-Ka))\,,$$

where the operator $\mathbf{S}_\alpha$ is defined component-wise by

$$\mathbf{S}_\alpha(x) = \sum_\lambda S_\alpha(x_\lambda)\psi_\lambda\,.$$

The proof follows from [16]. One can now define an iterative algorithm by repeated minimization of $\mathcal{F}^{\mathrm{sur}}_f$:

$$u^0 \ \textit{arbitrary};\ u^n = \arg\min_u \left(\mathcal{F}^{\mathrm{sur}}_f(\tilde{h}_\gamma(f,u),u;u^{n-1})\right)\,,\ n=1,2,\ldots\,. \tag{3.29}$$

The convergence results shown in [15, 17, 18] can be applied directly:

Theorem 3.11 *Suppose that K is a linear bounded operator, $\|K^*K\| < 1$, and that T_γ is defined as in* (3.26). *Then the sequence of iterates*

$$u^n_{\gamma,\alpha} = \mathbf{S}_\alpha(u^{n-1}_{\gamma,\alpha} + K^*T^2_\gamma(f - Ku^{n-1}_{\gamma,\alpha}))\ , \quad n = 1, 2, \ldots,$$

with arbitrarily chosen $u^0 \in L_2(\Omega)$, converges in norm to a minimizer $\tilde{u}_{\gamma,\alpha}$ of the functional

$$\mathcal{F}_f(\tilde{h}_\gamma(f,u), u) = \|T_\gamma(f - Ku)\|^2_{L_2(\Omega)} + 2\alpha|u|_{B^1_1(L_1(\Omega))}\,.$$

If $\mathcal{N}(T_\gamma K) = \{0\}$, then the minimizer $\tilde{u}_{\gamma,\alpha}$ is unique, and every sequence of iterates converges to $\tilde{u}_{\gamma,\alpha}$ in norm.

Combining the result of Theorem 3.11 and the representation for $\tilde{v}$, we summarize how the image can finally be decomposed in cartoon and oscillating components.

Corollary 3.12 *Assume that K is a linear bounded operator modeling the blur, with $\|K^*K\| < 1$. Moreover, if T_γ is defined as in* (3.26) *and if $\tilde{u}_{\gamma,\alpha}$ is the minimizing element of* (3.28), *obtained as a limit of $u^n_{\gamma,\alpha}$ (see Theorem 3.11), then the variational problem*

$$\inf_{(u,h)} \mathcal{F}_f(h,u)\,,\ \textit{with}$$

$$\mathcal{F}_f(h,u) = \|f - K(u + Th)\|^2_{L_2(\Omega)} + \gamma\|h\|^2_{L_2(\Omega)} + 2\alpha|u|_{B^1_1(L_1(\Omega))}$$

is minimized by the class

$$(\tilde{u}_{\gamma,\alpha}, (T^*K^*KT + \gamma)^{-1}T^*K^*(f - K\tilde{u}_{\gamma,\alpha}))\,,$$

where $\tilde{u}_{\gamma,\alpha}$ is the unique limit of the sequence

$$u^n_{\gamma,\alpha} = \mathbf{S}_\alpha(u^{n-1}_{\gamma,\alpha} + K^*T^2_\gamma(f - Ku^{n-1}_{\gamma,\alpha}))\ , \quad n = 1, 2, \ldots.$$

Numerical Experiments – Additional Redundancy and Adaptivity

The nonlinear filtering rule of Proposition 3.9 gives explicit descriptions of $\tilde{v}$ and $\tilde{u}$ that are computed by fast discrete wavelet schemes. However, nonredundant filtering very often creates artifacts in terms of undesirable oscillations, which manifest themselves as ringing and edge blurring, see Fig. 3.4. Poor directional selectivity of traditional tensor product wavelet bases likewise cause artifacts. Therefore, we apply various refinements on the basic algorithm that address this problem. In particular, we shall use redundant translation invariant schemes, see [12], complex wavelets, see e.g., [25, 40], and additional edge dependent penalty weights introduced in [16]. Here we limit ourselves to presenting the numerical results for the particular problem of simultaneously decomposing, deblurring and denoising a given image.

Fig. 3.4 An initial geometric image f (left), and two versions of f (the middle decomposed with the Haar wavelet basis and the right with the Db3 basis) where the soft-shrinkage operator with shrinkage parameter $\alpha = 0.5$ was applied

We start with the case where K is the identity operator. In order to show how the nonlinear (redundant) wavelet scheme acts on piecewise constant functions, we decompose a geometric image (representing cartoon components only) with sharp contours, see Fig. 3.5. We observe that $\tilde{u}$ represents the cartoon part very well. The texture component $\tilde{v}$ (plus a constant for illustration purposes) contains only some very weak contour structures.

Next, we demonstrate the performance of the Haar shrinkage algorithm successively incorporating redundancy (by cycle spinning) and local penalty weights. The local penalty weights are computed the following way: First, we apply the shrinkage operator $\mathbf{S}$ to f with a level dependent threshold α'. Second, for those λ according to the non-zero values of $S_{\alpha'}(f_\lambda)$ we put an extra weight $w_\lambda >> 1$ in the H^{-1} penalty. The coefficients $S_\xi(f_\lambda)$ for the first two scales of a segment of the image "Barbara" are visualized in Fig. 3.6. In Fig. 3.7, we present our numerical results. The upper row shows the original and the noisy image. The next row visualizes the results for non-redundant Haar shrinkage (Method A). The third row shows the same but incorporating cycle spinning (Method B), and the last row shows the incorporation of cycle spinning and local penalty weights. Each extension of the shrinkage method

Fig. 3.5 From left to right: initial geometric image f, $\tilde{u}$, $\tilde{v} + 150$, computed with Db3 in the translation invariant setting, $\alpha = 0.5$, $\gamma = 0.01$

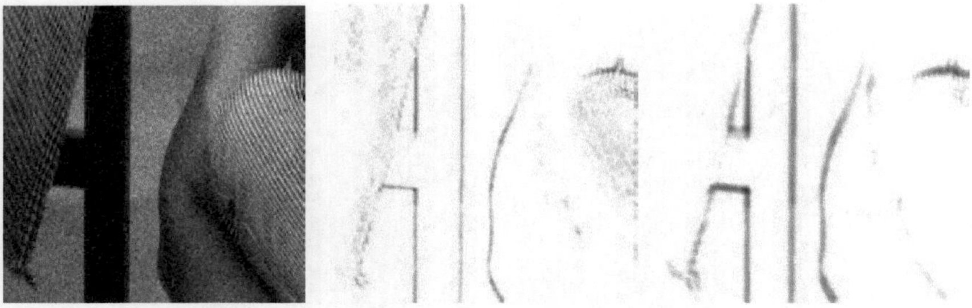

Fig. 3.6 Left: noisy segment of a woman image, middle and right: first two scales of $\mathbf{S}(f)$ inducing the weight function w

improves the results. This is also being confirmed by comparing the signal-to-noise-ratios (which is here defined as follows: $SNR(f,g) = 10\log_{10}(\|f\|^2/\|f-g\|^2)$), see Table 3.2.

In order to compare the performance with the Vese–Osher TV model and with the Vese–Solé–Osher H^{-1} model, we apply our scheme to a woman image (the same that was used in [35, 37]), see Fig. 3.8. We obtain very similar results as obtained with the TV model proposed in [35]. Compared with the results obtained with the H^{-1} model proposed in [37] we observe that our reconstruction of the texture component contains much less cartoon information. In terms of computational cost we have observed that even in the case of applying cycle spinning and edge enhancement our proposed wavelet shrinkage scheme is less time consuming than the Vese–Solé–Osher H^{-1} restoration scheme, see Table 3.3, even when the wavelet method is implemented in Matlab, which is slower than the compiled version for the Vese–Solé–Osher scheme.

We end this section with an experiment where K is not the identity operator. In our particular case K is a convolution operator with Gaussian kernel. The implementation is simply done in Fourier space. The upper row in Fig. 3.9 shows the original f and the blurred image Kf. The lower row visualizes the results: the cartoon component $\tilde{u}$, the texture component $\tilde{v}$, and the sum of both $\tilde{u}+\tilde{v}$. One may clearly see that the deblurred image $\tilde{u}+\tilde{v}$ contains (after a small number of iterations) more small scale details than Kf. This definitely shows the capabilities of the proposed iterative deblurring scheme (3.29).

3.3.2 Deconvolution of δ-Sequences

This section was inspired by discussions with signal and imaging experts in the field of preprocessing 1 D and 2 D mass spectroscopy data in proteomics. Both applications are mathematically modeled by a convolution operator, hence the data are some blurred and noisy signals or images. Accordingly, the classical approach for solving this inverse ill-posed problem would consist of a regularized deconvolution applied to the given data.

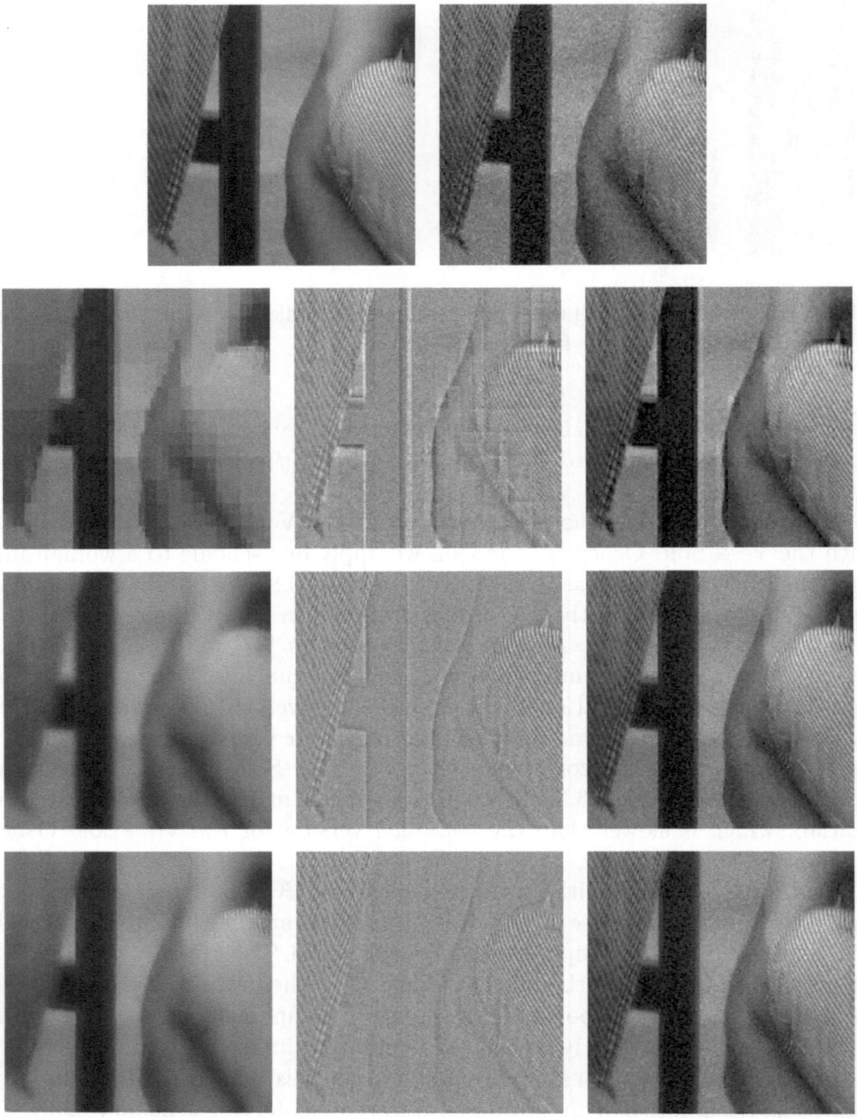

Fig. 3.7 Top: initial and noisy image, 2nd row: non-redundant Haar shrinkage (Method A), 3rd row: translation invariant Haar shrinkage (Method B), bottom: translation invariant Haar shrinkage with edge enhancement (Method C); 2nd-4th row from left to right: $\tilde{u}$, $\tilde{v} + 150$ and $\tilde{u} + \tilde{v}$, $\alpha = 0.5$, $\gamma = 0.0001$, computed with Haar wavelets and critical scale $j_e = -3$

Table 3.2 Signal-to-noise-ratios of the several decomposition methods (Haar shrinkage, translation invariant Haar shrinkage, translation invariant Haar shrinkage with edge enhancement)

Haar shrinkage	SNR(f, f_ε)	SNR(f,$u+v$)	SNR(f,u)
Method A	20,7203	18,3319	16,0680
Method B	20,7203	21,6672	16,5886
Method C	20,7203	23,8334	17,5070

However, the sought-after signals or images are mathematically modeled by finite sums of delta peaks, which are not captured by the classical theory. Moreover, numerical experiments in both fields indicate, that a somewhat "practical approach" yields better results. This "practical approach" proceeds by computing a wavelet-shrinkage on an appropriate wavelet decomposition followed by simply plotting the positions and amplitudes of the remaining coefficients. It has been shown [26], that this approach is indeed equivalent to a regularized deconvolution scheme in Besov scales. Besov scales are needed in order to obtain an appropriate mathematical model for the reconstruction of such sequences of delta peaks.

The aim of the present section is to summarize the mathematical justification of this approach as given in [26] and to present reconstruction results

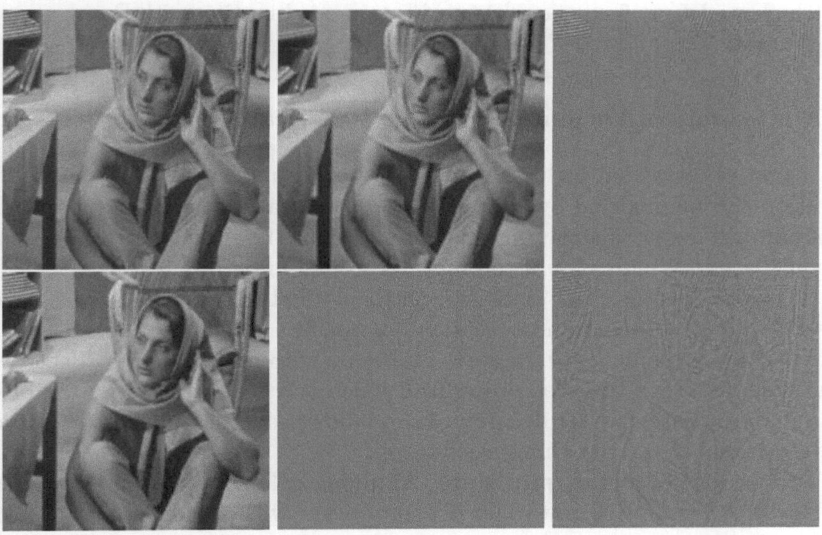

Fig. 3.8 Top from left to right: initial woman image f, $\tilde{u}$ and $\tilde{v}+150$, computed with Db10 (Method C), $\alpha = 0.5$, $\gamma = 0.002$; bottom from left to right: u and v obtained by the Vese–Osher TV model and the v component obtained by the Vese–Solé–Osher H^{-1} model

Table 3.3 Comparison of computational cost of the PDE- and the wavelet-based methods

Data basis	"Barbara" image (512×512 pixel)
Hardware architecture	PC
Operating system	linux
OS distribution	redhat7.3
Model	PC, AMD Athlon-XP
Memory size (MB)	1024
Processor speed (MHz)	1333
Number of CPUs	1
Computational cost	(Average over 10 runs)
PDE scheme in Fortran (compiler f77)	56,67 sec
Wavelet shrinkage Method A (Matlab)	4,20 sec
Wavelet shrinkage Method B (Matlab)	24,78 sec
Wavelet shrinkage Method C (Matlab)	26,56 sec

for deconvolving 1 D MALDI/TOF-data (Bruker Daltonics GmbH) and 2 D LCMS spectra (Hoffmann-La Roche AG).

This approach will require to measure the defect $\|Af - g^\delta\|_{B^s_p(L_p)}$ in an appropriate Besov space. Hence the resulting regularization method extends the recently proposed sparsity schemes for solving inverse problems [3, 17, 38], which treated L_2 defects, i.e., $\|Af - g^\delta\|_{L_2}$, in combination with Besov sparsity constraints.

A Mathematical Formulation of the Practical Approach

As a first observation, let us note that in both applications the sought-after function can be modeled by a finite set of delta peaks. Hence the sought-after function has a sparse structure and, in addition, the mass spectroscopy data is equispaced by multiples of the unitary atomic mass.

Therefore, let us now state the mathematical formulation which is the basis of the above mentioned practical approach:

We consider some noisy data g^δ, a function with finite support. The "practical approach" is a two-step procedure, which starts by a shrinkage operation on an appropriate wavelet decomposition followed by plotting the amplitudes and positions of the remaining coefficients.

Let us formalize this procedure. Applying a shrinkage operation S_λ to g^δ starts by computing a wavelet decomposition with a biorthogonal wavelet basis $\varphi, \psi, \tilde{\varphi}, \tilde{\psi}$,

$$g^\delta = \sum_{k\in\mathbb{Z}} c_k^J\, 2^{J/2}\varphi(2^J\cdot -k) + \sum_{j\geq J, k\in\mathbb{Z}} d_k^j\, 2^{j/2}\psi(2^j\cdot -k),$$

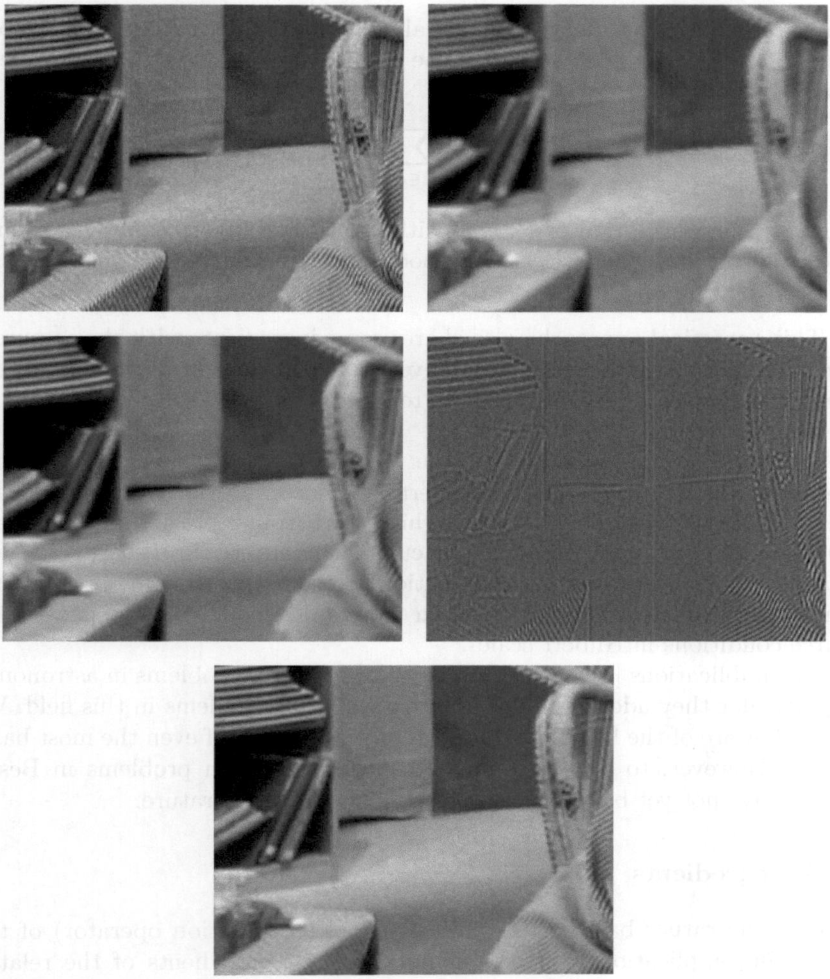

Fig. 3.9 Top from left to right: initial image f, blurred image Kf; middle from left to right: deblurred $\tilde{u}$, deblurred $\tilde{v} + 150$; **bottom**: deblurred $\tilde{u} + \tilde{v}$, computed with Db3 using the iterative approach, $\alpha = 0.2$, $\gamma = 0.001$

where the coefficients are given by $c_k = \langle g^\delta , 2^{J/2}\tilde{\varphi}(2^J \cdot -k)\rangle$ and $d_k^j = \langle g^\delta , 2^{j/2}\tilde{\psi}(2^j \cdot -k)\rangle$, see, e.g., [32, 33]. It follows a shrinkage of the coefficients, i.e., deleting all coefficients $|d_k^j| \leq \lambda$. This yields a function $S_\lambda g^\delta$ which is the finite sum of wavelet and scaling functions. We choose a "finest scale" $j_0 > J$ (for convenience of notation we set $j_0 = 0$) and delete all coefficients on scales finer than j_0. This amounts to a projection of $S_\lambda g^\delta$ and hence

$$P_0 S_\lambda g^\delta = \sum_{k\in\mathbb{Z}} c_k^J 2^{J/2}\varphi(2^J \cdot -k) + \sum_{0\geq j\geq J, k\in\mathbb{Z}} d_k^j 2^{j/2}\psi(2^j \cdot -k) = \sum_{k\in\mathbb{Z}} c_k\varphi(\cdot - k)$$

can be represented as a finite sum of scaling functions on this scale. The reconstruction by plotting the position of the coefficient sequence $\{c_k\}$ is equivalent to a reconstruction

$$R_\alpha P_0 S_\lambda g^\delta = \sum_{k\in\mathbb{Z}} c_k \, \delta(\cdot - k) \,,$$

which is a deconvolution of $P_0 S_\lambda g^\delta$ with the scaling function φ. This should give good results, whenever φ is a good approximation to the kernel of the true convolution operator.

This "practical approach" also shares some ingredients with the compressive sampling techniques [6, 7], however, they proceed in a different direction by analyzing achievable levels of resolution as well as deriving sampling theorems.

Finally, we want to mention some of the prominent papers in the vast literature which analyze specific properties of deconvolution problems. There are at least two fairly recent papers which start from a precise mathematical model for specific applications. In [22] cumulative spectra in Hilbert scales are treated, and [21] analyzes a deconvolution problem in astronomy in combination with an efficient CG solver. Both papers use models in L_2 spaces with source conditions in Hilbert scales.

The publications [4, 5] give an overview on inverse problems in astronomy, in particular they address the relevant convolution problems in this field. We are well aware of the fact that this is an incomplete list of even the most basic results. However, to our best knowledge, deconvolution problems in Besov spaces have not yet been addressed in the accessible literature.

Basic Ingredients

Despite the rather basic mathematical model (convolution operator) of the underlying application a precise definition of all ingredients of the related inverse problem (function spaces, convolution kernels, source conditions) requires some care. The most frequently used models use L_2-function spaces, mainly for convenience and in order to apply standard regularization theory. However, the nature of the specific convolution problems under consideration is characterized by

- a sparse structure of the solution, and
- a model which needs to capture spectral lines or point like objects, i.e., a chain of delta peaks.

Neither of these two requirements is captured by the standard theory.

In this section we will first introduce the convolution operators under consideration. They will be rather straightforward and classical. We then introduce the appropriate function spaces, which leads to Besov spaces and sparsity constraints.

The natural models for the applications described in the introduction is given by an operator $A : X \to Y$ which is an integral transform with a convolution kernel:

$$Af = \varphi * f = \int \varphi(\cdot - y) f(y)\, dy\,,$$

where φ approximates the point spread function of the measurement device. We will first address the case of a 1 D B-spline kernel φ, i.e.,

$$\varphi_k(x) = \underbrace{\varphi * \ldots * \varphi}_{k \text{ times}}(x), \text{ where } \varphi(x) = \frac{1}{2r}\chi_{[-r,r]}(x)\,. \tag{3.30}$$

These convolution kernels define standard operators A_k by

$$A_k f = \varphi_k * f\,. \tag{3.31}$$

They are continuous smoothing operators of the same order in Sobolev as well as in Besov scales

$$A_k : H^s \to H^{s+k}$$

$$A_k : B_p^s(L_p) \to B_p^{s+k}(L_p)\,.$$

The extension to higher dimensions by tensor products is straightforward, the case of general kernels k is addressed in Theorem 3.18. The appropriate mathematical model is a sum of delta peaks

$$f(x) = \sum_{k=1}^{N} f_k \delta(x - k)\,. \tag{3.32}$$

Suitable model spaces for point-like objects are defined via Besov norms: delta peaks in $\mathbb{R}$ are elements of any Besov space $B_p^s(L_p(\mathbb{R}^d))$ satisfying $(s+1)p < 1$. The most important cases are $p = 2$, i.e., the classical Sobolev case of negative order $(s < -1/2)$, and $p = 1$ which requires $s < 0$:

$$\delta \in B_1^{-\varepsilon}(L_1(\mathbb{R})) \quad \text{or} \quad \delta \in H^{-1/2-\varepsilon}(\mathbb{R}) \text{ for any } \varepsilon > 0\,.$$

Model Problems and Regularization Techniques

As we have described in the previous section, there are various meaningful choices for different model spaces as well as for defining convolution operators. Discussing the most general choice would involve a jungle of indices, which would obscure the main objective of the present section: to show the importance of Besov regularization schemes for solving inverse convolution problems.

We will therefore concentrate in the following on analyzing one model problem in detail and address the general cases in some remarks.

Following the discussion on the different modeling alternatives in the previous section we will now select some illustrative choices for the convolution kernel, the model space for source conditions, the noise model and a particular solution f^+. We will focus on the one-dimensional case ($d = 1$) in the following.

As convolution kernel we always choose a B-spline of order two in this section, the general case of an approximate kernel is discussed below.

The corresponding convolution operator A_2, defined by (3.31), is smoothing of order two in Sobolev- as well as in Besov scales:

$$A_2 \ : \ H^s(\mathbb{R}) \to \ H^{s+2}(\mathbb{R}) \ \text{ or } A_2 \ : \ B_1^s(L_1(\mathbb{R})) \to B_1^{s+2}(L_1(\mathbb{R}))\,.$$

As usual, the unavoidable data error will require to choose some weaker norms in the image space.

Problem 3.13 This model is the appropriate physical model for the above mentioned real life deconvolution problems in mass spectroscopy as well as in astronomy. It has no direct analogon in the classical regularization theory.

The problem of reconstructing sequences of Dirac peaks $f^+ = \sum f_k\delta(\cdot - k)$ requires a model space $X = B_1^{-\varepsilon}(L_1(\mathbb{R}))$ with an arbitrary small but fixed ε. We use a white noise model, hence $g^\delta = g^+ + \delta \mathrm{d}W$.

We exploit the smoothing properties of A_2, i.e., $Af^+ \in B_1^{2-\varepsilon}(L_1(\mathbb{R}))$. No additional source condition on the smoothness of f^+ is required.

In all cases, A_2 is smoothing images about two orders, i.e., $f \in H^s(\mathbb{R})$ implies $A_2 f \in H^{2+s}(\mathbb{R})$.

Convergence Analysis

Our primary objective concerns a mathematical analysis of the "practical approach" as explained in the introduction. To this end we will first analyze Problem 1 and compare the convergence results with other settings.

The approximation properties of wavelet shrinkage operators are well studied by now. We will use the results of [11, Theorem 4], which state the following.

Theorem 3.14 *Let s, σ, p, q, α denote real numbers s.t. $f \in B_q^s(L_q)$, $0 < \alpha \leq 1$, $q = (1-\alpha)p$, $s = \sigma p/q + (p/q - 1)/2$, $\sigma - 1/p > -\alpha/2$. Let further $f^\delta = f + \delta dW$ and denote with S_λ the hard shrinkage operator. Define the threshold λ and the cut-off scale J via*

$$\lambda \simeq \sqrt{2|\log\delta|}\delta, \qquad 2^J = -\frac{1}{2\delta^2 \log\delta} \tag{3.33}$$

and the projection P_J via

$$P_J f = \sum_{j \le J, k \in \mathbb{Z}} \langle f, \tilde{\psi}_{j,k} \rangle \psi_{j,k} \,. \tag{3.34}$$

Then the estimator

$$f_\lambda^\delta = P_J S_\lambda f^\delta = \sum_{j \le J, k \in \mathbb{Z}, |\langle f^\delta, \tilde{\psi}_{j,k} \rangle| > \lambda} \langle f^\delta, \tilde{\psi}_{j,k} \rangle \psi_{j,k} \tag{3.35}$$

fulfills

$$E(\|f_\lambda^\delta - f\|^p_{B_p^\sigma(L_p)}) \lesssim (\sqrt{|\log \delta|}\delta)^{\alpha p} \,. \tag{3.36}$$

Here and in the following "$a \simeq b$" means that both quantities can be uniformly bounded by constant multiples of each other. Likewise, "$\lesssim$" indicates inequalities up to constant factors. We will need this result only for measuring the approximation error on the scale $B_1^s(L_1)$ of Besov spaces.

Analysis of the Model Problem

This is the basic deconvolution problem for sparse, peak-like structures, e.g., mass spectroscopy data or certain astrophysical images.

The convolution operator A_2 maps a delta sequence $f^+ = \sum f_k \delta(\cdot - k)$ to a sum of hat functions $g^+ = A_2 f^+ = \sum f_k \varphi_2(\cdot - k)$ with φ_2 defined by (3.30). For the rest of this section we denote $\varphi = \varphi_2$. Hence, the exact inverse deconvolution operator is well defined on such finite sums $g = \sum c_k \varphi(\cdot - k)$ and yields a sequence of delta peaks: $A_2^{-1} g = \sum c_k \delta(\cdot - k)$.

We now exploit the denoising properties of wavelet shrinkage methods on the data side for given noisy data $g^\delta = g^+ + \delta\, dW$. The general result for the present situation is given by the following corollary, see [26].

Corollary 3.15 *Let $g^+ = \sum_{k \in \mathbb{Z}} f_k \varphi(\cdot - k)$ denote a finite sum of second order B-splines, i.e., $\{f_k\}$ is a finite set of non-zero indices. Then for every $\varepsilon > 0$*

$$g^+ \in B_q^{1/q+1-\varepsilon}(L_q) \,. \tag{3.37}$$

Assume $g^\delta = g^+ + \delta \mathrm{d}W$ and let λ and J be chosen according to (3.33).

For $0 < \varepsilon < 3/2$ and for every $3/2 > \tau \ge \varepsilon$ and $\alpha = 1 - (3 - 2\tau)/(3 - 2\varepsilon)$ we obtain the convergence rate

$$E(\|P_J S_\lambda g^\delta - g^+\|_{B_1^{2-\tau}(L_1)}) \lesssim (\sqrt{|\log \delta|}\delta)^\alpha \,. \tag{3.38}$$

This is an approximation result on the data side, which needs to be transferred to an estimate on the reconstruction side, see again [26].

Theorem 3.16 *Let $f^+ = \sum f_k \delta(\cdot - k)$, $g^+ = A_2 f^+$ and $g^\delta = g^+ + \delta \mathrm{d}W$. The "practical approach" as described in the introduction produces a regularized deconvolution of g^δ as*

$$f_\lambda^\delta = A_2^{-1} P_0 S_\lambda g^\delta \,. \tag{3.39}$$

If λ is chosen according to (3.33)*, the following convergence rate holds for every $3/2 \geq \tau > \varepsilon > 0$:*

$$E(\|f_\lambda^\delta - f^+\|_{B_1^{-\tau}(L_1)}) \lesssim (\sqrt{|\log\delta|}\delta)^{1-\frac{3-2\tau}{3-2\varepsilon}} \,. \tag{3.40}$$

We want to remark that this theorem justifies the "practical approach" of just plotting the remaining wavelet decomposition as described in the reconstruction. Hence, this approach, which is a pure shrinkage technique, can indeed be interpreted as a regularization method, which converges – arbitrarily slow – to the delta sequence of the exact solution. The convergence rate $(1-(3-2\tau)/(3-2\varepsilon))$ can be made better by choosing τ larger and ε smaller, but this weakens the norm in which we measure the convergence. Moreover, a generalization to infinite sequences of delta peaks $f^+ = \sum f_k \delta(\cdot - k)$ is obvious as long as $\sum_{k\in\mathbb{Z}} |f_k| < \infty$.

Remark 3.17 Implicitly, the practical approach makes use of the equivalent description of shrinkage methods via a variational approach. Minimizing

$$\|Af - g^\delta\|_{L_2} + \alpha\|f\|_{L_0}$$

for an operator which can be diagonalized by a wavelet basis also leads to a hard shrinkage approach. Hence, the practical approach can be regarded as some type of Tikhonov regularization in L_p spaces. However, we use a different noise model and measure the reconstruction error in a Besov space.

Approximate Kernels

The previous sections have analyzed regularization methods for reconstruction sequences of delta peaks from convolution data. These results were based on the assumption that the convolution kernel equals a B-spline. Obviously, they immediately extend to other wavelet kernels, i.e., we obtain the same convergence results for any convolution operator with kernel function φ whenever φ can be extended to a biorthogonal wavelet bases with a norm equivalence as stated in (3.20) and (3.21).

However, this is still very restrictive. In this section we address the case of a general kernel k, see again [26].

Theorem 3.18 *Let the assumptions of Theorem 3.16 be satisfied. Assume that the kernel k is approximated by the scaling function φ: $\|k-\varphi\|_{B_p^\kappa(L_p)} < \varepsilon$, and let $Af = k * f$. The "practical approach" with kernel φ applied to the noisy data g^δ yields an approximation*

$$E(\|f - A^{-1}P_0 S_\lambda g^\delta\|_{B_1^{-\tau}(L_1)}) \lesssim \|f\|_{B_1^{-\varepsilon}(L_1)} + \left(\delta\sqrt{|\log\delta|}\right)^{\frac{2(\tau-\varepsilon)}{3-2\varepsilon}} \,.$$

Numerical Simulations

For the numerical simulations we use an artificial example as well as a real world example from mass spectrography data. We start with an example where the convolution kernel coincides with a B-spline scaling function. We will discuss a kernel, which is only roughly approximated by the B-spline scaling function in the following example.

First we are going to illustrate the practical approach as described in the introduction for our model problem with simula data. We choose discrete data sets of 512 data-points. The solution f^+ is given by three delta-peaks of different height, and the convolution kernel is a hat function. In our case we used a biorthogonal wavelet base of the class `bior2.x`, and hence the reconstructing scaling function is a hat function. The noisy data g^δ was generated by adding white noise of variance δ. We chose the shrinkage parameter λ according to (3.33).

Figure 3.10 shows the true solution, the data, the reconstruction and an illustration of the convergence rate for $\delta \to 0$. Note that the convergence for $\delta \to 0$ shows very different behavior in different regions: slow convergence interrupted by jumps. The slow convergence is the behavior which is expected asymptotically (since we used $\tau = 0.1$, formula (3.40) predicts a convergence rate of 1/15 which is close to the results). The jumps have a simple explanation: As observable in Fig. 3.10 the reconstruction does not only show the delta peaks but also a number of smaller peaks, which are a result of the

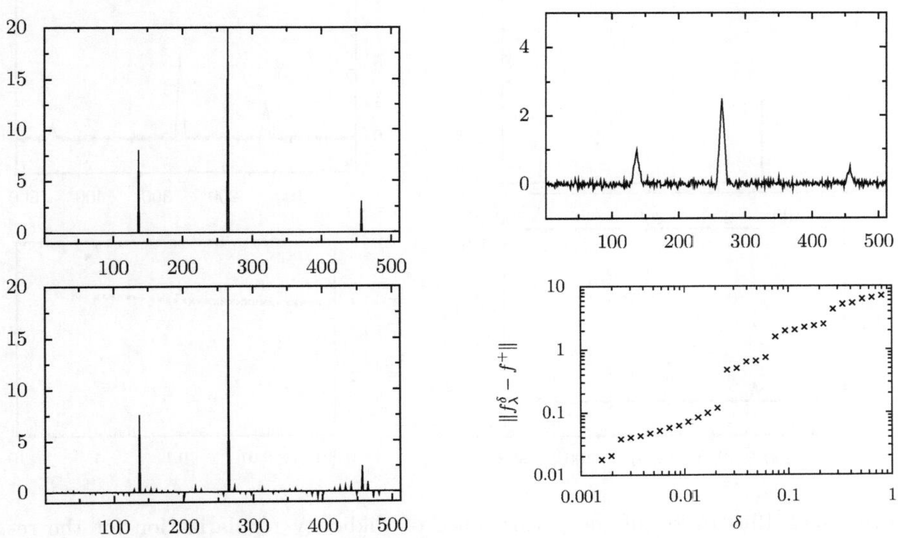

Fig. 3.10 Illustration of the performance of the practical approach. **Upper left**: The true solution f^+, **upper right**: the noisy data g^δ with approximately 12% relative error, **lower left**: the reconstruction by the practical approach, **lower right**: log–log plot of the reconstruction error measured in $B_1^{-\tau}(L_1)$ against the noise level

data errors which escape the shrinkage step. When noise level δ and shrinkage parameter λ tend to zero, it happens that more and more of these false peaks are wiped out, and a jump in the reconstruction error occurs every time this happens.

In a second experiment we compare the "practical approach" with standard Tikhonov regularization. We used the same data as for the first experiment and minimized the functional $\mathrm{J}_\alpha(f) = \|Af - g^\delta\|^2_{H^{-1/2}} - \alpha\|f\|^2_{H^{-1/2}}$. To illustrate the convergence rate for $\delta \to 0$, we used the optimal regularization parameter $\alpha \approx \delta^{1/2}$. The results shown in Fig. 3.11 show the expected behavior: oversmoothing of the regularized reconstruction and a very low convergence rate.

In the last experiment we used as convolution kernel a B-spline of order four but the same reconstruction scaling function (hat function) as in the other experiment: a B-spline of order two. The data consists of overlapping peaks. According to Theorem 3.18 the reconstruction by the practical approach leads to very good results. As Fig. 3.12 shows, the position of the major peaks is reconstructed perfectly. The height of the major peaks is slightly wrong, and there are some small sidepeaks in the reconstruction which are due to the fact the kernel and reconstruction scaling function do not fit together.

We now present reconstruction results for 1 D and 2 D mass spectroscopy data. Figure 3.13 shows the deconvolution of real world data from a

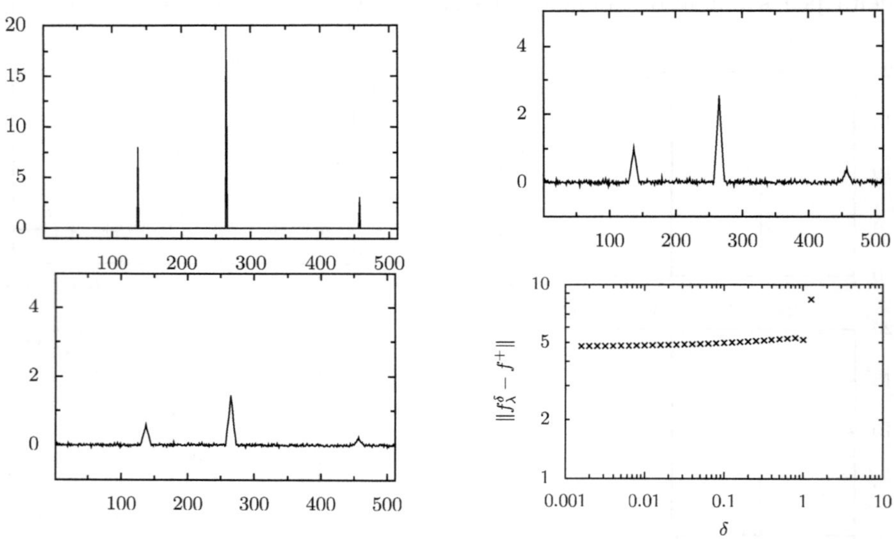

Fig. 3.11 Illustration of the performance of Tikhonov regularization for the reconstruction of delta peaks. **Upper left**: The true solution f^+, **upper right**: the noisy data g^δ with approximately 12% relative error, **lower left**: the reconstruction by Tikhonov regularization, **lower right**: log–log plot of the reconstruction error measured in $H^{-1/2}$ against the noise level

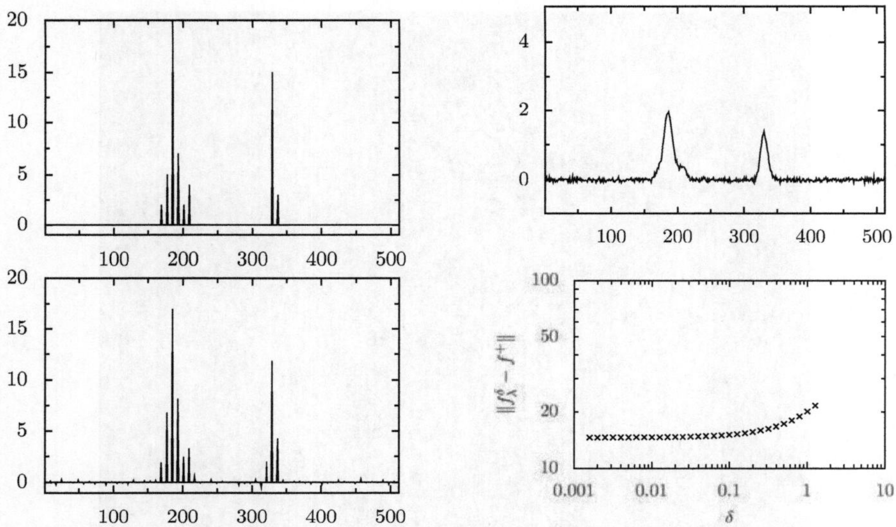

Fig. 3.12 Illustration of the performance of the practical approach for the reconstruction from overlapping peaks where the reconstruction scaling function does not fit to the convolution kernel. **Upper left**: The true solution f^+, **upper right**: the noisy data g^δ with approximately 12% relative error, **lower left**: the reconstruction by the practical approach, **lower right**: log–log plot of the reconstruction error measured in $B_1^{-\tau}(L_1)$ against the noise level

MALDI/SELDI-TOF mass spectrometer provided by an AutoFlex II by Broker Daltonics [31]. We have chosen a section of with a large peak consisting of different isotopes and two small peaks. We deconvolved the data by the practical approach with the `bior2.8` biorthogonal wavelet base and threshold and finest scale chosen by hand. The deconvolution of 2 D LCMS mass spectroscopy data follows the same outline. Figure 3.14 shows the practical approach applied to some real world data provided by Roche. The used data are from drug discovery metabolite identification and show metabolites of a drug produced by rat liver microsomes. The used spectrometer was a Sciex

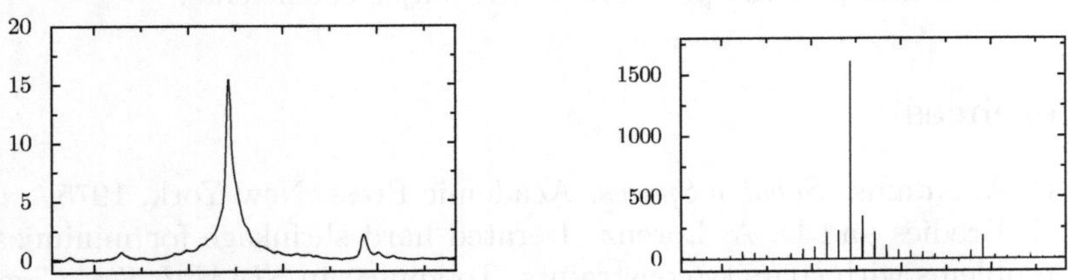

Fig. 3.13 Deconvolution of real world data. **Left**: A section of a spectrogram, right its deconvolution by the practical approach

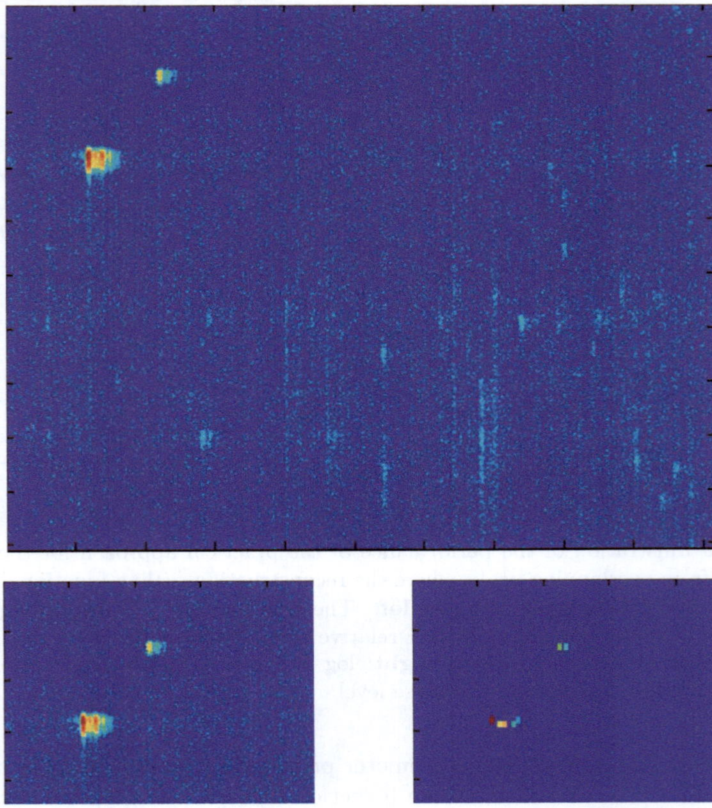

Fig. 3.14 Deconvolution of real world data. **Top**: 2D spectrum from drug metabolism, **bottom left**: part of the spectrum, **bottom right**: deconvolved spectrum part

API 365 Triple Quadrupole operated in positive ion electrospray mode. Applying the practical approach is to calculate the wavelet decomposition with the `bior6.8` biorthogonal wavelet, cut off 90% of the detail coefficients, choose a finest scale and plot the positions of the major coefficients.

References

[1] R. A. Adams. *Sobolev Spaces.* Academic Press, New York, 1975.

[2] K. Bredies and D. A. Lorenz. Iterated hard shrinkage for minimization problems with sparsity constraints. To appear in *SIAM J. Sci. Comput.*, 2007.

[3] K. Bredies, D. A. Lorenz, and P. Maass. A generalized conditional gradient method and its connection to an iterative shrinkage method. To appear in *Comput. Optim. Appl.*, 2008.

[4] J. C. Brown. Inverse problems in astrophysical spectrometry. *Inverse Probl.*, 11:783–794, 1995.

[5] J. C. Brown. Overview of topical issue on inverse problems in astronomy. *Inverse Probl.*, 11:635–638, 1995.

[6] E. J. Candès. Compressive sampling. Sanz-Solé, Marta (ed.) et al., *Proceedings of the International Congress of Mathematicians (ICM)*, Madrid, Spain, August 22–30, 2006. Volume III: Invited lectures. Zürich: European Mathematical Society (EMS), 2006.

[7] E. J. Candès, J.K. Romberg, and T. Tao. Stable signal recovery from incomplete and inaccurate measurements. *Commun. Pure Appl. Math.*, 59:1207–1223, 2005.

[8] C. Cabrelli, C. Heil, and U. Molter. Accuracy of lattice translates of several multidimensional refinable functions. *J. Approx. Theory*, 95:5–52, 1998.

[9] A. Cohen. *Numerical Analysis of Wavelet Methods.* Studies in Mathematics and its Applications. 32. Amsterdam: North-Holland. xviii, 336 p., 2003.

[10] A. Cohen, R. DeVore, P. Petrushev, and H. Xu. Nonlinear approximation and the space $BV(\mathbb{R}^2)$. *Am. J. Math.*, 121:587–628, 1999.

[11] A. Cohen, R. DeVore, G. Kerkyacharian, and D. Picard. Maximal spaces with given rate of convergence for thresholding algorithms. *Appl. Comput. Harmon. Anal.*, 11(2):167–191, 2001.

[12] R. R. Coifman and D. Donoho. Translation-invariant de-noising. *Wavelets and Statistics,* A. Antoniadis and G. Oppenheim, eds., Springer-Verlag, 21:125–150, 1995.

[13] S. Dahlke, E. Novak, and W. Sickel. Optimal approximation of elliptic problems by linear and nonlinear mappings II. *J. Complexity*, 22:549–603, 2006.

[14] I. Daubechies. *Ten Lectures on Wavelets.* CBMS-NSF Regional Conference Series in Applied Mathematics. 61. Philadelphia, PA: SIAM, Society for Industrial and Applied Mathematics. xix, 357 p., 1992.

[15] I. Daubechies and G. Teschke. Wavelet-based image decomposition by variational functionals. *Wavelet Applications in Industrial Processing; Frederic Truchetet; Ed.*, 5266:94–105, 2004.

[16] I. Daubechies and G. Teschke. Variational image restoration by means of wavelets: simultaneous decomposition, deblurring and denoising. *Appl. Comput. Harmon. Anal.*, 19(1):1–16, 2005.

[17] I. Daubechies, M. Defrise, and C. DeMol. An iterative thresholding algorithm for linear inverse problems with a sparsity constraint. *Commun. Pure Appl. Math.*, 57(11):1413–1457, 2004.

[18] I. Daubechies, G. Teschke, and L. Vese. Iteratively solving linear inverse problems with general convex constraints. *Inverse Probl. Imaging*, 1(1): 29–46, 2007.

[19] B. Han. Symmetry property and construction of wavelets with a general dilation matrix. *Linear Algebra Appl.*, 353(1–3):207–225, 2002.

[20] B. Han. Symmetric multivariate orthogonal refinable functions. *Appl. Comput. Harmon. Anal.*, 17(3):277–292, 2004.
[21] M. Hanke and J. G. Nagy. Restoration of atmospherically blurred images by symmetric indefinite conjugate gradient techniques. *Inverse Probl.*, 12 (2):157–173, 1996.
[22] M. Hegland and R. S. Anderssen. Resolution enhancement of spectra using differentiation. *Inverse Probl.*, 21(3):915–934, 2005.
[23] R. Q. Jia. Shift-invariant spaces and linear operator equations. *Isr. J. Math.*, 103:259–288, 1998.
[24] Q. T. Jiang. Multivariate matrix refinable functions with arbitrary matrix dilation. *Trans. Am. Math. Soc.*, 351:2407–2438, 1999.
[25] N. Kinsbury. Image processing with complex wavelets. *Philos. Trans. R. Soc. Lond. Ser. A, Math. Phys. Eng. Sci.* 357(1760):2543–2560, 1999.
[26] E. Klann, M. Kuhn, D. A. Lorenz, P. Maass, and H. Thiele. Shrinkage versus deconvolution. *Inverse Probl.*, 23(5):2231–2248, 2007.
[27] K. Koch. Interpolating scaling vectors. *Int. J. Wavelets Multiresolut. Inf. Process.*, 3(3):389–416, 2005.
[28] K. Koch. Nonseparable orthonormal interpolating scaling vectors. *Appl. Comput. Harmon. Anal.*, 22(2):198–216, 2007.
[29] K. Koch. Multivariate symmetric interpolating scaling vectors with duals. Preprint Nr. 145, SPP 1114, to appear in *J. Fourier Anal. Appl.*, 2006.
[30] K. Koch. *Interpolating Scaling Vectors and Multiwavelets in* $\mathbb{R}^d$. PhD thesis, Philipps-University of Marburg, 2006.
[31] M. Lindemann, M. Diaz, P. Maass, F.-M. Schleif, J. Decker, T. Elssner, M. Kuhn, and H. Thiele. Wavelet based feature extraction in the analysis of clinical proteomics mass spectra. In *54th Conference of the American Society for Mass Spectrometry*, 2006.
[32] A. Louis, P. Maass, and A. Rieder. *Wavelets: Theory and Applications.* Wiley, New York, 1997.
[33] S. Mallat. Multiresolution approximation and wavelet orthonormal bases for $L^2(\mathbb{R})$. *Trans. Am. Math. Soc.*, 315:69–88, 1989.
[34] Y. Meyer. Oscillating patterns in image processing and nonlinear evolution equations. *University Lecture Series*, 22, 2002.
[35] S. Osher and L. Vese. Modeling textures with total variation minimization and oscillating patterns in image processing. *J. Sci. Comput.*, 19 (1–3):553–572, 2003.
[36] S. Osher and L. Vese. Image denoising and decomposition with total variation minimization and oscillatory functions. *J. Math. Imaging Vis.*, 20(1–2):7–18, 2004.
[37] S. Osher, A. Sole, and L. Vese. Image decomposition and restoration using total variation minimization and the H^{-1} norm. *SIAM J. Multiscale Model. Simul.*, 1(3):349–370, 2003.
[38] R. Ramlau and G. Teschke. Tikhonov replacement functionals for iteratively solving nonlinear operator equations. *Inverse Probl.*, 21:1571–1592, 2005.

[39] L. Rudin, S. Osher, and E. Fatemi. Nonlinear total variations based noise removal algorithms. *Physica D*, 60:259–268, 1992.
[40] I. W. Selesnick. Hilbert transform pairs of wavelet bases. *IEEE Signal Process. Lett.*, 8(6):170–173, 2001.
[41] P. Wojtaszczyk. *A Mathematical Introduction to Wavelets.* Cambridge University Press, Cambridge 1997.
[42] X.-G. Xia and Z. Zhang. On sampling theorem, wavelets and wavelet transforms. *IEEE Trans. Signal Process.*, 41:3524–3535, 1993.

4

Inverse Problems and Parameter Identification in Image Processing

Jens F. Acker[5], Benjamin Berkels[2], Kristian Bredies[3], Mamadou S. Diallo[6], Marc Droske[2], Christoph S. Garbe[4], Matthias Holschneider[1], Jaroslav Hron[5], Claudia Kondermann[4], Michail Kulesh[1], Peter Maass[3], Nadine Olischläger[2], Heinz-Otto Peitgen[7], Tobias Preusser[7], Martin Rumpf[2], Karl Schaller[9], Frank Scherbaum[8], and Stefan Turek[5]

[1] Institute for Mathematics, University of Potsdam, Am Neuen Palais 10, 14469 Potsdam, Germany
{mkulesh,hols}@math.uni-potsdam.de

[2] Institute for Numerical Simulation, University of Bonn, Wegelerstr. 6, 53115 Bonn, Germany
{benjamin.berkels,nadine.olischlaeger,martin.rumpf}@ins.uni-bonn.de

[3] Center of Industrial Mathematics (ZeTeM), University of Bremen, Postfach 33 04 40, 28334 Bremen, Germany
{kbredies,pmaass}@math.uni-bremen.de

[4] Interdisciplinary Center for Scientific Computing, University of Heidelberg, Im Neuenheimer Feld 368, 69120 Heidelberg, Germany
{Christoph.Garbe,Claudia.Nieuwenhuis}@iwr.uni-heidelberg.de

[5] Institute for Applied Mathematics, University of Dortmund, Vogelpothsweg 87, 44227 Dortmund, Germany
{jens.acker,jaroslav.hron,ture}@math.uni-dortmund.de

[6] Now at ExxonMobil Upstream Research company, Houston, Texas, USA
mamadou.s.diallo@exxonmobil.com

[7] Center for Complex Systems and Visualization, University of Bremen, Universitätsallee 29, 28359 Bremen, Germany
{heinz-otto.peitgen,preusser}@cevis.uni-bremen.de

[8] Institute for Geosciences, University of Potsdam, Karl-Liebnecht-Strasse 24, 14476 Potsdam, Germany
fs@geo.uni-potsdam.de

[9] Hôpitaux Universitaires de Genève, Rue Micheli-du-Crest 24, 1211 Genève, Switzerland
karl.Schaller@hcuge.ch

4.1 Introduction

Many problems in imaging are actually inverse problems. One reason for this is that conditions and parameters of the physical processes underlying the actual image acquisition are usually not known. Examples for this are the inhomogeneities of the magnetic field in magnetic resonance imaging (MRI) leading to nonlinear deformations of the anatomic structures in the recorded images, material parameters in geological structures as unknown parameters for the simulation of seismic wave propagation with sparse measurement on the surface, or temporal changes in movie sequences given by intensity changes or moving image edges and resulting from deformation, growth and transport processes with unknown fluxes. The underlying physics is mathematically described in terms of variational problems or evolution processes. Hence, solutions of the forward problem are naturally described by partial differential equations. These forward models are reflected by the corresponding inverse problems as well. Beyond these concrete, direct modeling links to continuum mechanics abstract concepts from physical modeling are successfully picked up to solve general perceptual problems in imaging. Examples are visually intuitive methods to blend between images showing multiscale structures at different resolution or methods for the analysis of flow fields.

This chapter is organized as follows. In Sect. 4.2 wavelet based method for the identification of parameters describing heterogeneous media in subsurface structures from sparse seismic measurements on the surface are investigated by Kulesh, Holschneider, Scherbaum and Diallo. It is shown how recent wavelet methodology gives further insight and outperforms classical Fourier techniques for these applications.

In Sect. 4.3 close links between surface matching and morphological image matching are established. Berkels, Droske, Olischläger, Rumpf and Schaller describe how to encode image morphology in terms of the map of regular level set normals (the Gauss map of an image) and the singular normal field on edges. Variational methods are presented to match these geometric quantities of images in a joint Mumford Shah type approach. These techniques are complemented by a related approach for explicit surface matching in geometric modeling.

In Sect. 4.4 anisotropic diffusion models with a control parameter on the right hand side are investigated by Bredies, Maass and Peitgen. The aim is a visually natural blending between image representations on different scales. The method is applied for the morphing between medical images of different detail granularity. Here the transition between different scales is captured by the diffusion, whereas the right hand side of the corresponding parabolic initial value problem is considered as a control parameter to ensure that the coarse scale image is actually meet at time 1 starting from the fine scale image at time 0. Existence of solution for this type of control problem is established.

The inverse problem of optical flow is investigated in Sect. 4.5. Here, the focus is in particular on restoration methods for dense optical flow and the underlying image sequence. Garbe, Kondermann, Preusser and Rumpf describe confidence measure for local flow estimation and flow inpainting based on variational techniques. Furthermore, Mumford Shah type approaches for joint motion estimation and image segmentation as well as motion deblurring are presented. Finally, Acker, Hron, Preusser and Rumpf consider in Sect. 4.6 multiscale visualization methods for fluid flow based on anisotropic diffusion methods from image processing. Here, efficient finite element methodology is investigated to resolve temporal flow patterns in a perceptually intuitive way based on time dependent texture mapping. In addition algebraic multigrid methods are applied for a hierarchical clustering of flow pattern.

4.2 Inverse Problems and Parameter Identification in Geophysical Signal Processing

Surface wave propagation in heterogeneous media can provide a valuable source of information about the subsurface structure and its elastic properties. For example, surface waves can be used to obtain subsurface rigidity through inversion of the shear wave velocity. The processing of experimental seismic data sets related to the surface waves is computationally expensive and requires sophisticated techniques in order to infer the physical properties and structure of the subsurface from the bulk of available information.

Most of the previous studies related to these problems are based on Fourier analysis. However, the frequency-dependent measurements, or time-frequency analysis (TFR) offer additional insight and performance in any applications where Fourier techniques have been used. This analysis consists of examining the variation of the frequency content of a signal with time and is particularly suitable in geophysical applications.

The continuous wavelet transform (CWT) of a real or complex signal $S(t) \in L^2(\mathbb{R})$ with respect to a real or complex mother wavelet is the set of L^2–scalar products of all dilated and translated wavelets with this signal [37]:

$$\begin{aligned} \mathcal{W}_g S(t,a) &= \langle T_t D_a g, S\rangle = \int\limits_{-\infty}^{+\infty} \frac{1}{a} g^*\Big(\frac{\tau - t}{a}\Big) S(\tau)\,\mathrm{d}\tau\,, \\ S(t) &= \mathcal{M}_h \mathcal{W}_g S(t,a) = \frac{1}{C_{g,h}} \int\limits_{-\infty}^{+\infty}\int\limits_{-\infty}^{+\infty} h\left(\frac{t-\tau}{a}\right) \mathcal{W}_g S(\tau,a)\,\frac{\mathrm{d}\tau\,\mathrm{d}a}{a^2}\,, \end{aligned} \tag{4.1}$$

where g, h are wavelets used for the direct and inverse wavelet transforms, $D_a : g(\tau) \mapsto g(\tau/a)/a$ and $T_t : g(\tau) \mapsto g(\tau - t)$ define the dilation $a \in \mathbb{R}$ and translation $t \in \mathbb{R}$ operations correspondingly. If we select a wavelet with a unit central frequency, it is possible to obtain the physical frequency directly by taking the inverse of the scale: $f = 1/a$.

This approach is powerful and elegant, but is not the only one available for practical applications. Other TFR methods such as the Gabor transform, the S-transform [71] or bilinear transforms like the Wigner-Ville [64] or smoothed Wigner-Ville transform can be used as well. The relative performance of time-frequency analysis from different TFR approaches is primarily controlled by the frequency resolution capability that motivated the use of CWT in the present work.

With multicomponent data, one is usually confronted with the issue of separating seismic signals of different polarization characteristics. For instance, one would like to distinguish between the body waves (P- and S- waves) that are linearly polarized from elliptically polarized Rayleigh waves. Polarization analysis is also used to identify shear wave splitting. Unfortunately, there is no mathematically exact a priori definition for the instantaneous polarization attributes of a multicomponent signal. Therefore any attempts to produce one are usually arbitrary.

Time-frequency representations can be incorporated in polarization analysis [67, 71, 73]. We proposed several different wavelet based methods for the polarization analysis and filtering.

4.2.1 Polarization Properties for Two-Component Data

Given a signal from three-component record, with $S_x(t)$, $S_y(t)$, and $S_z(t)$ representing the seismic traces recorded in three orthogonal directions, any combination of two orthogonal components can be selected for the polarization analysis: $Z(t) = S_k(t) + iS_m(t)$. Let us consider the instantaneous angular frequency defined as the derivative of the complex spectrum's phase: $\Omega^{\pm}(t,f) = \pm\partial \arg \mathcal{W}_g^{\pm} Z(t,f)/\partial t$. Then, near time instant t, the wavelet spectrum can be represented as follows:

$$\mathcal{W}_g Z(t+\tau, f) \simeq \mathcal{W}_g^{+} Z(t,f) \mathrm{e}^{\mathrm{i}\Omega^{+}(t,f)\tau} + \mathcal{W}_g^{-} Z(t,f) \mathrm{e}^{-\mathrm{i}\Omega^{-}(t,f)\tau} ,$$

which yields the time-frequency spectrum for each of the parameters (see [26, 52]):

$$\begin{aligned} R(t,f) &= |\mathcal{W}_g^{+} Z(t,f)| + |\mathcal{W}_g^{-} Z(t,f)|/2 , \\ r(t,f) &= ||\mathcal{W}_g^{+} Z(t,f)| - |\mathcal{W}_g^{-} Z(t,f)||/2 , \\ \theta(t,f) &= \arg[\mathcal{W}_g^{+} Z(t,f) \mathcal{W}_g^{-} Z(t,f)]/2 , \\ \Delta\phi(t,f) &= \arg\left(\frac{\mathcal{W}_g^{+} Z(t,f) + \mathcal{W}_g^{-} Z(t,f)^*}{\mathcal{W}_g^{+} Z(t,f) - \mathcal{W}_g^{-} Z(t,f)^*}\right) \bmod \pi , \end{aligned} \tag{4.2}$$

where R is the semi-major axis $R \geq 0$, r is the semi-minor axis $R \geq r \geq 0$, θ is the tilt angle, which is the angle of the semi-major axis with the horizontal axis, $\theta \in (-\pi/2, \pi/2]$ and $\Delta\phi$ is the phase difference between $S_k(t)$ and $S_m(t)$ components.

If we analyze seismic data, an advantage of the method (4.2) is the possibility to perform the complete wave-mode separation/filtering process in the wavelet domain and the ability to provide the frequency dependence of ellipticity, which contains important information on the subsurface structure. With the extension of the polarization analysis to the wavelet domain, we can construct filtering algorithms to separate different wave types based on the instantaneous attributes by a combination of constraints posed on the range of the reciprocal ellipticity $\rho(t,f) = r(t,f)/R(t,f)$ and the tilt angle $\theta(t,f)$ [26].

4.2.2 Polarization Properties for Three-Component Data

Reference [61] proposed a method based on a variational principle that allows generalization to any number of components, and they briefly addressed the possibility of using the instantaneous polarization attributes for wavefield separation and shear-wave splitting identification. In [24], we extended the method of [61] to the wavelet domain in order to use the instantaneous attributes for filtering and wavefield separation for any number of components. As an example, reference [63], used this method for spectral analysis and multicomponent polarization analyses on the Gubbio Piana (central Italy) recordings to identify the frequency content of the different phases composing the recorded wavefield and to highlight the importance of basin-induced surface waves in modifying the main strong ground-motion parameters.

In more general terms, particle motions captured with three-component recordings can be characterized by a polarization ellipsoid. Several methods are proposed in the literature to introduce such an approximation. They are based on the analysis of the covariance matrix of multicomponent recordings and principal components analysis using singular value decomposition [43]. In [50], we extended the covariance method to the time-frequency domain. Following the method, proposed by [25], we use an approximate analytical formula to compute the elements of the covariance matrix $\mathbf{M}(t,f)$ for a time window which is derived from an averaged instantaneous frequency of the multicomponent record:

$$\begin{aligned}
M_{km}(t,f) &= |\mathcal{W}_g S_k(t,f)|\,|\mathcal{W}_g S_m(t,f)|\{\operatorname{sinc}(\varGamma^-_{km}(t,f))\cos(A^-_{km}(t,f)) \\
&\qquad + \operatorname{sinc}(\varGamma^+_{km}(t,f))\cos(A^+_{km}(t,f))\} - \mu_{km}\mu_{mk}\,, \\
\varGamma^{\pm}_{km}(t,f) &= \tfrac{\Delta t_{km}(t,f)}{2}\left(\varOmega_k(t,f) \pm \varOmega_m(t,f)\right), \\
A^{\pm}_{km}(t,f) &= \arg \mathcal{W}_g S_k(t,f) \pm \arg \mathcal{W}_g S_m(t,f)\,, \\
\Delta t_{km}(t,f) &= \tfrac{4\pi n}{\varOmega_k(t,f)+\varOmega_m(t,f)}\,, \quad n \in \mathbb{N}\,, \\
\mu_{km} &= \Re\left[\mathcal{W}_g S_k(t,f)\right] \operatorname{sinc}\left(\tfrac{\Delta t_{km}(t,f)\varOmega_k(t,f)}{2}\right), \quad k,m = x,y,z,
\end{aligned} \tag{4.3}$$

where $\operatorname{sinc}(x)$ indicates the sine cardinal function.

The eigenanalysis performed on $\mathbf{M}(t,f)$ yields the principal component decomposition of the energy. Such a decomposition produces three eigenvalues $\lambda_1(t,f) \geq \lambda_2(t,f) \geq \lambda_3(t,f)$ and three corresponding eigenvectors $\mathbf{v}_k(t,f)$ that fully characterize the magnitudes and directions of the principal components of the ellipsoid that approximates the particle motion in the considered time window $\Delta t_{km}(t,f)$:

- the major half-axis $\mathbf{R}(t,f) = \sqrt{\lambda_1(t,f)}\mathbf{v}_1(t,f)/\|\mathbf{v}_1(t,f)\|$;
- the minor half-axis $\mathbf{r}(t,f) = \sqrt{\lambda_3(t,f)}\mathbf{v}_3(t,f)/\|\mathbf{v}_3(t,f)\|$;
- the intermediate half-axis $\mathbf{r}_s(t,f) = \sqrt{\lambda_2(t,f)}\mathbf{v}_2(t,f)/\|\mathbf{v}_2(t,f)\|$;

- the reciprocal ellipticity $\rho(t,f) = \|\mathbf{r}_s(t,f)\| / \|\mathbf{R}(t,f)\|$;
- the minor reciprocal ellipticity $\rho_1(t,f) = \|\mathbf{r}(t,f)\| / \|\mathbf{r}_s(t,f)\|$;
- the dip angle $\delta(t,f) = \arctan(\sqrt{v_{1,x}(t,f)^2 + v_{1,y}(t,f)^2} / v_{1,z}(t,f))$;
- the azimuth $\alpha(t,f) = \arctan(v_{1,y}(t,f)/v_{1,x}(t,f))$.

Note, when the instantaneous frequencies are the same for all components, this method produces the same results as those by [61] in terms of polarization parameters.

4.2.3 Modeling a Wave Dispersion Using a Wavelet Deformation Operator

The second problem in the context of surface wave analysis (especially with high frequency signals) is the robust determination of dispersion curves from multivariate signals. Wave dispersion expresses the phenomenon by which the phase and group velocities are functions of the frequency. The cause of dispersion may be either geometric or intrinsic. For seismic surface waves, the cause of dispersion is of a geometrical nature. Geometric dispersion results from the constructive interferences of waves in bounded or heterogeneous media. Intrinsic dispersion arises from the causality constraint imposed by the Kramers–Krönig relation or from the microstructure properties of material. If the dispersive and dissipative characteristics of the medium are represented by the frequency-dependent wavenumber $k(f)$ and attenuation coefficient $\alpha(f)$, the relation between the Fourier transforms of two propagated signals reads

$$O[\mathcal{D}_{\mathcal{F}}] : \hat{S}(f) \mapsto \mathrm{e}^{-\mathrm{i}\mathbb{K}(f)D - 2\pi \mathrm{i} n} \hat{S}(f) ,$$

where D is the propagation distance, $n \in \mathbb{N}$ is any integer number and $\mathbb{K}(f)$ is the complex wavenumber, which can be defined by real functions $k(f)$ and $\alpha(f)$ as $\mathbb{K}(f) = 2\pi k(f) - \mathrm{i}\alpha(f)$.

In order to analyze the dynamical behavior of multivariate signals using the continuous wavelet transforms, one may be interested in investigating a diffeomorphic deformation of the wavelet space. These deformations establish an algebra of wavelet pseudodifferential operators acting on signals [85]. In the most general case, a wavelet deformation operator can be defined as

$$\begin{aligned} &O[\mathcal{D}] : S(t) \mapsto \mathcal{M}_h \mathcal{D} \mathcal{W}_g S(t,f) , \quad \mathcal{D} : \mathbb{H} \to \mathbb{H} , \\ &\mathbb{H} := \{(t,f) : \ t \in \mathbb{R}, f > 0\} . \end{aligned}$$

We investigated some practical models that give concrete expressions of this deformation operator related to the used dispersion parameters of the medium. Reference [49] has shown how the wavelet transform of the source and the propagated signals are related through a transformation operator that explicitly incorporates the wavenumber as well as the attenuation factor of the medium:

$$\mathcal{D}_{\mathcal{W}} : \mathcal{W}_g S(t,f) \mapsto \mathrm{e}^{-\alpha(f)D} \mathrm{e}^{-\mathrm{i}\psi_1(f)} \mathcal{W}_g S\left(t - k'(f)D, f\right) , \tag{4.4}$$

where $\psi_1(f) = 2\pi[k(f) - fk'(f)]D + 2\pi n$.

In a special case, with the assumption that the analyzing wavelet has a linear phase (with time-derivative approximately equal to 2π, as it is the case for the Morlet wavelet, the approximation (4.4) can be written in terms of the phase $C_p(f) = f/k(f)$ and group $C_g(f) = 1/k'(f)$ velocities as [52]:

$$\begin{aligned}\mathcal{D}_{\mathcal{W}} : \mathcal{W}_g S(t,f) \mapsto \mathrm{e}^{-\alpha(f)D} \left| \mathcal{W}_g S\left(t - \tfrac{D}{C_g(f)}, f\right) \right| \cdot \\ \exp\left[\mathrm{i} \arg \mathcal{W}_g S\left(t - \tfrac{D}{C_p(f)} - \tfrac{n}{f}, f\right)\right] .\end{aligned} \tag{4.5}$$

The relationship (4.5) has the following interpretation. The group velocity is a function that "deforms" the image of the absolute value of the source signal's wavelet spectrum, the phase velocity "deforms" the image of the wavelet spectrum phase, and the attenuation function determines the frequency-dependent real coefficient by which the spectrum is multiplied.

4.2.4 How to Extract the Dispersion Properties from the Wavelet Coeffitients?

Equation (4.5) allows us to formulate the ideas how the frequency-dependent dispersion properties can be obtained using the wavelet spectra' phases of source and propagated signals. To obtain the phase velocities of multi-mode and multivariate signals, we can perform "frequency-velocity" analysis on the analogy of the frequency-wavenumber method [17] for a seismogram $S_k(t)\,,\, k = 1, N$. The main part of this analysis consists of calculating of correlation spectrum $\mathbf{M}(f,c)$ as follows (see [51]):

$$\begin{aligned}\mathbf{M}(f,c) &= \int_{t_{\min}}^{t_{\max}} \left| \sum_{k,m} A_k(\tau,f) A_m^*\left(\tau - \frac{D_{mk}}{c}, f\right) \right| \mathrm{d}\tau \\ &= \int_{t_{\min}}^{t_{\max}} \left| \sum_{k,m} \mathrm{e}^{\mathrm{i}B_k(\tau,f)} \exp\left(-\mathrm{i}B_m\left(\tau - \frac{D_{mk}}{c}, f\right)\right) \right| \mathrm{d}\tau , \\ A_k(\tau,f) &= \mathcal{W}_g S_k(\tau,f)/|\mathcal{W}_g S_k(\tau,f)|, \quad B_k(\tau,f) = \arg \mathcal{W}_g S_k(\tau,f) ,\end{aligned} \tag{4.6}$$

where $[t_{\min}, t_{\max}]$ indicates the total time range for which the wavelet spectrum was calculated, $c \in [C_p^{\min}, C_p^{\max}]$ is an unbounded variable corresponding to the phase velocity, A_k is a complex-valued wavelet phase and B_k is a real-valued wavelet phase.

For a given parametrization of wavenumber and attenuation functions, finding an acceptable set of parameters can be thought of as an optimization problem that seeks to minimize a cost function χ^2 and can be formulated as follows:

$$\chi^2(\alpha(f,\mathbf{p}), k(f,\mathbf{q})) \to \min, \ \mathbf{p} \in \mathbb{R}^P, \ \mathbf{q} \in \mathbb{R}^Q ,$$

where P is the number of parameters used to model the attenuation $\alpha(f)$ and Q is the number of parameters used to model the wavenumber $k(f)$. $\mathbf{p}$ and $\mathbf{q}$ represent the vectors of parameters describing $\alpha(f)$ and $k(f)$ respectively. This cost function involves a propagator described above.

At this stage, we need to distinguish between the case where the analyzed signal consists only of one coherent arrival from the case where it consists of several coherent arrivals. In the former case, the derived functions are meaningful and characterize those analyzed event. However in the latter, these functions cannot be easily interpreted since the signals involved consist of many overlapping arrivals.

If only one single phase is observed in all the traces $S_k(t)$, it will be enough to minimize a cost function that involves some selected seismic traces in order to estimate the attenuation and phase velocity using the modulus and the phase of the wavelet transforms correspondingly, see [38]:

$$\chi^2(\mathbf{p},\mathbf{q}) = \sum_{m,k} \iint \left| |\mathcal{W}_g S_k(t,f)| - |\mathcal{D}_{\mathcal{W}}(\mathbf{p},\mathbf{q}) \mathcal{W}_g S_m(t,f)| \right|^2 \,\mathrm{d}t\,\mathrm{d}f\,,$$
$$\chi^2(\mathbf{p},\mathbf{q}) = \sum_{m,k} \iint \left| \arg \mathcal{W}_g S_k(t,f) - \arg \mathcal{D}_{\mathcal{W}}(\mathbf{p},\mathbf{q}) \mathcal{W}_g S_m(t,f) \right|^2 \,\mathrm{d}t\,\mathrm{d}f\,. \tag{4.7}$$

The first step will consist of seeking a good initial condition by performing an image matching using the modulus of the wavelet transforms of a pair of traces. The optimization is carried out over the whole frequency range of the signal. In order to reduce the effect of uncorrelated noise in our estimates, it is preferable to use a propagator based on the cross-correlations, see [38].

In the case where the observed signals consist of a mixture of different wave types and modes, a cascade of optimizations in the wavelet domain will be necessary in order to fully determine the dispersion and attenuation characteristics specific to each coherent arrival.

Since the dependence of the cost functions (4.7) on the parameters $\mathbf{p}$ and $\mathbf{q}$ is highly non-linear, each function may have several local minima. To obtain the global minimum that corresponds to the true parameters, a non-linear least-squares minimization method that proceeds iteratively from a reasonable set of initial parameters is required. In the present contribution, we use the Levenberg–Marquardt algorithm [69].

Finally, the obtained dispersion curves (especially phase and group velocities) for defined wave types can be used for the determination of physical and geometrical properties of the subsurface structure. Because of the non-uniqueness of earth models that can be fitted to a given dispersion curve, the inversion for the average shear velocity profile is usually treated as an optimization problem where one tries to minimize the misfit between experimental and theoretical dispersion curves computed for a given earth model that is assumed to best represent the subsurface under investigation.

4.3 The Interplay of Image Registration and Geometry Matching

Image registration is one of the fundamental tools in image processing. It deals with the identification of structural correspondences in different images of the same or of similar objects acquired at different times or with different image devices. For instance, the revolutionary advances in the development of imaging modalities has enabled clinical researchers to perform precise studies of the immense *variability* of human anatomy. As described in the excellent review by Miller et al. [60] and the overview article of Grenander and Miller [33], this field aims at automatic detection of anatomical structures and their evaluation and comparison. Different images show corresponding structures at usually nonlinearly transformed positions.

In image processing, registration is often approached as a variational problem. One asks for a deformation ϕ on an image domain Ω which maps structures in the reference image u_R onto corresponding structures in the template image u_T. This leads ill-posed minimization problem if one considers the infinite dimensional space of deformations [12]. A iterative, multilevel regularization of the descent direction has been investigated in [22]. Alternatively, motivated by models from continuum mechanics, the deformation can additionally be controlled by *elastic stresses* on images regarded as elastic sheets. For example see the early work of Bajcsy and Broit [5] and more recent, significant extensions by Grenander and Miller [33]. In [29] nonlinear elasticity based on polyconvex energy functionals is investigated to ensure a one-to-one image matching. As the image modality differs there is usually no correlation of image intensities at corresponding positions. What still remains, at least partially, is the local geometric image structure or "morphology" of corresponding objects. Viola and Wells [82] and Collignon et al. [23] presented an information theoretic approach for the registration of multi-modal images. Here, we consider "morphology" as a geometric entity and will review registration approaches presented in [28, 29, 30].

Obviously, geometry matching is also a widespread problem in computer graphics and geometric modeling [35]. E.g. motivated by the ability to scan geometry with high fidelity, a number of approaches have been developed in the graphics literature to bring such scans into correspondence [9, 56]. Given a reference surface $\mathcal{M}_R$ and a template surface $\mathcal{M}_T$ a particular emphasize is on the proper alignment of curved features and the algorithmic issues associated with the management of irregular meshes and their effective overlay. Here, we will describe an image processing approach to the nonlinear elastic matching of surface patches [57]. It is based on a proper variational parametrization method [21] and on the matching of surface characteristics encoded as images u_R and u_T on flat parameter domains ω_R and ω_T, respectively. Here, it is particularly important to take into account of the metric distortion, to ensure a physically reasonable matching of the actual surfaces $\mathcal{M}_R$ and $\mathcal{M}_T$.

4.3.1 The Geometry of Images

In mathematical terms, two images $u, v : \Omega \to \mathbb{R}$ with $\Omega \subset \mathbb{R}^d$ for $d = 2, 3$ are called morphologically equivalent, if they only differ by a change of contrast, i.e., if $u(x) = (\beta \circ v)(x)$ for all $x \in \Omega$ and for some monotone function $\beta : \mathbb{R} \to \mathbb{R}$. Obviously, such a contrast modulation does not change the order and the shape of super level sets $l_c^+[u] = \{x : u(x) \geq c\}$. Thus, image morphology can be defined as the upper topographic map, defined as the set of all these sets $\text{morph}[u] := \{l_c^+[u] : c \in \mathbb{R}\}$. Unfortunately, this set based definition is not feasible for a variational approach and it does not distinguish between edges and level sets in smooth image regions. Hence, in what follows, we derive an alternative notion and consider image functions $u : \Omega \to \mathbb{R}$ in SBV [3] – by definition L^1 functions, whose derivative Du is a vector-valued Radon measure with vanishing Cantor part. We consider the usual splitting $Du = D^{ac}u + D^j u$ [3], where $D^{ac}u$ is the regular part, which is the usual image gradient apart from edges and absolutely continuous with respect to the Lebesgue measure $\mathcal{L}$, and a singular part $D^j u$, which represents the jump and is defined on the jump set J, which consists of the edges of the image. We denote by n^j the vector valued measure representing the normal field on J. Obviously, n^j is a morphological invariant. For the regular part of the derivative we adopt the classical gradient notion $\nabla^{ac}u$ for the $\mathcal{L}$ density of $D^{ac}u$, i.e., $D^{ac}u = \nabla^{ac}u\,\mathcal{L}$ [3]. As long as it is defined, the normalized gradient $\nabla^{ac}u(x) / \|\nabla^{ac}u(x)\|$ is the outer normal on the upper topographic set $l_{u(x)}^+[u]$ and thus again a morphological quantity. It is undefined on the flat image region $F[u] := \{x \in \Omega : \nabla^{ac}u(x) = 0\}$. We introduce n^{ac} as the normalized regular part of the gradient $n^{ac} = \chi_{\Omega \setminus F[u]} \nabla^{ac}u / \|\nabla^{ac}u\|$. We are now able to redefine the morphology $\text{morph}[u]$ of an image u as a unit length vector valued Radon measure on Ω with $\text{morph}[u] = n^{ac}\mathcal{L} + n^s$. We call $n^{ac}\mathcal{L}$ the regular morphology or Gauss map (GM) and n^s the singular morphology. In the next section, we aim to measure congruence of two image morphologies with respect to a matching deformation making explicit use of this decomposition.

4.3.2 Matching Image Morphology

Let us suppose that an initial template image u_T^0 and an initial reference image u_R^0 are given on an image domain Ω. Both images are assumed to be noisy. We aim for a simultaneous robust identification of smoothed and structural enhanced representations $u_T, u_R \in SBV$ and a deformation ϕ, which properly matches the underlying image morphologies, such that $\text{morph}[u_T \circ \phi] = \text{morph}[u_R]$. To phrase this in terms a variational approach we treat the two different components of the morphology separately.

Matching the Singular Morphology

We aim for a deformation ϕ a proper matching of the singular morphologies requesting that $\phi(J_R) = J_T$ for the edge sets $J_R := J_{u_R}$ and $J_T := J_{u_T}$. Now, we ask for a simultaneous edge segmentation, denoising and matching of images in terms of a Mumford Shah approach jointly applied to both images and linked via the unknown elastic deformation. I.e., we consider as set of unknowns u_T, u_R, J_T and ϕ. For the template image we take into account the usual Mumford Shah approach and define the energy

$$E^T_{\mathrm{MS}}[u_T, J_T] = \frac{1}{2}\int_\Omega (u_T - u_T^0)^2 \,\mathrm{d}\mathcal{L} + \frac{\mu}{2}\int_{\Omega\setminus J_T} \|\nabla u_T\|^2 \,\mathrm{d}\mathcal{L} + \eta\mathcal{H}^{d-1}(S_T)$$

with μ, $\eta > 0$. For the reference image we make use of our correspondence assumption and define

$$E^R_{\mathrm{MS}}[u_R, J_T, \phi] = \frac{1}{2}\int_\Omega (u_R - u_R^0)^2 \,\mathrm{d}\mathcal{L} + \frac{\mu}{2}\int_{\Omega\setminus\phi^{-1}(J_T)} \|\nabla u_R\|^2 \,\mathrm{d}\mathcal{L}\,,$$

where the $\mathcal{H}^{d-1}$ -measure of J_R is supposed to be implicitly controlled by the $\mathcal{H}^{d-1}$ -measure of J_T and a smooth deformation ϕ. Hence, we omit the corresponding energy term here. Finally, the energy for the joint Mumford Shah segmentation and matching model in the reference and the template image is given by $E_{\mathrm{MS}}[u_R, u_T, S_T, \phi] = E^T_{\mathrm{MS}}[u_T, J_T] + E^R_{\mathrm{MS}}[u_R, J_T, \phi]$. So far, the deformation ϕ is needed only on the singularity set S_T and thus it is highly under determined.

Matching the Regular Morphology

The regular image morphology consists of the normal field n^{ac}. Given regularized representations u_T and u_R of noisy initial images we observe a perfect match of the corresponding regular morphologies, if the deformation of the reference normal field $n^{ac}_R := \nabla^{ac} u_R \,/\, \|\nabla^{ac} u_R\|$ coincides with the template normals field $n^{ac}_T := \nabla^{ac} u_T \,/\, \|\nabla^{ac} u_T\|$ at the deformed position. In fact, all level sets of the pull back template image $u_T \circ \phi$ and the reference image u_R would then be nicely aligned. In the context of a linear mapping A normals deformed with the inverse transpose A^{-T}, Thus, we obtain the deformed reference normal $n^{ac,\phi}_R = \mathrm{Cof}\, D\phi\, \nabla^{ac} u_R \,/\, \|\mathrm{Cof}\, D\phi\, \nabla^{ac} u_R\|$, where $\mathrm{Cof}\, A := \det A\, A^{-T}$ and ask for a deformation $\phi : \Omega \to \mathbb{R}^d$, such that $n^{ac}_T \circ \phi = n^{ac,\phi}_R$. This can be phrased in terms of an energy integrand $g_0 : R^d \times R^d \times \mathbb{R}^{d,d} \to \mathbb{R}^+_0$, which is zero-homogeneous in the first two arguments as long as they both do not vanish and zero elsewhere. It measures the misalignment of directions of vectors on R^d. For instance we might define

$$g_0(w, z, A) := \gamma \left\| \left(\;-\frac{w}{\|w\|} \otimes \frac{w}{\|w\|}\right) \frac{\mathrm{Cof}\, Az}{\|\mathrm{Cof}\, Az\|} \right\|^m$$

for $w, z \neq 0$, with $\gamma > 0$ and $m \geq 2$, $a \otimes b = ab^T$. Based on this integrand we finally define a Gauss map registration energy

$$E_{GM}[u_T, u_R, \phi] = \int_\Omega g_0(D^{ac}u_T \circ \phi, D^{ac}u_R, \operatorname{Cof} D\phi)\, \mathrm{d}\mathcal{L}\,.$$

For the analytical treatment of the corresponding variational problem we refer to [29].

In a variational setting neither the matching energy for the singular morphology nor the one for the regular morphology uniquely identify the deformation ϕ. Indeed, the problem is still ill-posed. For instance, arbitrary reparametrizations of the level sets ∂l_c^+ or the edge set J, and an exchange of level sets induced by the deformation do not change the energy. Thus, we have to regularize the variational problem. On the background of elasticity theory [20], we aim to model the image domain as an elastic body responding to forces induced by the matching energy. Let us consider the deformation of length, volume and for $d = 3$ also area under a deformation ϕ, which is controlled by $D\phi \,/\, \|D\phi\|$, $\det D\phi$, and $\operatorname{Cof} D\phi \,/\, \|\operatorname{Cof} D\phi\|$, respectively. In general, we consider a so called polyconvex energy functional

$$E_{\text{reg}}[\phi] := \int_\Omega W(D\phi, \operatorname{Cof} D\phi, \det \mathrm{d}\phi)\, \mathrm{d}\mathcal{L}\,, \tag{4.8}$$

where $W : \mathbb{R}^{d,d} \times \mathbb{R}^{d,d} \times \mathbb{R} \to \mathbb{R}$ is supposed to be convex. In particular, a suitable built-in penalization of volume shrinkage, i. e., $W(A, C, D) \overset{D\to 0}{\longrightarrow} \infty$, enables us to ensure bijectivity of the deformation (cf. [6]) and one-to-one image matches. For details we refer to [29]. With respect to the algorithmical realization we take into account a phase field approximation of the Mumford Shah energy E_{MS} picking up the approach by Ambrosio and Tortorelli [2]. Thereby, the edge set J_T in the template image will be represented by a phase field function v, hence $v \circ \phi$ can regarded as the phase field edge representation in the reference image [30]. As an alternative a shape optimization approach based on level sets can be used [28]. Results of the morphological matching algorithm are depicted in Figs. 4.1–4.3.

4.3.3 Images Encoding Geometry

So far, we have extensively discussed the importance of geometry encoded in images for the purpose of morphological image matching. Now, we will discuss how surface geometry can be encoded in images and how to make use of this encoding for surface matching purposes. Consider a smooth surface $\mathcal{M} \subset \mathbb{R}^3$, and suppose $x : \omega \to \mathcal{M};\ \xi \mapsto x(\xi)$ is a parameterization of $\mathcal{M}$ on a parameter domain ω. The metric $g = Dx^T Dx$ is defined on ω, where $Dx \in \mathbb{R}^{3,2}$ is the Jacobian of the parameterization x. It acts on tangent vectors v, w on the parameter domain ω with $(g\, v) \cdot w = Dx\, v \cdot Dx\, w$ and describes how length, area and angles are distorted under the parameterization x. This distortion is measured by the inverse metric $g^{-1} \in \mathbb{R}^{2,2}$. In fact, $\sqrt{\operatorname{tr} g^{-1}}$ measures the average

Fig. 4.1 The morphological registration is demonstrated for a test case. From left to right the reference image u_R, the contrast modulated and artificially deformed template image u_T, the jump set J_T in the template image u_T (represented by a phase field function), the deformation ϕ^{-1} of the template image visualized with a deformed underlying checkerboard, and finally the registered template image $u_T \circ \phi$ are displayed

change of length of tangent vectors under the mapping from the surface onto the parameter plane, whereas $\sqrt{\det g^{-1}}$ measures the corresponding *change of area*. As a surface classifier the mean curvature on $\mathcal{M}$ can be considered as a function h on the parameter domain ω. Similarly a feature set $\mathcal{F}_\mathcal{M}$ on the surface $\mathcal{M}$ can be represented by a set $\mathcal{F}$ on ω. Examples for feature sets for instance on facial surfaces are particularly interesting sets such as the eye holes, the center part of the mouth, or the symmetry line of a suitable width between the left and the right part of the face. Finally, surface textures $\mathcal{T}$ usually live on the parameter space. Hence, the quadruple $(x, h, \mathcal{F}, \mathcal{T})$ can be regarded as an encoding of surface geometry in a geometry image on the parameter domain ω. The quality of a parameterization can be described via a suitable distortion energy $E_{param}[x] = \int_{x^{-1}(\mathcal{M})} W(\mathrm{tr}\,(g^{-1}), \det g^{-1})\,\mathrm{d}x$. For details on the optimization of the parametrization based on this variational approach we refer to [21].

4.3.4 Matching Geometry Images

Let us now consider a reference surface patch $\mathcal{M}_R$ and a template patch $\mathcal{M}_T$ to be matched, where geometric information is encoded via two initially fixed parameter maps x_R and x_T on parameter domains ω_R and ω_T. In what follows we

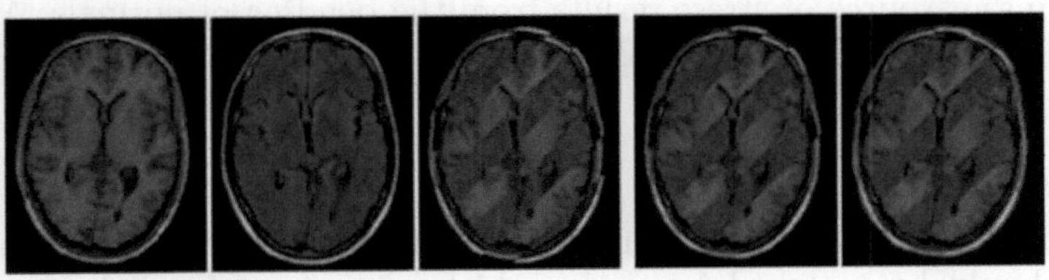

Fig. 4.2 The registration of FLAIR and T1-weighted magnetic resonance brain images is considered. From left to right: the reference T1 weighted MR image u_R, the template FLAIR image u_T, the initial mismatch (with alternating stripes from u_T and u_R), and in the same fashion results for a registration only of the regular morphology and finally for the complete energy are shown

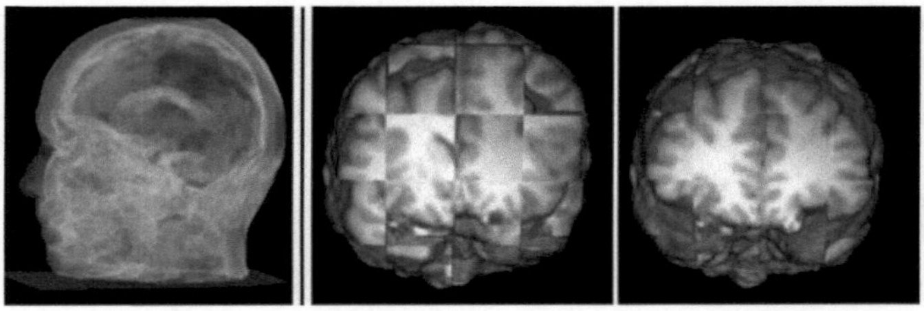

Fig. 4.3 On the left the 3 D phasefield corresponding to the edge set in the an MR image is shown. Furthermore, the matching of two MR brain images of different patients is depicted. We use a volume renderer based on ray casting (VTK) for a 3 D checkerboard with alternating boxes of the reference and the pull back of the template image to show the initial mismatch of MR brain images of two different patients (**middle**) and the results of our matching algorithm (**right**)

always use indices R and T to distinguish quantities on the reference and the template parameter domain. First, let us consider a one-to-one deformation $\phi : \omega_R \to \omega_T$ between the two parameter domains. This induces a deformation between the surface patches $\phi_{\mathcal{M}} : \mathcal{M}_R \to \mathcal{M}_T$ defined by $\phi_{\mathcal{M}} := x_T \circ \phi \circ x_R^{-1}$. Now let us focus on the distortion from the surface $\mathcal{M}_R$ onto the surface $\mathcal{M}_T$. In elasticity, the distortion under an elastic deformation ϕ is measured by the Cauchy-Green strain tensor $D\phi^T D\phi$. Properly incorporating the metrics g_R and g_T we can adapt this notion and obtain the Cauchy Green tangential distortion tensor $\mathcal{G}[\phi] = g_R^{-1}\, D\phi^T\, (g_T \circ \phi)\, D\phi$, which acts on tangent vectors on the parameter domain ω_R. As in the parameterization case, one observes that $\sqrt{\operatorname{tr}\mathcal{G}[\phi]}$ measures the average change of length of tangent vectors from $\mathcal{M}_R$ when being mapped to tangent vectors on $\mathcal{M}_T$ and $\sqrt{\det \mathcal{G}[\phi]}$ measures the change of area under the deformation $\phi_{\mathcal{M}}$. Thus, $\operatorname{tr}\mathcal{G}[\phi]$ and $\det \mathcal{G}[\phi]$ are natural variables for an energy density in a variational approach measuring the tangential distortion, i.-e. we define an energy of the type

$$E_{\text{reg}}[\phi] = \int_{\omega_R} W(\operatorname{tr}\mathcal{G}[\phi], \det \mathcal{G}[\phi]) \sqrt{\det g_R}\, \mathrm{d}\xi\, .$$

When we press a given surface $\mathcal{M}_R$ into the thin mould of the surface $\mathcal{M}_T$, a second major source of stress results from the bending of normals. A simple thin shell energy reflecting this is given by

$$E_{\text{bend}}[\phi] = \int_{\omega_R} (h_T \circ \phi - h_R)^2 \sqrt{\det g_R}\, \mathrm{d}\xi\, .$$

Frequently, surfaces are characterized by similar geometric or texture features, which should be matched in a way which minimizes the difference of the deformed reference set $\phi_{\mathcal{M}}(\mathcal{F}_{\mathcal{M}_R})$ and the corresponding template set $\mathcal{F}_{\mathcal{M}_T}$. Hence, we consider a third energy

$$E_{\mathcal{F}}[\phi] = \mu \int_{\omega_R} \chi_{\mathcal{F}_R} \chi_{\phi^{-1}(F_T)} \sqrt{\det g_R} + \mu \int_{\omega_T} \chi_{\phi(\mathcal{F}_R)} \chi_{(F_T)} \sqrt{\det g_T}\, .$$

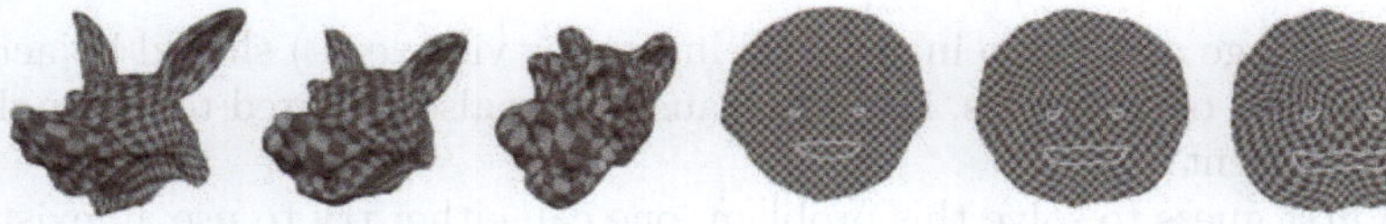

Fig. 4.4 Large deformations are often needed to match surfaces that have very different shapes. A checkerboard is texture mapped onto the first surface as it morphs to the second surface (**top**). The matching deformation shown in the parameter domain (**bottom**) is smooth and regular, even where the distortion is high (e.g., around the outlines of the mouth and eyes)

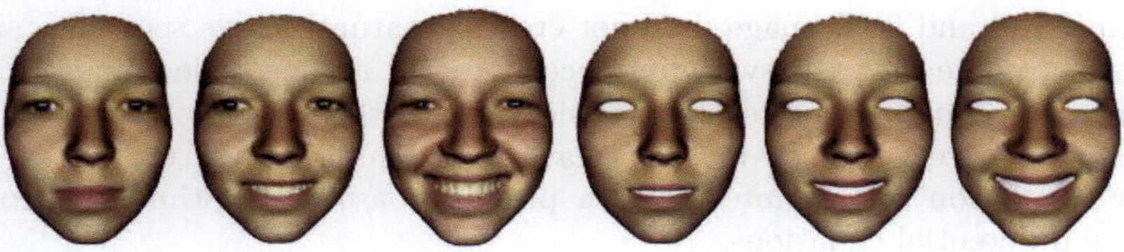

Fig. 4.5 Morphing through keyframe poses A, B, C is accomplished through pairwise matches $A \to B$ and $B \to C$. The skin texture from A is used throughout. Because of the close similarity in the poses, one can expect the intermediate blends A', B', C' to correspond very well with the original keyframes A, B, C, respectively

Usually, we cannot expect that $\phi_{\mathcal{M}}(\mathcal{M}_R) = \mathcal{M}_T$. Therefore, we must allow for a partial matching. For details on this and on the numerical approximation we refer to [57]. Figs. 4.4–4.5 show two different application of the variational surface matching method.

4.4 An Optimal Control Problem in Medical Image Processing

In this section we consider the problem of creating a "natural" movie which interpolates two given images showing essentially the same objects. In many situations, these objects are not at the same position or – more importantly - may be out-of-focus and blurred in one image while being in focus and sharp in the other. This description may be appropriate for frames in movies but also for different versions of a mammogram emphasizing coarse and fine details, respectively. The problem is to create an interpolating movie from these images which is perceived as "natural". In this context, we specify "natural" according to the following requirements. On the one hand, objects from the initial image should move smoothly to the corresponding object in the final image. On the other hand, the interpolation of an object which is blurred in the initial image and sharp in the final image (or vice versa) should be across different stages of sharpness, i.e. , the transition is also required to interpolate between different scales.

As a first guess to solve this problem, one can either try to use an existing morphing algorithm or to interpolate linearly between the two images. However, morphing methods are based on detecting matching landmarks in both images. They are not applicable here, since we are particularly interested in images containing objects, which are not present or heavily diffused in the initial image but appear with a detailed structure in the final image. Hence, there are no common landmarks for those objects. Mathematically speaking, it is difficult or impossible to match landmark points for an object which is given on a coarse and fine scale, respectively. Also linear interpolation between initial and final image does not create a natural image sequence, since it does not take the scale sweep into account, i.e. , all fine scale are appearing immediately rather than developing one after another.

Hence, more advanced methods have to be employed. In this article we show a solution of this interpolation problem based on optimal control of partial differential equations.

To put the problem in mathematical terms, we start with a given image y_0 assumed to be a function on $\Omega =]0,1[^2$. Under the natural assumption of finite-energy images, we model them as functions in $L^2(\Omega)$. The goal is to produce a movie (i.e. a time-dependent function) $y : [0,1] \to L^2(\Omega)$ such that appropriate mathematical implementations of the above conditions are satisfied.

4.4.1 Modeling as an Optimal Control Problem

Parabolic partial differential equations are a widely used tool in image processing. Diffusion equations like the heat equation [84], the Perona-Malik equation [66] or anisotropic equations [83] are used for smoothing, denoising and edge enhancing.

A smoothing of a given image $y_0 \in L^2(\Omega)$ can for example be done by solving the heat equation

$$\begin{aligned} y_t - \Delta y &= 0 \quad \text{in }]0,1[\times \Omega \\ y_\nu &= 0 \quad \text{on }]0,1[\times \partial\Omega \\ y(0) &= y_0 \,, \end{aligned}$$

where y_ν stands for the normal derivative, i.e. we impose homogeneous Neumann boundary conditions. The solution $y : [0,1] \to L^2(\Omega)$ gives a movie which starts at the image y_0 and becomes smoother with time t. This evolution is also called scale space and is analyzed by the image processing community in detail since the 1980s. Especially the heat equation does not create new features with increasing time, see e.g. [32] and the references therein. Thus, it is suitable for fading from fine to coarse scales.

The opposite direction, the sweep from coarse to fine scales, however, is not modeled by the heat equation. Another drawback of this PDE is that generally, all edges of the initial image will be blurred. To overcome this problem, the equation is modified such that it accounts for the edges and allows the formation of new structures. The isotropic diffusion is replaced with the degenerate diffusion tensor given by

$$D_p^2 = \left(I - \sigma(|p|)\frac{p}{|p|} \otimes \frac{p}{|p|}\right), \tag{4.9}$$

where the vector field $p :]0,1[\times \Omega \to \mathbb{R}^d$ with $|p| \leq 1$ describes the edges of the interpolating sequence and $\sigma : [0,1] \to [0,1]$ is an edge-intensity function. The special feature of this tensor is that it is allowed to degenerate for $|p| = 1$, blocking the diffusion in the direction of p completely.

Consequently, the degenerate diffusion tensor D_p^2 can be used for the preservation of edges. Additionally, in order to allow brightness changes and to create fine-scale structures, a source term u is introduced. The model under consideration then reads as:

$$\begin{aligned} y_t - \operatorname{div}\left(D_p^2 \nabla y\right) &= u \quad \text{in }]0,1[\times \Omega \\ \nu \cdot D_p^2 \nabla y &= 0 \quad \text{on }]0,1[\times \partial\Omega \\ y(0) &= y_0 \,. \end{aligned} \tag{4.10}$$

The above equation is well-suited to model a sweep from an image y_0 to an image y_1 representing objects on different scales. Hence, we take the image y_0 as initial value. To make the movie y end at a certain coarse scale image y_1 instead of the endpoint $y(1)$ which is already determined through (y_0, u, p), we propose the following optimal control problem:

$$\text{Minimize } J(y,u,p) = \frac{1}{2}\int_\Omega |y(1) - y_1|^2 \, \mathrm{d}x + \int_0^1 \int_\Omega \frac{\lambda_1}{2}|u|^2 + \lambda_2 \sigma(|p|) \, \mathrm{d}x \, \mathrm{d}t$$

$$\text{subject to } \begin{cases} y_t - \operatorname{div}\left(D_p^2 \nabla y\right) = u & \text{in }]0,1[\times \Omega \\ \nu \cdot D_p^2 \nabla y = 0 & \text{on }]0,1[\times \partial\Omega \\ y(0) = y_0 \,. \end{cases} \tag{4.11}$$

In other words, the degenerate diffusion process is forced to end in y_0 with the help of a heat source u and the edge field p and such that the energy for u and the edge-intensity $\sigma(|p|)$ becomes minimal.

4.4.2 Solution of the Optimal Control Problem

The minimization of the functional (4.11) is not straightforward. An analytical treatment of the minimization problem involves a variety of mathematical tasks. First, an appropriate weak formulation for (4.10) has to be found for which existence and uniqueness of solutions can be proven. Second, we have to ensure that a minimizer of a possibly regularized version of (4.11) exists. The main difficulty in these two points is to describe the influence of the parameter p in the underlying degenerate parabolic equation which control where the position and evolution of the edges in the solution. A general approach for minimizing Tikhonov functionals such as (4.11) by a generalized gradient method can be found in [11].

The Solution of the PDE

The solution of diffusion equations which are uniformly elliptic is a classical task. The situation changes when degenerate diffusion tensors like (4.9) are considered. In the following we fix an edge field p and examine the PDE (4.10) only with respect to (u, y_0) which is now linear. Here, when considering weak solutions, the choice of $L^2\big(0,1;H^1(\Omega)\big)$ for the basis of a solution space is not sufficient. This has its origin in one of the desired features of the equation: In order to preserve and create edges, which correspond to discontinuities in y with respect to the space variable, the diffusion tensor is allowed to degenerate. Such functions cannot be an element of $L^2\big(0,1;H^1(\Omega)\big)$. Hence, spaces adapted to the degeneracies have to be constructed by the formal closure with respect to a special norm (also see [62] for a similar approach):

$$\mathcal{V}_p = L^2\big(0,1;H^1(\Omega)\big)\big|^{\sim}_{\|\cdot\|_{\mathcal{V}_p}}, \quad \|y\|_{\mathcal{V}_p} = \Big(\int_0^1 \int_\Omega |y|^2 + |D_p \nabla y|^2 \,\mathrm{d}x\,\mathrm{d}t\Big)^{1/2}.$$

Elements y can be thought of square-integrable functions for which formally $D_p \nabla y \in L^2\big(0,1;L^2(\Omega)\big)$. One can moreover see that functions which admit discontinuities where $|p| = 1$ are indeed contained in $\mathcal{V}_p$. In the same manner, the solution space

$$W_p(0,1) = \big\{y \in \mathcal{AC}\big(0,1;H^1(\Omega)\big) \;\big|\; \|y\|_{W_p} < \infty\big\}\big|^{\sim}_{\|\cdot\|_{W_p}},$$
$$\|y\|_{W_p} = \big(\|y\|^2_{\mathcal{V}_p} + \|y_t\|^2_{\mathcal{V}_p^*}\big)^{1/2}.$$

A weak formulation of (4.10) then reads as: Find $y \in \mathcal{V}_p$ such that

$$-\langle z_t\,, y\rangle_{\mathcal{V}_p^* \times \mathcal{V}_p} + \langle D_p \nabla y,\; D_p \nabla z\rangle_{L^2} = \langle y_0,\; z(0)\rangle_{L^2} + \langle u,\; z\rangle_{L^2} \tag{4.12}$$

for all $z \in W_p(0,1)$ with $z(T) = 0$. One can prove that a unique solution exists in this sense.

Theorem 4.1 *For $p \in L^\infty(]0,1[\times \Omega, \mathbb{R}^d)$ with $\|p\|_\infty \le 1$, $u \in L^2\big(0,1;L^2(\Omega)\big)$ and $y_0 \in L^2(\Omega)$, there exists a unique solution of (4.12) in $W_p(0,T)$ with*

$$\|y\|^2_{W_p} \le C\big(\|u\|_2^2 + \|y_0\|_2^2\big)$$

where C is also independent of p.

Proof A solution can be obtained, for example with Lions' projection theorem or with Galerkin approximations. Both approaches yield the same solution in $W_p(0,1)$, whose uniqueness can be seen by a monotonicity argument. However, other solutions may exist in the slightly larger space

$$\bar{W}_p(0,1) = \{y \in \mathcal{V}_p \mid y_t \in \mathcal{V}_p^*\}, \quad \|y\|_{\bar{W}_p} = \|y\|_{W_p},$$

see [10] for details. □

Unfortunately, for each p, the solution space may be different and, in general, no inclusion relation holds. This complicates the analysis of the solution operator with respect to p in a profound way.

But fortunately, the spaces $W_p(0,1)$ still possess the convenient property that each $W_p(0,1) \hookrightarrow \mathcal{C}\big([0,1];L^2(\Omega)\big)$ with embedding constant independent of p. So, the solution operator

$$\mathcal{S} : L^2\big(0,1;L^2(\Omega)\big) \times \{\|p\|_\infty \leq 1\} \to \mathcal{C}\big([0,1];L^2(\Omega)\big), \quad (u,p) \mapsto y$$

is well-defined and bounded on bounded sets.

Examining the continuity of $\mathcal{S}$, a bounded sequence $\{u_l\}$ and arbitrary $\{p_l\}$ have, up to a subsequence, weak- and weak*-limits u and p. Since $\mathcal{C}\big([0,1];L^2(\Omega)\big)$ is not reflexive, we cannot assure weak convergence of the bounded sequence $\{y_k\}$, but it is possible to show that a weak limit y exists in the slightly larger space $\mathcal{C}^*\big([0,1];L^2(\Omega)\big)$ in which point-evaluation is still possible, again see [10] for details. The problem now is to show that the solution operator is closed in the sense that $\mathcal{S}(u,p) = y$.

Characterization of the Solution Spaces

One difficulty in examining the varying solution spaces $\mathcal{V}_p$ is the definition as a closure with respect to a norm which depends on p, resulting in equivalence classes of Cauchy sequences. A more intuitive description of the $\mathcal{V}_p$ is given in terms of special weak differentiation notions, as it is demonstrated in the following. In particular, this allows to describe the behavior of the solution operator $\mathcal{S}$ with respect to p.

For $w \in H^{1,\infty}(\Omega)$ and $q \in H^{1,\infty}(\Omega,\mathbb{R}^d)$, the *weak weighted derivative* and *weak directional derivative* of y are the functions, denoted by $w\nabla y$ and $\partial_q y$, respectively, satisfying

$$\int_\Omega (w\nabla y)\cdot z \,\mathrm{d}x = -\textstyle\int_\Omega y(w \operatorname{div} z + \nabla w \cdot z)\,\mathrm{d}x \quad \text{for all } z \in \mathcal{C}_0^\infty(\Omega,\mathbb{R}^d)$$

$$\int_\Omega \partial_q y z \,\mathrm{d}x = -\textstyle\int_\Omega y(z \operatorname{div} q + \nabla z \cdot q)\,\mathrm{d}x \quad \text{for all } z \in \mathcal{C}_0^\infty(\Omega).$$

With the help of these notions, a generalization of the well-known weighted Sobolev spaces [48] can be introduced, the *weighted and directional Sobolev spaces* associated with a weight $w \in H^{1,\infty}(\Omega)$ and directions $q_1, \ldots, q_K \in H^{1,\infty}(\Omega, \mathbb{R}^d)$:

$$H^2_{w,\partial q_1,\ldots,\partial q_K}(\Omega) = \{y \in L^2(\Omega) \mid w\nabla y \in L^2(\Omega, \mathbb{R}^d), \partial_{q_1} y, \ldots, \partial_{q_K} y \in L^2(\Omega)\}$$
$$\|y\|_{H^2_{w,\partial q_1,\ldots,\partial q_K}} = \Big(\|y\|_2^2 + \|w\nabla y\|_2^2 + \sum_{k=1}^{K} \|\partial_{q_k} y\|_2^2\Big)^{1/2}.$$

These spaces generalize weighted Sobolev spaces in the sense that ∇y does not necessarily exist for elements in $H^2_w(\Omega)$ and that $w = 0$ is allowed on non-null subsets of Ω.

The gain now is that the following weak closedness properties can be established:

$$\begin{matrix} y_l \rightharpoonup y \\ w_l \nabla y_l \rightharpoonup \theta \\ \partial_{q_{k,l}} y_l \rightharpoonup v_k \end{matrix} \quad \text{and} \quad \left.\begin{matrix} w_l \overset{*}{\rightharpoonup} w \\ q_{k,l} \overset{*}{\rightharpoonup} q_k \\ \operatorname{div} q_{k,l} \to \operatorname{div} q_k \\ \text{pointwise a.e.} \end{matrix}\right\} \Rightarrow \begin{cases} w\nabla y = \theta \\ \partial_{q_k} y = v_k . \end{cases} \tag{4.13}$$

Such a result is the key to prove that the solution operator $\mathcal{S}$ possesses appropriate closedness properties.

The construction of the weighted and directional weak derivative as well as the associated spaces can also be carried out for the time-variant case, resulting in spaces $\mathcal{H}^2_{w,\partial q_1,\ldots,\partial q_K}$. Splitting the diffusion tensor (4.9) then into a weight and direction as follows

$$w = \sqrt{1 - \sigma(|p|)}\,, \quad q = \begin{pmatrix} 0 & -1 \\ 1 & 0 \end{pmatrix} \frac{\sqrt{\sigma(|p|)}}{|p|} p$$

yields $D_p^2 = w^2 I + q \otimes q$, so $\nabla z \cdot D_p^2 \nabla y = w\nabla y \cdot w\nabla z + \partial_q y \partial_q z$. This gives an equivalent weak formulation in terms of weak weighted and directional derivatives.

Theorem 4.2 *For $\|p(t)\|_{H^{1,\infty}} \leq C$ a.e. and $|p| \leq C < 1$ on $]0,1[\times\partial\Omega$ follows that $\mathcal{V}_p = \mathcal{H}^2_{w,q}$ and $W_p(0,1) = \bar{W}_p(0,1)$. A $y \in \mathcal{V}_p$ is the unique solution of (4.12) if and only if*

$$-\langle z_t,\, y\rangle_{\mathcal{H}^{2*}_{w,q}\times\mathcal{H}^2_{w,q}} + \langle w\nabla y,\, w\nabla z\rangle_{L^2} + \langle \partial_q y,\, \partial_q z\rangle_{L^2} = \langle y_0,\, z(0)\rangle_{L^2} + \langle u,\, z\rangle_{L^2} \tag{4.14}$$

for each $z \in W_p(0,1)$, $z_t \in L^2\big(0,1; L^2(\Omega)\big)$ and $z(T) = 0$.

Proof For the proof and further details we again refer to [10]. □

Existence of Optimal Solutions

The characterization result of Theorem 4.2 as well as time-variant versions of the closedness property (4.13) are the crucial ingredients to obtain existence of solutions for a regularized version of (4.11).

Theorem 4.3 *Let $\mathcal{P}$ a weak*-compact set such that each $p \in \mathcal{P}$ satisfies the prerequisites of Theorem 4.2. The control problem*

$$\min_{\substack{u \in L^2(0,1;L^2(\Omega)) \\ p \in \mathcal{P}}} \frac{\|y(1) - y_1\|_2^2}{2} + \lambda_1 \|u\|_2^2 + \lambda_2 \int_0^T \int_\Omega \sigma(|p|) \,\mathrm{d}x \,\mathrm{d}t$$
$$+ \mu_1 \mathrm{tv}^*(p) + \mu_2 \operatorname*{ess\,sup}_{t \in [0,1]} \mathrm{TV}\big(\nabla p(t)\big)$$
$$\text{subject to} \quad \begin{cases} y_t - \mathrm{div}\big(D_p^2 \nabla y\big) = u & \text{in }]0,1[\times \Omega \\ \nu \cdot D_p^2 \nabla y = 0 & \text{on }]0,1[\times \partial\Omega \\ y(0) = y_0 \,. \end{cases}$$

possesses at least one solution (u^*, p^*). Here, tv^* and TV denote the semi-variation with respect to t and the total variation, respectively.

Proof The proof can roughly be sketched as follows, see [10] for a rigorous version. For a minimizing sequence (y_l, u_l, p_l), one obtains weak- and weak*-limits (y^*, u^*, p^*) according to the compactness stated above. Theorem 4.2 gives weakly convergent sequences $w_l \nabla y_l$ and $\partial_{q_l} y_l$ as well as the alternative weak formulation (4.14). The total-variation regularization terms then ensure the applicability of closedness properties analog to (4.13), so passing to the limit in (4.14) yields that $y^* \in \bar{W}_{p^*}(0,1)$ is the unique solution associated with (u^*, p^*). Finally, with a lower-semicontinuity argument, the optimality is verified. □

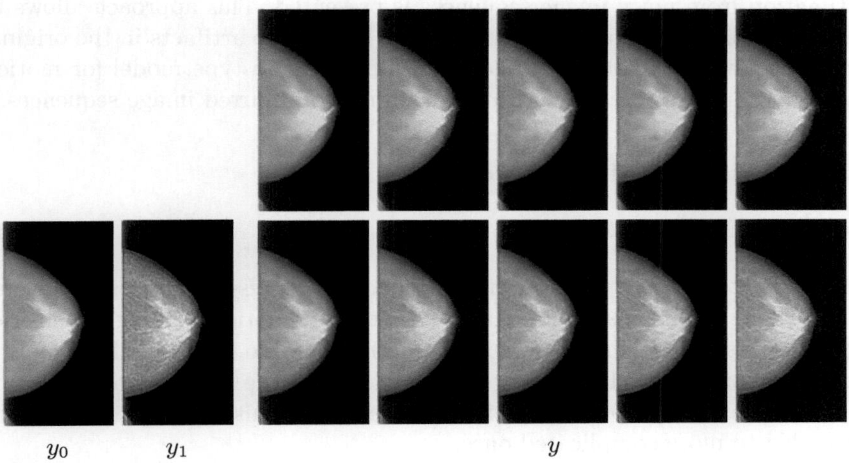

Fig. 4.6 Illustration of an interpolating sequence generated by solving the proposed optimal control problem. The two leftmost images depict y_0 and y_1, respectively (a coarse- and fine-scale version of a mammography image), while some frames of the optimized image sequence can be seen on the right

Having established the existence of at least one minimizing element, one can proceed to derive an optimality system based on first-order necessary conditions (which is possible for $||p||_\infty < 1$). Furthermore, numerical algorithms for the optimization of the discrete version of (4.11) can be implemented, see Fig. 4.6 for an illustration of the proposed model.

4.5 Restoration and Post Processing of Optical Flows

The estimation of motion in image sequences has gained wide spread importance in a number of scientific applications stemming from diverse fields such as environmental and life-sciences. Optical imaging systems acquire imaging data non-invasively, a prerequisite for accurate measurements. For analyzing transport processes, the estimation of motion or optical flow plays a central role. Equally, in engineering applications the estimation of motion from image sequences is not only important in fluid dynamics but can also be used in novel products such as driver assisting systems or in robot navigation. However, frequently the image data is corrupted by noise and artifacts. In infrared thermography, temperature fluctuations due to reflections are often impossible to fully eliminate. In this paper, novel techniques will be presented which detect artifacts or problematic regions in image sequences. Optical flow computations based on local approaches such as those presented in Chap. 7 can then be enhanced by rejecting wrong estimates and inpainting the flow fields from neighboring areas. Furthermore, a joint Mumford Shah type approach for image restoration, image and motion edge detection and motion estimation from noisy image sequences is presented. This approach allows to restore missing information, which may be lost due to artifacts in the original image sequence. Finally, we discuss a Mumford Shah type model for motion estimation and restoration of frames from motion-blurred image sequences.

4.5.1 Modeling and Preprocessing

Standard Motion Model

The estimation of motion from image sequences represents a classical inverse problem. As such, constraint equations that relate motion to image intensities and changes thereof are required. In Chap. 7, a wide range of these motion models is presented. Here we will just introduce the simplest one, keeping in mind that the proposed algorithms based upon this model can readily be extended to more complicated ones.

For a finite time interval $[0,T]$ and a spatial domain $\Omega \subset \mathbb{R}^d$ with $d = 1,2,3$ the image sequence $u : D \to \mathbb{R}$ is defined on the space time domain $D = [0,T] \times \Omega$. If $x : [0,T] \to \mathbb{R}^d$ describes the trajectory of a point of an object such that the velocity $w = (1, v)$ is given by $\dot{x} = w$ we can model a constant brightness intensity u as $u(t, x(t)) = \text{const}$. A first order approximation yields

$$\frac{du}{dt} = 0 \quad \Leftrightarrow \quad \frac{\partial u}{\partial t} + v \cdot \nabla_{(x)} u = 0 \quad \Leftrightarrow \quad w \cdot \nabla_{(t,x)} u = 0\,, \tag{4.15}$$

where ∇ is the gradient operator with respect to parameters given as indices. Models based on this equation are called *differential models* since they are based on derivatives.

The parameters w of the motion model (4.15) can be solved by incorporating additional constraints such as local constancy of parameters or global smoothness (a more refined approach of assuming global piecewise smoothness will be presented in Sect. 4.5.2). Refined techniques for local estimates extending the common structure tensor approach have been outlined in Chap. 7 and will not be repeated here.

Comparison of Confidence and Situation Measures and Their Optimality for Optical Flows

In order to detect artifacts in image sequences, one can analyze confidence and situation measures. Confidence measures are used to estimate the correctness of flow fields, based on information derived from the image sequence and/or the displacement field. Since no extensive analysis of proposed confidence measures has been carried out so far, in [45] we compare a comprehensive selection of previously proposed confidence measures based on the theory of intrinsic dimensions [86], which have been applied to analyze optical flow methods in [42]. We find that there are two kinds of confidence measures, which we distinguish into situation and confidence measures. Situation measures are used to detect locations, where the optical flow cannot be estimated unambiguously. This is contrasted by confidence measures, which are suited for evaluating the degree of accuracy of the flow field. Situation measures can be applied, e.g., in image reconstruction [58], to derive dense reliable flow fields [74] or to choose the strength of the smoothness parameter in global methods (e.g., indirectly mentioned in [54]). Confidence measures are important for quantifying the accuracy of the estimated optical flow fields. A successful way to obtain robustness to noise in situation and confidence measures is also discussed in [45].

Previously, confidence measures employed were always chosen as innate to the flow estimation technique. By combining flow methods with non-inherent confidence measures we were able to show considerable improvements for confidence and situation measures. Altogether the results of the known measures are only partially satisfactory as many errors remain undetected and a large number of false positive error detections have been observed. Based on a derived optimal confidence map we obtain the results in Fig. 4.7 for Lynn Quam's Yosemite sequence [36], and the Street [59] test sequences. For situation measures we conclude by presenting the best measure for each intrinsic dimension. Quantitative results can be found in [45].

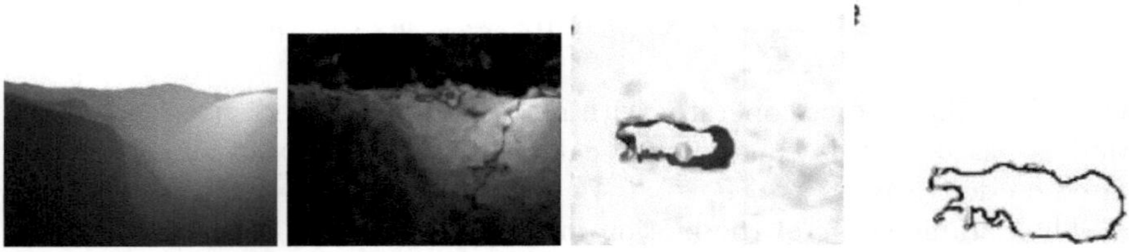

Fig. 4.7 Comparison of optimal confidence measure (**left**) to best known confidence measure (**right**) for Yosemite and Street sequences

An Adaptive Confidence Measure Based on Linear Subspace Projections

For variational methods, the inverse of the energy after optimization has been proposed as a general confidence measure in [13]. For methods not relying on global smoothness assumptions, e.g. local methods, we propose a new confidence measure based on linear subspace projections in [46]. The idea is to derive a spatio-temporal model of typical flow field patches using e.g. principal component analysis (PCA). Using temporal information the resulting eigenflows can represent complex temporal phenomena such as a direction change, a moving motion discontinuity or a moving divergence. Then the reconstruction error of the flow vector is used to define a confidence measure.

Quantitative analysis shows that using the proposed measure we are able to improve the previously best results by up to 31%. A comparison between the optimal, the obtained confidence and the previously often applied gradient measure [4, 13] is shown in Fig. 4.8.

Surface Situation Measures

In [47] we present a new type of situation measure for the detection of positions in the image sequence, where the full optical flow cannot be estimated reliably (e.g. in the case of occlusions, intensity changes, severe noise, transparent structures, aperture problems or homogeneous regions), that is in unoccluded situations of intrinsic dimension two. The idea is based on the concept of surface functions. A surface function for a given flow vector v reflects the

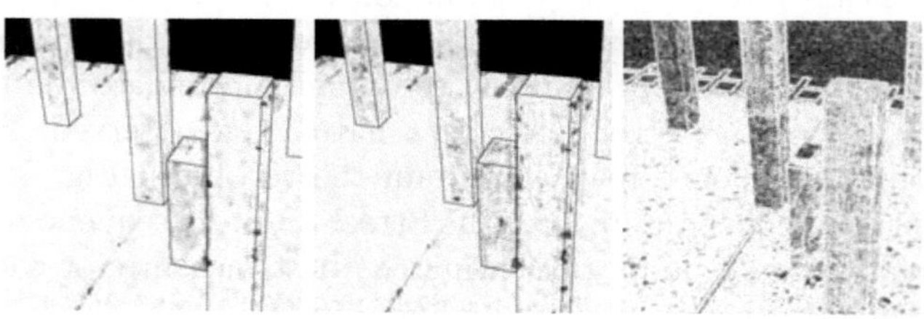

Fig. 4.8 Comparison to optimal confidence, **left**: optimal confidence map, center: pcaReconstruction confidence map, **right**: previously often used gradient confidence measure

variation of a confidence measure c over the set of variations of the current displacement vector.

$$S_{x,v,c} : \mathbb{R}^2 \to [0,1]\,, \;\; S_{x,v,c}(d) := c(x, v+d)\,. \tag{4.16}$$

By analyzing the curvature of a given surface function statements on the intrinsic dimension and possible occlusions can be made. The surface situation measures have proven superior to all previously proposed measures and are robust to noise as well.

4.5.2 Restoration of Optical Flows

Optical Flows via Flow Inpainting Using Surface Situation Measures

Based on the surface situation measures introduced in Sect. 4.5.1, in [47] we suggest a postprocessing technique for optical flow methods, a flow inpainting algorithm, which integrates the information provided by these measures and obtains significantly reduced angular errors. We demonstrate that 100% dense flow fields obtained from sparse fields via flow inpainting are superior to dense flow fields obtained by local and global methods. Table 4.1 shows the reduction of the angular error of four flow fields computed by the local structure tensor (ST) [8] and the global combined local global (CLG) method [14] by means of flow inpainting.

Comparing the angular error obtained by the derived flow inpainting algorithm to the angular error of the original flow fields computed with two state of the art methods (the fast local structure tensor method and the highly accurate combined local global method) we could achieve up to 38% lower angular errors and an improvement of the accuracy in all cases. We conclude that both local and global methods can be used alike to obtain dense optical flow fields with lower angular errors than state of the art methods by means of the proposed flow inpainting algorithm. The algorithm was also used to compute accurate flow fields on real world applications. In Fig. 4.9 two examples

Table 4.1 Original and inpainting angular error for surface measures and inpainting error based on the best previously known situation measure [45] on average for ten frames of the test sequences for the combined local global and the structure tensor method

	Combined local global		Structure tensor	
	Original	Inpainting	Original	Inpainting
Marble	3.88 ± 3.39	3.87 ± 3.38	4.49 ± 6.49	3.40 ± 3.56
Yosemite	4.13 ± 3.36	3.85 ± 3.00	4.52 ± 10.10	2.76 ± 3.94
Street	8.01 ± 15.47	7.73 ± 16.23	5.97 ± 16.92	4.95 ± 13.23
Office	3.74 ± 3.93	3.59 ± 3.93	7.21 ± 11.82	4.48 ± 4.49

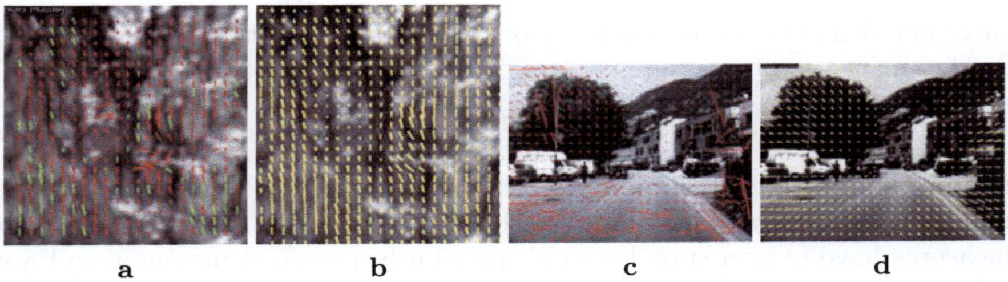

Fig. 4.9 In (**a**) the estimated flow field based on the structure tensor is shown for an infrared sequence of the air–water interface. Reflections lead to wrong estimates. The post processed motion field is shown in (**b**). In (**c**) and (**d**) the same is shown for a traffic scene.

for typical applications are presented. The inpainting algorithm signifiacntly reduces errors due to reflections in thermographic image sequences of the air–water interface and errors in different situations in traffic scenes.

Joint Estimation of Optical Flow, Segmentation and Denoising

In the previous section, separate techniques for detecting artifacts were presented, followed by an algorithm to inpaint parts of the flow field corrupted by the artifacts. In this section we will outline a technique for jointly denoising an image sequence, estimating optical flow and segmenting the objects at the same time [75]. Our approach is based on an extension of the well known Mumford Shah functional which originally was proposed for the joint denoising and segmentation of still images. Given a noisy initial image sequence $u_0 : D \to \mathbb{R}$ we consider the energy

$$\begin{aligned} E_{\mathrm{MSopt}}[u,w,S] = &\int_D \frac{\lambda_u}{2}(u-u_0)^2 \,\mathrm{d}\mathcal{L} + \int_{D\setminus S} \frac{\lambda_w}{2}\left(w\cdot\nabla_{(t,x)}u\right)^2 \mathrm{d}\mathcal{L} \\ &+ \int_{D\setminus S} \frac{\mu_u}{2}\left|\nabla_{(t,x)}u\right|^2 \mathrm{d}\mathcal{L} \\ &+ \int_{D\setminus S} \frac{\mu_w}{2}\left|P_\delta[\zeta]\nabla_{(t,x)}w\right|^2 \mathrm{d}\mathcal{L} + \nu\mathcal{H}^d(S) \end{aligned}$$

for a piecewise smooth denoised image sequence $u : D \to \mathbb{R}$, and a piecewise smooth motion field $w = (1, v)$ and a set $S \subset D$ of discontinuities of u and w. The first term models the fidelity of the denoised image-sequence u, the second term represents the fidelity of the flow field w in terms of the optical flow (4.15). The smoothness of u and w is required on $D \setminus S$ and finally, the last term is the Hausdorff measure of the set S. A suitable choice of the projection $P_\delta[\zeta]$ leads to an anisotropic smoothing of the flow field along the edges indicated by ζ.

The model is implemented using a phase field approximation in the spirit of Ambrosio and Tortorelli's approach [2]. Thereby the edge set S is replaced by a phase field function $\zeta : D \to \mathbb{R}$ such that $\zeta = 0$ on S and $\zeta \approx 1$ far from S. Taking into account the Euler-Lagrange equations of the corresponding yields a system of three partial differential equations for the image-sequence u, the optical flow field v and the phase field ζ:

$$
\begin{aligned}
-\mathrm{div}_{(t,x)}\Big(\frac{\mu_u}{\lambda_u}(\zeta^2+k_\epsilon)\nabla_{(t,x)}u + \frac{\lambda_w}{\lambda_u}w(\nabla_{(t,x)}u\cdot w)\Big) + u = u_0 \\
-\epsilon\Delta_{(t,x)}\zeta + \left(\frac{1}{4\epsilon} + \frac{\mu_u}{2\nu}\left|\nabla_{(t,x)}u\right|^2\right)\zeta = \frac{1}{4\epsilon} \qquad (4.17)\\
-\frac{\mu_w}{\lambda_w}\mathrm{div}_{(t,x)}\left(P_\delta[\zeta]\nabla_{(t,x)}v\right) + (\nabla_{(t,x)}u\cdot v)\nabla_{(x)}u = 0\,.
\end{aligned}
$$

For details on this approximation and its discretization we refer to [31].

In Fig. 4.10 we show results from this model on a noisy test-sequence where one frame is completely missing. But this does not hamper the restoration of the correct optical flow field shown in the fourth column, because of the anisotropic smoothing of information from the surrounding frames into the destroyed frame.

Furthermore, in Fig. 4.11 we consider a complex, higher resolution video sequence showing a group of walking pedestrians. The human silhouettes are well extracted and captured by the phase field. The color-coded optical flow plot shows how the method is able to extract the moving limbs of the pedestrians.

Joint Motion Estimation and Restoration of Motion Blur

Considering video footage from a standard video camera, it is quite noticeable that relatively fast moving objects appear blurred. This effect is called *motion blur*, and it is linked to the aperture time of the camera, which roughly speaking integrates information in time. The actual motion estimation suffers from motion blur and on the other hand given the motion the blur can be removed by "deconvolution". Hence, these two problems are intertwined, which motivates the development of a method that tackles both problems at once. In [7] a corresponding joint motion estimation and deblurring model has

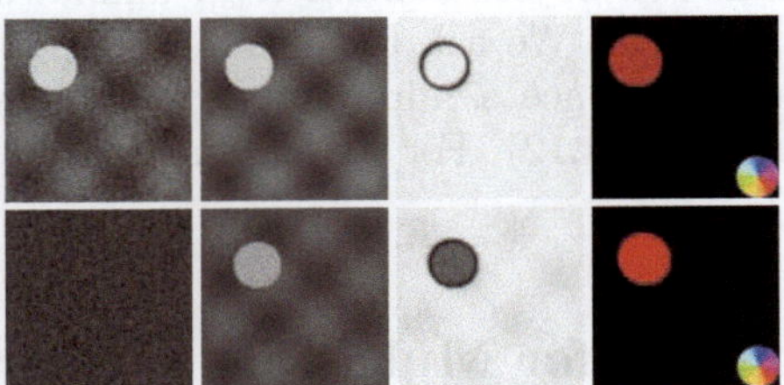

Fig. 4.10 Noisy test sequence: From top to bottom frames 9 and 10 are shown. (**a**) original image sequence, (**b**) smoothed images, (**c**) phase field, (**d**) estimated motion (color coded)

Fig. 4.11 Pedestrian video: frames from original sequence (**left**); phase field (**middle**); optical flow, color coded (**right**)

been presented. For simplicity let us assume that an object is moving with constant velocity v in front of a still background and we observe m frames $g_1, \cdots u_m$ at times $t_1, \cdots, t_m$. From the object and background intensity functions f_{obj} and f_{bg}, respectively, one assembles the actual scene intensity function $f(t,x) = f_{\mathrm{obj}}(x-tv)\chi_{\mathrm{obj}}(x-vt)+f_{\mathrm{bg}}(x)(1-\chi_{\mathrm{obj}}(x-vt))$. Now, it turns out to be crucial close to motion edges to observe that the theoretically observed motion blur at time t is a properly chosen average of background intensity and motion blurred object intensity. Indeed, the expected intensity is given by $G_i[\Omega_{\mathrm{obj}}, v, f_{\mathrm{obj}}, f_{\mathrm{bg}}](x) := ((f_{\mathrm{obj}}\chi_{\mathrm{obj}}) * h_v)(x - t_i v) + f_{\mathrm{bg}}(x)(1 - (\chi_{\mathrm{obj}} * h_v)(x - t_i v))$, where χ_{obj} is the characteristic function of the object domain Ω_{obj} and $h_v := \delta_0((v^\perp / |v|) \cdot y)h((v / |v|) \cdot y)$ a one dimensional filter kernel with filter width $\tau|v|$ in the direction of the motion trajectory $\{y = x + sv : s \in \mathbb{R}\}$. Here $v^\perp$ denotes v rotated by 90 degrees, δ_0 is the usual 1D Dirac distribution and h the 1D block filter with $h(s) = 1 / (\tau|v|)$ for $s \in [-(\tau|v|) / 2, (\tau|v|) / 2]$ and $h(s) = 0$, else. Hence, a Mumford Shah type approach for joint motion estimation and deblurring comes along with the energy

$$E[\Omega_{\mathrm{obj}}, v, f_{\mathrm{obj}}] = \sum_{i=1,2} \int_\Omega \left(G_i[\Omega_{\mathrm{obj}}, v, f_{\mathrm{obj}}, f_{\mathrm{bg}}] - g_i\right)^2 \, \mathrm{d}\mathcal{L} + \int_\Omega \mu |\nabla f_{\mathrm{obj}}| \, \mathrm{d}\mathcal{L} + \nu |\partial \Omega_{\mathrm{obj}}|$$

depending on the unknown object domain Ω_{obj}, unknown velocity v, object intensity f_{obj} to be restored. We ask for a minimizing set of the degrees of freedom Ω_{obj}, v, and f_{obj}. Once a minimizer is known, we can retrieve the deblurred images (see Fig. 4.12). For details on this approach and further results we refer to [7].

4.6 FEM Techniques for Multiscale Visualization of Time-Dependent Flow Fields

The analysis and post-processing of flow fields is one of the fundamental tasks in scientific visualization. Sophisticated multiscale methods are needed to visualize and analyze the structure of especially nonstationary flow fields for which

Fig. 4.12 From two real blurred frames (**left**), we automatically and simultaneously estimate the motion region, the motion vector, and the image intensity of the foreground (**middle**). Based on this and the background intensity we reconstruct the two frames (**right**)

the standard tools may fail. A huge variety of techniques for the visualization of steady as well as time-dependent flow fields in 2 D and 3 D has been presented during the last years. The methods currently available range from particle tracing approaches [79, 81] over texture based methods [16, 27, 40, 72, 80] to feature extraction for 3 D flow fields [19, 39, 41, 76]. An overview is given by Laramee et al. [55].

In this section we discuss the application of an anisotropic transport diffusion method to complex flow fields resulting from CFD computations on arbitrary grids. For general unstructured meshes, we apply the discretization of the arising transport diffusion problems by the streamline-diffusion (SD) FEM scheme, and we discuss iterative solvers of type Krylov-space or multigrid schemes for the arising nonsymmetric auxiliary problems. We analyze a corresponding balancing of the involved operators and blending strategies. The application to several test examples shows that the approaches are excellent candidates for efficient visualization methods of highly nonstationary flow with complex multiscale behavior in space and time.

Moreover we show a technique for multiscale visualization of static flow fields which is based on an algebraic multigrid method. Starting from a standard finite element discretization of the anisotropic diffusion operator, the algebraic multigrid yields a hierarchy of inter-grid prolongation operators. These prolongations can be used to define coarse grid finite element basis functions whose support is aligned with the flow field.

4.6.1 The Anisotropic Transport Diffusion Method

In [15, 70] special methods which are based on anisotropic diffusion and anisotropic transport diffusion for the visualization of static and time-dependent vector fields have been presented. In this section we briefly review these models, the according parameters and a blending strategy which is needed to produce a visualization of time-dependent flow fields.

The Transport Diffusion Operator

We consider a time-dependent vector field $v : I \times \Omega \to \mathbb{R}^d$, $(s, x) \mapsto v(s, x)$ given on a finite time-space cylinder $I \times \Omega$ where $I = [0, T]$ and $\Omega \subset \mathbb{R}^d$ for

$d = 2, 3$. Here, we restrict to $d = 2$. If the vector field v is constant in time, i.e., $v(s, x) = v_0(x)$ for all $s \in I$, we can create a multiscale visualization of the flow field in form of a family of textures $\{u(t)\}_{t\in\mathbb{R}^+}$ by the following anisotropic diffusion equation:
Find $u : \mathbb{R}^+ \times \Omega \to \mathbb{R}$ *such that*

$$\begin{aligned} \partial_t u - \operatorname{div}(A(v,\nabla u)\nabla u) &= f(u) && \text{in } \mathbb{R}^+ \times \Omega\,, \\ A(v,\nabla u)\partial_n u &= 0 && \text{on } \mathbb{R}^+ \times \partial\Omega\,, \\ u(0,\cdot) &= u_0(\cdot) && \text{in } \Omega\,. \end{aligned} \tag{4.18}$$

We start this evolution with an initial image u_0 showing random white noise. Since we have assumed the vector field to be continuous, there exists a family of orthogonal mappings $B(v) \in SO(d)$ such that $B(v)e_1 = v$. And denoting the identity matrix of dimension d with Id_d, the diffusion tensor reads

$$A(v,\nabla u) = B(v) \begin{pmatrix} \alpha(\|v\|) & 0 \\ 0 & G(\|\nabla u\|)\mathrm{Id}_{d-1}\,, \end{pmatrix} B(v)^T$$

where α is a monotone increasing function which prescribes a linear diffusion in direction of v for $\|v\| > 0$. We will choose α appropriately below. During the evolution, patterns are generated which are aligned with the flow field. The function $G(s) := \varepsilon/(1 + c\,s^2)$ – well known in image processing [65] – controls the diffusion in the directions orthogonal to the flow. It is modeled such that the evolution performs a clustering of streamlines and thus generates coarser representations of the vector field with increasing scale t. The definition of the diffusion tensor G depends on the gradient of a regularized image $u^\sigma = u * \chi^\sigma$. This regularization is theoretically important for the well-posedness of the presented approach [18, 44]. To our experience, in the implementation this regularization can be neglected or can be replaced by a lower bound for the value of $G(\cdot)$. For $\|v\| = 0$ we use an isotropic diffusion operator. The role of the right hand side $f(u)$ (4.18) is to strengthen the contrast of the image during the evolution, because for $f \equiv 0$ the asymptotic limit would be an image of constant gray value. We set $f(u) = \rho \times \left((2\,u - 1) - (2\,u - 1)^3\right)$ with $\rho = 80$ to increase the set of asymptotic states of the evolution. An example[1] of the multiscale evolution is shown in Fig. 4.13, where the multiscale visualization of a flow field is displayed for the Venturi pipe problem in 2 D [1].

Let us now suppose that the vector field varies smoothly in time. If we would consider the evolution equation separately for each fixed time $s \in I$, the resulting textures at a fixed scale $t_0 \in \mathbb{R}^+$ would not give a smooth animation of the flow in time. This is due to a lack of correlation between the line-structures of the separate textures. However, if there would be a correlation between the structure of separate textures, the resulting animation would only give an Eulerian type representation of the flow.

[1] This example was computed with a time step of $\Delta t = 0.005$ on a mesh with 82753 nodes.

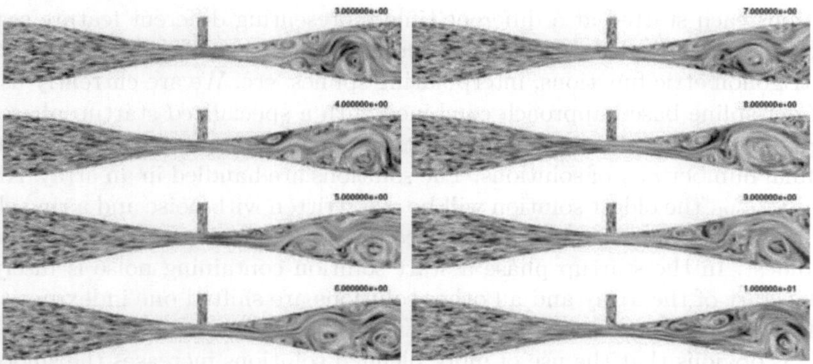

Fig. 4.13 Multiscale visualization of the Venturi pipe example (with transport)

To obtain a Lagrangian type representation, we consider the following anisotropic transport diffusion operator for the multiscale representation $u : \mathbb{R}^+ \times \Omega \to \mathbb{R}$ and the corresponding inhomogeneous transport diffusion equation

$$\begin{aligned} \partial_t u + v \cdot \nabla u - \operatorname{div}(A(v, \nabla u)\nabla u) &= f(u) \quad \text{in } \mathbb{R}^+ \times \Omega\,, \\ A(v)\,\partial_n u &= 0 \qquad \text{on } \mathbb{R}^+ \times \partial\Omega\,, \\ u(0,\cdot) &= u_0(\cdot) \quad \text{in } \Omega\,. \end{aligned} \tag{4.19}$$

In this equation we have identified the time s of the vector field with the scale t of the multiscale visualization. Indeed the resulting texture shows again structures aligned with streamlines which are now transported with the flow. But due to the coupling of s and t the feature scale gets coarser with increasing time, i.e., we are not able to fix a scale t_0 and typically an animation is created at this scale showing always patterns of the same size. This fact makes the use of an appropriate blending strategy unavoidable.

Balancing the Parameters

In general the transport and the diffusion of the patterns of the texture u are opposite processes. Denoting a time-step of the transport diffusion equation with Δt and introducing the balance parameter $\beta > 0$ we have [15]

$$\alpha(\|v\|)(x) = \frac{\beta^2 \max(\|v(x)\|\,,\, \|v\|_{min})^2 \Delta t}{2}\,.$$

In our applications we use the setting $\beta = 10$ and $\|v\|_{min} = 0.05$.

Blending Strategies

Blending strategies have to be used to get a visual representation of a given flow inside of a desired feature scale range. This means we have a set of solutions each started at a different time representing different feature scales which will be blended together. Different blending strategies are possible, e.g. trigonometric functions, interpolating splines, etc. We are currently using a Bézier-spline based approach combined with a specialized startup phase.

At the startup phase we will bootstrap our blending from one solution to the final number n_{tot} of solutions. The solutions are handled in an array. After time Δt_{blend}, the oldest solution will be overwritten with noise and a ring shift will be carried out that brings the second oldest solution to the position of the oldest. In the startup phase a start solution containing noise is inserted at the start of the array and all other solutions are shifted one index position higher.

Is is obvious that the use of more blended solutions increases the smoothness of the transitions between visible feature scales. However, the computational time increases linearly with the number of used solutions which comes down to a tradeoff between quality and time. For preview purposes, two blended solutions are sufficient. High quality visualizations will need more.

4.6.2 Discretization

A Finite Element Discretization for Static Flow Fields

For static flow fields and the scheme (4.18) we can use a standard finite element method on a given discretizational grid of the domain. A semi-implicit Backward Euler scheme with time step width Δt is applied, which results in the evaluation of the diffusion tensor A and the right hand side f at the previous time steps. This leads to the semi-discrete scheme

$$\frac{u_{n+1} - u_n}{\Delta t} - \operatorname{div}\left(A(v_{n+1}, \nabla u_n)\nabla u_{n+1}\right) = f(u_n)\,, \tag{4.20}$$

where u_n denotes the evaluation of u at time $n\Delta t$. Denoting the finite element basis functions with ϕ_i the spatial discretization yields the well known mass matrix M with entries $M_{ij} = \int_\Omega \phi_i\, \phi_j\, dx$ and the stiffness matrix L^n at time step n with entries $L^n_{ij} = \int_\Omega A(v_n)\nabla\phi_i \cdot \nabla\phi_j$. In summary we get a system of equations $(M + (\Delta t)L^{n-1})U^n = MU^{n-1} + (\Delta t)F^{n-1}$ for the vector U^n of nodal values of u^n. This system can be solved with e.g. a conjugate gradient method.

The SD Finite Element Method for Time-Dependent Flow Fields

In [15] a characteristic-upwinding algorithm due to Pironneau [68] is used to discretize the transport diffusion scheme (4.19) on quadtree/octtree grids for $\Omega = [0,1]^d$. For the diffusive parts and the right hand side again a semi-implicit Backward Euler scheme with time step Δt is applied (cf. (4.20)):

$$\frac{u_{n+1} - u_n}{\Delta t} + v_{n+1} \cdot \nabla u_{n+1} - \operatorname{div}\left(A(v_{n+1}, \nabla u_n)\nabla u_{n+1}\right) = f(u_n)\,. \tag{4.21}$$

However the application of the anisotropic diffusion visualization method on rectangular or cubical domains is often unrealistic in complex CFD applications. Moreover, vector field data typically coming from CFD simulations is rarely given on structured quadtree/octtree grids. Furthermore, the scheme introduces some numerical diffusion which decreases the quality of the final animation. In this section we discuss a higher order discretization scheme on general meshes which leads to high quality animations, showing sharp patterns moving with the flow field.

The variational formulation of (4.21) reads

$$(u_{n+1}, \psi) + \Delta t(v_{n+1} \cdot \nabla u_{n+1}, \psi) + \Delta t(A(v_{n+1}, \nabla u_n)\nabla u_{n+1}, \nabla \psi) = \tag{4.22}$$

$$\Delta t(f(u_n), \psi) + (u_n, \psi) \quad \forall \psi \in \mathcal{V} \tag{4.23}$$

with the space of test-functions $\mathcal{V}$ and test functions $\psi \in \mathcal{V}$.

The convection part of our equation demands some kind of additional stabilization. Since the diffusion operator A is already decomposed, in a way that allows to control the diffusion in flow direction, we replace A with a slightly modified version $\tilde{A}$:

$$\tilde{A}(v, \nabla u) = B(v) \begin{pmatrix} \alpha(\|v\|) + \mathrm{sd} & 0 \\ 0 & G(\|\nabla u\|)\mathrm{Id}_{d-1} \end{pmatrix} B(v)^T .$$

This modification allows an easy implementation of the streamline-diffusion scheme. The scalar function *sd* is the necessary *streamline diffusion* added in flow direction and is computed by

$$Re_{\mathrm{loc}} := \frac{\|v\|_{\mathrm{loc}}\, h_{\mathrm{loc}}}{\alpha(\|v\|)}, \qquad \mathrm{sd} := \mathrm{sd}_{\mathrm{par}}\, h_{\mathrm{loc}} \frac{Re_{\mathrm{loc}}}{1 + Re_{\mathrm{loc}}} .$$

The parameter $\mathrm{sd}_{\mathrm{par}} \in (0, 2)$ is user-specified and h_{loc} is the local mesh width, that means defined on each mesh cell, analogously to $\|v\|_{\mathrm{loc}}$ as local flow speed (see [77] for more details). The advantage of this scheme is that it can be easily applied on general unstructured meshes, giving sufficient robustness for treating the convection dominated parts while at the same time the amount of numerical diffusion is not too big. Moreover, since it is a *linear scheme* - in contrast to TVD methods – the resulting subproblems are linear and can be efficiently treated via standard iterative solvers. However, being a linear scheme, the SD scheme suffers potentially from spurious numerical oscillations, due to over and undershooting, and the choice for the user-specific parameter $\mathrm{sd}_{\mathrm{par}}$ can be critical. In a forthcoming paper, we plan to analyze the influence of the parameter $\mathrm{sd}_{\mathrm{par}}$ onto the behavior of accuracy, robustness and efficiency of the described numerical approaches.

4.6.3 Multiscale Analysis with Algebraic Multigrid (AMG)

In [34] we use the fact that the structure of the flow is completely encoded in the diffusion operator $-\mathrm{div}\,(A(v, \nabla u)\nabla u)$ to create a multiscale representation of the flow field. Let us assume that we have discretized the diffusion operator by standard finite elements on a regular grid yielding the stiffness-matrix L introduced in Subsect. 4.6.2. The algebraic multigrid method (AMG) finds a hierarchy of finite element basis functions which leads to optimal convergence of a multigrid solver of the linear system of equations determined by L. Thereby it generates a set of inter-grid prolongation matrices P^k which define the coarse grid basis.

Since the structure of the flow is encoded into the discretized diffusion operator, the AMG aligns the support of coarse grid basis functions to the diffusion of mass along the vector field. Consequently the prolongation matrices can be used for a multiscale visualization of flow fields. In Fig. 4.14 we show the AMG multiscale representation of the vector field of a convective flow.

Fig. 4.14 Multiscale visualization of a convective flow using AMG. The left column shows flow field clusters which are obtained from the supports of basis functions at different grid levels. The right column shows a representation of the clusters with arrow icons. The grid level increases from top to bottom

4.6.4 Conclusions and Outlook

We have discussed multiscale visualization techniques for time-dependent and static flow fields coming from CFD simulations on general 2 D and 3 D domains. The proposed models are based on PDEs with anisotropic transport and diffusion operators which are linearized in time by a semi-implicit approach. The simple diffusion problem can be discretized by a standard FEM scheme, for the transport diffusion scheme the resulting problem in each time step is discretized by a sophisticated streamline-diffusion FEM scheme on unstructured quadrilateral grids. The main features of the proposed numerical methods together with improved blending strategies and a discussion of the involved parameters have been tested via numerical examples.

In a next step, the use of the Crank–Nicholson or a related 2nd order time stepping scheme, for instance fractional-step-θ-methods (see [78]), will be analyzed which we expect to yield better accuracy results and hence enables for the use of larger time steps. Another aspect is the improvement of the iterative solvers, particularly of special multigrid schemes which are able to cope with the very anisotropic differential operators and the related very ill-conditioned linear systems. These fast solvers and improved variants of the streamline-diffusion or monotone and oscillation-free FEM-TVD techniques (cf. [53]) will be the key ingredients for efficient visualization tools for complex 3 D flows.

References

[1] J. F. Acker. *PDE basierte Visualisierungsmethoden für instationäre Strömungen auf unstrukturierten Gittern.* PhD thesis, Universität Dortmund, to appear in 2008.

[2] L. Ambrosio and V. M. Tortorelli. On the approximation of free discontinuity problems. *Bollettino de la Unione Matematica Italiana* 6(7): 105–123, 1992.

[3] L. Ambrosio, N. Fusco, and D. Pallara. *Functions of Bounded Variation and Free Discontinuity Problems.* Oxford Mathematical Monographs. The Clarendon Press, New York, 2000.

[4] M. Arredondo, K. Lebart, and D. Lane. Optical flow using textures. *Pattern Recognition Letters*, 25(4):449–457, 2004.

[5] R. Bajcsy and C Broit. Matching of deformed images. In *Proceedings of the 6th International Conference on Pattern Recognition*, pages 351–353, 1982.

[6] J. M. Ball. Global invertibility of sobolev functions and the interpenetration of matter. *Proceedings of the Royal Society of Edinburgh*, 88A: 315–328, 1981.

[7] L. Bar, B. Berkels, M. Rumpf, and G. Sapiro. A variational framework for simulateneous motion estimation and restoration of motion-blurred video. In *Proceedings ICCV, to appear*, 2007.

[8] J. Bigün, G.H. Granlund, and J.Wiklund. Multidimensional orientation estimation with applications to texture analysis and optical flow. IEEE Transactions on Pattern Analysis and Machine Intelligence 13(8): 775–790, 1991.

[9] V. Blanz and T. Vetter. A morphable model for the synthesis of 3d faces. In *Proceedings of SIGGRAPH 99*, Computer Graphics Proceedings, Annual Conference Series, pages 187–194, 1999.

[10] K. Bredies. *Optimal control of degenerate parabolic equations in image processing.* PhD thesis, University of Bremen, 2007.

[11] K. Bredies, D. A. Lorenz, and P. Maass. A generalized conditional gradient method and its connection to an iterative shrinkage method. To appear in *Computational Optimization and Applications*, 2008.

[12] L. Gottesfeld Brown. A survey of image registration techniques. *ACM Computing Surveys*, 24(4):325–376, 1992.

[13] A. Bruhn and J. Weickert. A confidence measure for variational optic flow methods. In R. Klette, R. Kozera, L. Noakes, J. Weickert, editors, *Geometric Properties for Incomplete data Series: Computational Imaging and Vision*, vol. 31, pages 283–298. Springer-Verlag, 2006.

[14] A. Bruhn, J. Weickert, and C. Schnörr. Lucas/kanade meets horn/schunck: Combining local and global optic flow methods. *International Journal of Computer Vision*, 61(3):211–231, 2005.

[15] D. Bürkle, T. Preusser, and M. Rumpf. Transport and anisotropic diffusion in time-dependent flow visualization. In *Proceedings Visualization '01*, 2001.

[16] B. Cabral and L. Leedom. Imaging vector fields using line integral convolution. In James T. Kajiya, editor, *Computer Graphics (SIGGRAPH '93 Proceedings)*, volume 27, pages 263–272, August 1993.

[17] J. Capon. High-resolution frequency-wavenumber spectrum analysis. *Proceedings of the IEEE*, 57(8):1408–1418, 1969.

[18] F. Catté, P.-L. Lions, J.-M. Morel, and T. Coll. Image selective smoothing and edge detection by nonlinear diffusion. *SIAM Journal On Numerical Analysis*, 29(1):182–193, 1992.

[19] M. S. Chong, A. E. Perry, and B. J. Cantwell. A general classification of three-dimensional flow fields. *Physics of Fluids A*, 2(5):765–777, 1990.

[20] P. G. Ciarlet. *Three-Dimensional Elasticity.* Elsevier, New York, 1988.

[21] U. Clarenz, N. Litke, and M. Rumpf. Axioms and variational problems in surface parameterization. *Computer Aided Geometric Design*, 21(8): 727–749, 2004.

[22] U. Clarenz, M. Droske, S. Henn, M. Rumpf, and K. Witsch. Computational methods for nonlinear image registration. In O. Scherzer, editor, *Mathematical Models for Registration and Applications to Medical Imaging, Mathematics in Industry*, volume 10, Springer, 2006.

[23] A. Collignon et al. Automated multi-modality image registration based on information theory. In Y. Bizais, C. Barillot and R. Di Paola, editors, Proceedings of the XIVth international conference on information

processing in medical imaging - IPMI'95, computational imaging and vision, vol. 3, pp. 263–274, June 26–30, Ile de Berder, France, 1995. Kluwer Academic Publishers.

[24] M. S. Diallo, M. Kulesh, M. Holschneider, and F. Scherbaum. Instantaneous polarization attributes in the time-frequency domain and wavefield separation. *Geophysical Prospecting*, 53(5):723–731, 2005.

[25] M. S. Diallo, M. Kulesh, M. Holschneider, K. Kurennaya, and F. Scherbaum. Instantaneous polarization attributes based on an adaptive approximate covariance method. *Geophysics*, 71(5):V99–V104, 2006.

[26] M. S. Diallo, M. Kulesh, M. Holschneider, F. Scherbaum, and F. Adler. Characterization of polarization attributes of seismic waves using continuous wavelet transforms. *Geophysics*, 71(3):V67–V77, 2006.

[27] U. Diewald, T. Preusser, and M. Rumpf. Anisotropic diffusion in vector field visualization on euclidean domains and surfaces. *IEEE Transactions on Visualization and Computer Graphics*, 6(2):139–149, 2000.

[28] M. Droske and W. Ring. A Mumford-Shah level-set approach for geometric image registration. *SIAM Journal on Applied Mathematics*, 66(6):2127–2148, 2006.

[29] M. Droske and M. Rumpf. A variational approach to non-rigid morphological registration. *SIAM Applied Mathematics*, 64(2):668–687, 2004.

[30] M. Droske and M. Rumpf. Multi scale joint segmentation and registration of image morphology. *IEEE Transaction on Pattern Recognition and Machine Intelligence*, 29(12):2181–2194, December 2007.

[31] M. Droske, C. Garbe, T. Preusser, M. Rumpf, and A. Telea. A phase field method for joint denoising, edge detection and motion estimation. *SIAM Applied Mathematics, Revised Version Submitted*, 2007.

[32] L. Florack and A. Kuijper. The topological structure of scale-space images. *Journal of Mathematical Imaging and Vision*, 12(1):65–79, 2000. ISSN 0924-9907.

[33] U. Grenander and M. I. Miller. Computational anatomy: An emerging discipline. *Quarterly Applied Mathematics*, 56(4):617–694, 1998.

[34] M. Griebel, T. Preusser, M. Rumpf, M.A. Schweitzer, and A. Telea. Flow field clustering via algebraic multigrid. In *Proceedings IEEE Visualization*, pages 35–42, 2004.

[35] X. Gu and B. C. Vemuri. Matching 3D shapes using 2D conformal representations. In *MICCAI 2004*, LNCS 3216, pages 771–780, 2004.

[36] D. Heeger. Model for the extraction of image flow. *Journal of the Optical Society of America*, 4(8):1455–1471, 1987.

[37] M. Holschneider. *Wavelets: An Analysis Tool.* Clarendon Press, Oxford, 1995.

[38] M. Holschneider, M. S. Diallo, M. Kulesh, M. Ohrnberger, E. Lück, and F. Scherbaum. Characterization of dispersive surface waves using continuous wavelet transforms. *Geophysical Journal International*, 163(2): 463–478, 2005.

[39] J. C. R. Hunt, A. A. Wray, and P. Moin. Eddies, stream and convergence zones in turbulent flow fields. Technical Report CTR-S88, Center for turbulence research, 1988.

[40] V. Interrante and C. Grosch. Stragegies for effectively visualizing 3D flow with volume LIC. In *Proceedings Visualization '97*, pages 285–292, 1997.

[41] J. Jeong and F. Hussain. On the identification of a vortex. *Journal of Fluid Mechanics*, 285:69–94, 1995.

[42] S. Kalkan, D. Calow, M. Felsberg, F. Worgotter, M. Lappe, and N. Kruger. Optic flow statistics and intrinsic dimensionality, 2004.

[43] E. R. Kanasewich. *Time Sequence Analysis in Geophysics.* University of Alberta Press, Edmonton, Alberta, 1981.

[44] B. Kawohl and N. Kutev. Maximum and comparison principle for one-dimensional anisotropic diffusion. *Mathematische Annalen*, 311(1):107–123, 1998.

[45] C. Kondermann, D. Kondermann, B. Jähne, and C. Garbe. Comparison of confidence and situation measures and their optimality for optical flows. *submitted to International Journal of Computer Vision*, February 2007.

[46] C. Kondermann, D. Kondermann, B. Jähne, and C. Garbe. An adaptive confidence measure for optical flows based on linear subspace projections. In *Proceedings of the DAGM-Symposium*, pages 132–141, 2007. http://dx.doi.org/10.1007/978-3-540-74936-3_14.

[47] C. Kondermann, D. Kondermann, B. Jähne, and C. Garbe. Optical flow estimation via flow inpainting using surface situation measures. *submitted*, 2007.

[48] A. Kufner. Weighted sobolev spaces, 1980. Teubner-Texte zur Mathematik, volume 31.

[49] M. Kulesh, M. Holschneider, M. S. Diallo, Q. Xie, and F. Scherbaum. Modeling of wave dispersion using continuous wavelet transforms. *Pure and Applied Geophysics*, 162(5):843–855, 2005.

[50] M. Kulesh, M. S. Diallo, M. Holschneider, K. Kurennaya, F. Krüger, M. Ohrnberger, and F. Scherbaum. Polarization analysis in the wavelet domain based on the adaptive covariance method. *Geophysical Journal International*, 170(2):667–678, 2007.

[51] M. Kulesh, M. Holschneider, M. Ohrnberger, and E. Lück. Modeling of wave dispersion using continuous wavelet transforms II: wavelet based frequency-velocity analysis. Technical Report 154, Preprint series of the DFG priority program 1114 "Mathematical methods for time series analysis and digital image processing", January 2007.

[52] M. A. Kulesh, M. S. Diallo, and M. Holschneider. Wavelet analysis of ellipticity, dispersion, and dissipation properties of Rayleigh waves. *Acoustical Physics*, 51(4):425–434, 2005.

[53] D. Kuzmin and S. Turek. High-resolution FEM-TVD schemes based on a fully multidimensional flux limiter. *Journal of Computational Physics*, 198:131–158, 2004.

[54] S. H. Lai and B.C. Vemuri. Robust and efficient algorithms for optical flow computation. In *Proceedings of the International Symposium on Computer Vision*, pages 455–460, November 1995.

[55] R. S. Laramee, H. Hausser, H. Doleisch, B. Vrolijk, F.H. Post, and D. Weiskopf. The state of the art in flow visualization: Dense and texture-based techniques. *Computer Graphics Forum*, 23(2):203–221, 2004.

[56] A. Lee, D. Dobkin, W. Sweldens, and P. Schröder. Multiresolution mesh morphing. In *Proceedings of SIGGRAPH 99*, Computer Graphics Proceedings, Annual Conference Series, pages 343–350, August 1999.

[57] N. Litke, M. Droske, M. Rumpf, and P. Schröder. An image processing approach to surface matching. In M. Desbrun and H. Pottmann, editors, *Third Eurographics Symposium on Geometry Processing*, Eurographics Association, pages 207–216, 2005.

[58] S. Masnou and J. Morel. Level lines based disocclusion. In *Proceedings of ICIP*, volume 3, pages 259–263, 1998.

[59] B. McCane, K. Novins, D. Crannitch, and B. Galvin. On benchmarking optical flow. http://of-eval.sourceforge.net/, 2001.

[60] M. I. Miller, A. Trouvé, and L. Younes. On the metrics and euler-lagrange equations of computational anatomy. *Annual Review of Biomedical Enginieering*, 4:375–405, 2002.

[61] I. B. Morozov and S. B. Smithson. Instantaneous polarization attributes and directional filtering. *Geophysics*, 61(3):872–881, 1996.

[62] O. A. Oleĭnik and E. V. Radkevič. Second order equations with nonnegative characteristic form. *American Mathematical Society*, Providence, Rhode Island and Plenum Press, New York, 1973.

[63] F. Pacor, D. Bindi, L. Luzi, S. Parolai, S. Marzorati, and G. Monachesi. Characteristics of strong ground motion data recorded in the Gubbio sedimentary basin (Central Italy). *Bulletin of Earthquake Engineering*, 5(1):27–43, 2007.

[64] H. A. Pedersen, J. I. Mars, and P.-O. Amblard. Improving surface-wave group velocity measurements by energy reassignment. *Geophysics*, 68(2): 677–684, 2003.

[65] P. Perona and J. Malik. Scale space and edge detection using anisotropic diffusion. In *IEEE Computer Society Workshop on Computer Vision*, 1987.

[66] P. Perona and J. Malik. Scale-space and edge detection using anisotropic diffusion. Technical Report UCB/CSD-88-483, EECS Department, University of California, Berkeley, December 1988.

[67] C. R. Pinnegar. Polarization analysis and polarization filtering of three-component signals with the time-frequency S transform. *Geophysical Journal International*, 165(2):596–606, 2006.

[68] O. Pironneau. On the transport-diffusion algorithm and its applications to the Navier-Stokes equations. *Numerische Mathematics*, 38:309–332, 1982.

[69] W. H. Press, S. A. Teukolsky, W. T. Vetterling, and B. P. Flannery. *Numerical Recipe in C: The Art of Scientific Computing.* Cambridge University Press, 1992.
[70] T. Preusser and M. Rumpf. An adaptive finite element method for large scale image processing. *Journal of Visual Communication and Image Representation*, 11:183–195, 2000.
[71] M. Schimmel and J. Gallart. The inverse S-transform in filters with time-frequency localization. *IEEE Transaction on Signal Processing*, 53(11): 4417–4422, 2005.
[72] H.-W. Shen and D. L. Kao. Uflic: A line integral convolution algorithm for visualizing unsteady flows. In *Proceedings Visualization '97*, pages 317–322, 1997.
[73] N. Soma, H. Niitsuma, and R. Baria. Reflection technique in time-frequency domain using multicomponent acoustic emission signals and application to geothermal reservoirs. *Geophysics*, 67(3):928–938, 2002.
[74] H. Spies and C. Garbe. Dense parameter fields from total least squares. In L. Van Gool, editor, *Pattern Recognition*, volume LNCS 2449 of *Lecture Notes in Computer Science*, pages 379–386, Zurich, CH, 2002. Springer-Verlag.
[75] A. Telea, T. Preusser, C. Garbe, M. Droske, and M. Rumpf. A variational approach to joint denoising, edge detection and motion estimation. In *Proceedings of DAGM 2006*, pages 525–535, 2006.
[76] M. Tobak and D. J. Peake. Topology of 3D separated flow. *Annual Review of Fluid Mechanics*, 14:61–85, 1982.
[77] S. Turek. *Efficient Solvers for Incompressible Flow Problems: An Algorithmic and Computational Approach*, volume 6 of *LNCSE.* Springer Verlag Berlin Heidelberg New York, 1999.
[78] S. Turek, L. Rivkind, J. Hron, and R. Glowinski. Numerical analysis of a new time-stepping θ-scheme for incompressible flow simulations, *Journal of Scientific Computing*, 28(2–3):533–547, September 2006.
[79] G. Turk and D. Banks. Image-guided streamline placement. In *Proc. 23rd annual conference on Computer graphics, August 4–9, 1996, New Orleans, LA USA*. ACM Press, 1996.
[80] J. J. van Wijk. Spot noise-texture synthesis for data visualization. In T.W̃. Sederberg, editor, *Computer Graphics (SIGGRAPH '91 Proceedings)*, volume 25, pages 309–318, Addison Wesley July 1991.
[81] J. J. van Wijk. Flow visualization with surface particles. *IEEE Computer Graphics and Applications*, 13(4):18–24, July 1993.
[82] P. Viola and W. M. Wells. Alignment by maximization of mutual information. *International Journal of Computer Vision*, 24(2):137–154, 1997.
[83] J. Weickert. *Anisotropic Diffusion in Image Processing.* European Consortium for Mathematics in Industry. Teubner, Stuttgart, Leipzig, 1998.
[84] A. P. Witkin. Scale-space filtering. In *Proceedings of the 8th IJCAI*, pages 1019–1022, Karlsruhe, Germany, 1983.

[85] Q. Xie, M. Holschneider, and M. Kulesh. Some remarks on linear diffeomorphisms in wavelet space. Technical Report 37, Preprint series of the DFG priority program 1114 "Mathematical methods for time series analysis and digital image processing", July 2003.
[86] C. Zetzsche and E. Barth. Fundamental limits of linear filters in the visual processing of two dimensional signals. *Vision Research*, 30(7):1111–1117, 1990.

5

Analysis of Bivariate Coupling by Means of Recurrence

Christoph Bandt[1], Andreas Groth[1], Norbert Marwan[2], M. Carmen Romano[2,3], Marco Thiel[2,3], Michael Rosenblum[2], and Jürgen Kurths[2]

[1] Institute of Mathematics, University of Greifswald, Jahnstrasse 15 a, D-17487 Greifswald, Germany
{bandt,groth}@uni-greifswald.de

[2] Institute of Physics, University of Potsdam, Am Neuen Palais 10, D-14469 Potsdam, Germany
{marwan,romano,thiel,mros,jkurths}@agnld.uni-potsdam.de

[3] Department of Physics, University of Aberdeen, UK-Aberdeen AB24 3UE, UK

5.1 Introduction

In the analysis of coupled systems, various techniques have been developed to model and detect dependencies from observed bivariate time series. Most well-founded methods, like Granger-causality and partial coherence, are based on the theory of linear systems: on correlation functions, spectra and vector autoregressive processes. In this paper we discuss a nonlinear approach using recurrence.

Recurrence, which intuitively means the repeated occurrence of a very similar situation, is a basic notion in dynamical systems. The classical theorem of Poincaré says that for every dynamical system with an invariant probability measure P, almost every point in a set B will eventually return to B. Moreover, for ergodic systems the mean recurrence time is $1/P(B)$ [23]. Details of recurrence patterns were studied when chaotic systems came into the focus of research, and it turned out that they are linked to Lyapunov exponents, generalized entropies, the correlation sum, and generalized dimensions [20, 38].

Our goal here is to develop methods for time series which typically contain a few hundreds or thousands of values and which need not come from a stationary source. While Poincaré's theorem holds for stationary stochastic processes, and linear methods require stationarity at least for sufficiently large windows, recurrence methods need less stationarity. We outline different concepts of recurrence by specifying different classes of sets B. Then we visualize recurrence and define recurrence parameters similar to autocorrelation.

We are going to apply recurrence to the analysis of bivariate data. The basic idea is that coupled systems show similar recurrence patterns. We can study joint recurrences as well as cross-recurrence. We shall see that both approaches have their benefits and drawbacks.

Model systems of coupled oscillators form a test bed for analysis of bivariate time series since the corresponding differential equations involve a parameter which precisely defines the degree of coupling. Changing the parameter we can switch to phase synchronization and generalized synchronization. The approaches of cross- and joint recurrence are compared for several models. In view of possible experimental requirements, recurrence is studied on ordinal scale as well as on metric scale. Several quantities for the description of synchronization are derived and illustrated. Finally, two different applications to EEG data will be presented.

5.2 Recurrence on Different Scales

5.2.1 Nominal Scale

We start with an ordinary time series of numbers $x_1, x_2, \ldots, x_N$. Recurrence basically means that certain numbers will repeat: $x_i = x_j$. This is the proper concept when the values x_i form a nominal scale – they are just symbols from a finite or countable alphabet. A typical example is the nucleotide sequence of a DNA segment, with values A,C,G and T (which we can code 1, 2, 3, 4). Since letters will repeat very often, we usually prescribe a length d for the word which should repeat:

$$x_{i+n} = x_{j+n}\ , \qquad n = 0, \ldots, d-1\,.$$

Here d is a parameter which indicates the *strength* of recurrence. Finding occurrences of words in large data is a basic algorithmic task in bioinformatics. The statistical structure of such sequences is modeled by Hidden Markov Models, also called probabilistic automata [7].

5.2.2 Metric Scale

If the x_i are real numbers, instead of $x_i = x_j$ we require that x_j is in the vicinity or *neighborhood* of x_i :

$$|x_i - x_j| \leq \varepsilon\,,$$

where ε is a predefined threshold. According to the ergodic theorem mentioned above, the mean recurrence time is of order $1/\varepsilon$ which gives a clue on how to choose ε.

Due to different density of the values, different x_i will have different numbers of neighbors. This can be mended by taking *rank numbers*

$$r_i = \#\{k | 1 \leq k \leq N, x_k < x_i\}$$

instead of the x_i, and integer ε. Then each x_i (except for the ε largest and smallest values) has 2ε recurrences. Eckmann et al. [8] used constant number of neighbors when they introduced recurrence plots.

However, it makes little sense to require that only single values repeat. For the function $\sin t$, $t \geq 0$ the value 0 repeats at $t = \pi$, but this is a false neighbor, proper recurrence (in fact periodicity) appears at 2π. Thus we shall again choose a strength parameter d and require

$$|x_{i+n} - x_{j+n}| \leq \varepsilon \ , \qquad n = 0, ..., d-1 \, .$$

5.2.3 Vector Recurrence

The last condition can also be interpreted in a different way. We take the d-dimensional vectors $\boldsymbol{x}_i = (x_i, x_{i+1}, ..., x_{i+d-1}) \in \mathbb{R}^d$ and consider their approximate equality

$$\|\boldsymbol{x}_i - \boldsymbol{x}_j\| \leq \varepsilon \, , \tag{5.1}$$

with respect to the maximum norm in $\mathbb{R}^d$. However, it is also possible to consider any other norm on $\mathbb{R}^d$, like the Euclidean norm, or similarity indices like cosine similarity and the Mahalanobis distance. The choice of the distance function and the threshold (e.g. fixed, time-dependent, fixed amount of nearest neighbors) depends on the particular problem under consideration. For an overview we refer to [20].

Vector recurrence is certainly the appropriate concept when our time series does not consist of numbers but of vectors. This is the case for multivariate time series treated below, in particular for d-dimensional time series obtained numerically from a model system of d differential equations. For such systems, studied in Sects. 5 and 6, we need a slightly different notation.

5.2.4 Differential Equations and Delay Embedding

In the formalism of differentiable dynamical systems, the state of a system at time t is described by a vector

$$\boldsymbol{x}(t) = [x_1(t), x_2(t), \dots, x_d(t)] \in \mathbb{R}^d \, , \tag{5.2}$$

where $x_n(t)$ denotes the n-th component at time t. The evolution of the state of the system in time, i.e., its trajectory, is determined by a flow $F(\cdot)$, such that $\dot{\boldsymbol{x}}(t) = F(\boldsymbol{x}(t))$. The components $x_n(t)$ of the state vector $\boldsymbol{x}(t)$ are observable physical variables, such as the position and velocity of a particle. However, in an experimental setting typically not all relevant components are known or can be measured. If certain conditions are fulfilled, it is possible to reconstruct the trajectory of the system from a scalar measurement $u(t) = f[\boldsymbol{x}(t)]$, e.g., by means of its delay embedding [31, 35]

$$\boldsymbol{u}(t) = [u(t), u(t+\vartheta), \ldots, u(t+(m-1)\vartheta)] \in \mathbb{R}^m \,, \tag{5.3}$$

where ϑ denotes the time delay and m the embedding dimension. In the ideal case there is a functional relationship (strictly speaking, a diffeomorphism) between the original unknown components and those of the delay embedding.

Although the underlying system evolves continuously in time, we measure the system at discrete time points $i\Delta t$, where $i = 1, \ldots, N$ and Δt is the sampling rate. When confusion is possible, we denote by $\boldsymbol{x}_i = \boldsymbol{x}(i\Delta t)$ and $\boldsymbol{u}_i = \boldsymbol{u}(i\Delta t)$ the points of the original and reconstructed trajectory, respectively. Otherwise we use $\boldsymbol{x}_i$ for both, as in (5.1).

5.2.5 Remarks on Dynamical Systems

Periodicity is an extreme case of recurrence. And in deterministic systems, an exact recurrence to a state $\boldsymbol{x}_i$ at a later time point j is only possible in the case of periodic dynamics. Otherwise, the required uniqueness of the solution of a dynamical system is not fulfilled.

In our definition (5.1) of vector recurrence, the recurrence set B is a ball of radius ε around $\boldsymbol{x}_i$ with respect to the given norm on $\mathbb{R}^d$. In case of the maximum norm it is a cube of side length 2ε. These sets are not disjoint for different i. It may happen that j realizes a recurrence to both i_1 and i_2, but i_2 does not represent a recurrence to i_1. Thus sometimes one might wish the recurrence sets to form a disjoint partition of phase space, $\mathbb{R}^d = B_1 \cup \ldots \cup B_k$.

When we decompose the phase space into regions and assign the same symbol to all states within one region, our metric vector data become nominal data (symbolization), and we neglect all further information about distances. This coarse-graining leads to a lower level of resolution, but on the other hand also to a weaker stationarity condition on the measurement. Note that for recurrence on a metric scale, stationarity is only required up to a threshold ε. By varying this threshold, and in case of $\boldsymbol{u}_i$ also the embedding dimension, we are able to balance between resolution and stationarity requirements.

In the sequel, we give a partial answer to the question for an appropriate decomposition of phase space.

5.2.6 Ordinal Scale

The ordinal scale of numbers is between the metric and the nominal one: the order structure of the states is known, but no meaningful distance of the values is defined. The analysis on an ordinal scale, in contrast to the one on a metric scale, is invariant with respect to a strictly monotonic transformation. The classification of levels of measurement resolution into metric, ordinal and nominal, was originally proposed in statistics [34]. Here, we suggest this approach for the analysis of dynamical systems.

We consider two states $\boldsymbol{u}_i$ and $\boldsymbol{u}_j$ in the reconstructed phase space. Order patterns are related to the time-delayed embedding (3). They will not be applied to the systems (5.2). We define *recurrence on the ordinal scale* if both states exhibit the same order structure

$$\pi(\boldsymbol{u}_i) = \pi(\boldsymbol{u}_j)\,, \tag{5.4}$$

where π is a mapping function that encodes the order structure.

To illustrate this idea, suppose that the reconstructed trajectory has embedding dimension $m = 2$ and time delay ϑ. In this case, two relationships between u_i and $u_{i+\vartheta}$ are possible, apart from equality.[1] We encode the order structure as a new symbol

$$\pi(\boldsymbol{u}_i) = \begin{cases} 0: & u_i < u_{i+\vartheta} \\ 1: & u_i > u_{i+\vartheta}\,, \end{cases} \tag{5.5}$$

where π is called *order pattern* of $\boldsymbol{u}_i$. Thus the phase space is divided by the identity into two areas (Fig. 5.1). This way of generating a new symbol sequence is common in statistics (e.g. [10, 11, 13]). Our approach was originally motivated by Kendall's tau-correlation [16], which was modified to an autocorrelation function for time series [3, 9]. In classical time series analysis there are practically no methods which use order patterns of higher dimensions. Here we can use order patterns of length d, so that we have again a parameter for the strength of recurrence, as well as more complicated order patterns [5].

5.2.7 Order Patterns of Length 3

Let us consider embedding dimension $m = 3$, which is related to the phase of an oscillator [12], discussed in Sect. 5.5. Here the phase space is nicely

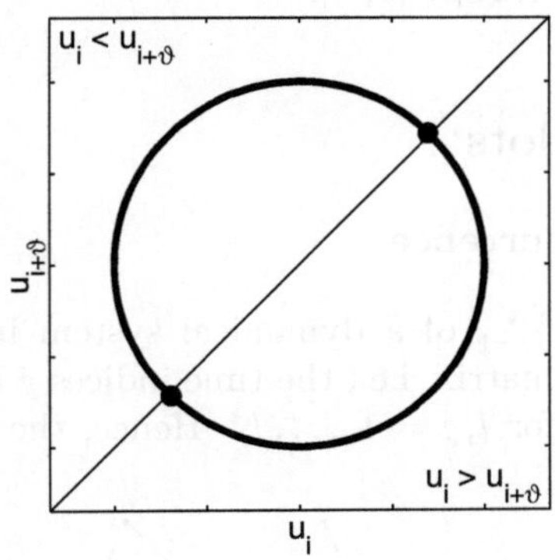

Fig. 5.1 Periodic trajectory in phase space and decomposition of the phase space $(u_i, u_{i+\vartheta})$ by order patterns

[1] In general we neglect the equality of values. This is reasonable if we consider systems with continuous distribution of the values, where equality has measure zero.

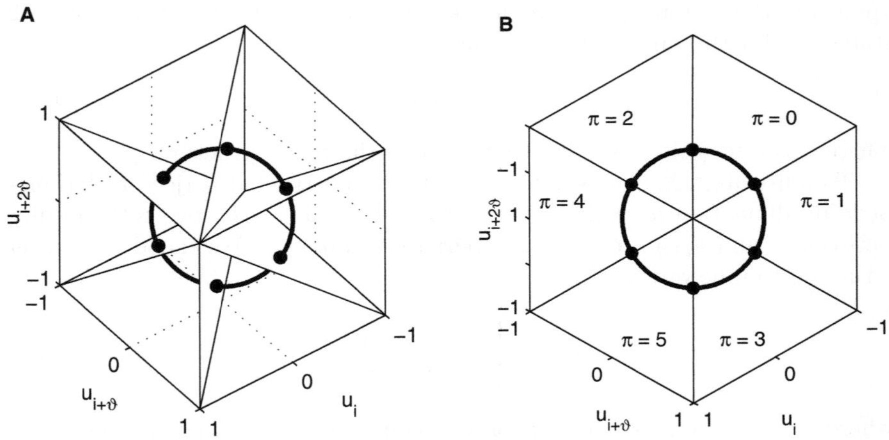

Fig. 5.2 (**A**) Decomposition of the phase space $(u_i, u_{i+\vartheta}, u_{it+2\,\vartheta})$ by order patterns π and possible trajectory of a sine function. (**B**) Same plot with viewing angle in direction of the main diagonal

decomposed into $m! = 6$ regions. These regions are separated by planes of pairwise equalities ($u_i = u_{i+\vartheta}$, $u_{i+\vartheta} = u_{i+2\vartheta}$, $u_i = u_{i+2\vartheta}$) and are arranged around the main diagonal $u_i = u_{i+\vartheta} = u_{i+2\vartheta}$ (Fig. 5.2). All states $\boldsymbol{u}_i$ within a single region of the phase space have the same structure of order relations. Hence, they are associated to the same symbol $\pi(\boldsymbol{u}_i)$ (Fig. 5.3).

This scheme of mapping states $\boldsymbol{u}_i$ to symbols $\pi(\boldsymbol{u}_i)$ works for arbitrary dimension m. The phase space decomposition into $m!$ regions is related to the concept of *permutation entropy* [4] which for various dynamical systems agrees with the metric entropy [1, 6].

5.3 Recurrence Plots

5.3.1 Univariate Recurrence

Given a trajectory $\{\boldsymbol{x}_i\}_{i=1}^N$ of a dynamical system in phase space, we can compute its recurrence matrix, i.e., the time indices j at which the trajectory recurs to the state $\boldsymbol{x}_i$, for $i, j = 1, \ldots, N$. Hence, the *recurrence matrix* is a

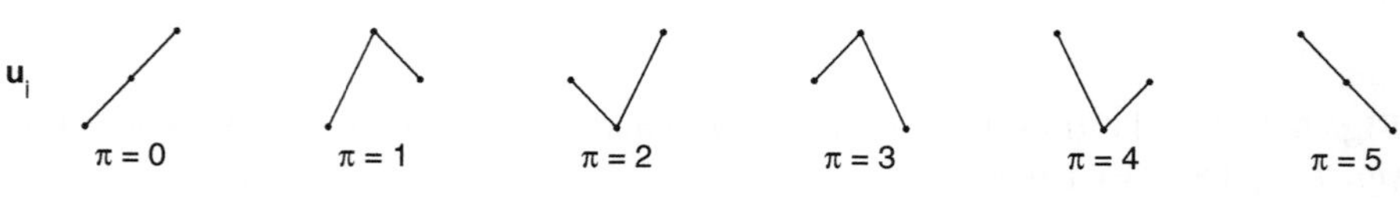

Fig. 5.3 The vector $\boldsymbol{u}_i = (u_i, u_{i+\vartheta}, u_{it+2\,\vartheta})$ in reconstructed phase space can form six different order patterns. The labeling is not important for the analysis and is added just for illustration

binary $N \times N$ matrix with entry $\mathbf{R}_{i,j} = 1$ if the trajectory at time j recurs to the state $\boldsymbol{x}_i$ and entry $\mathbf{R}_{i,j} = 0$, otherwise.

As mentioned above, recurrence can be defined on a metric, nominal or ordinal scale. Accordingly, the recurrence matrix on a metric scale is

$$\mathbf{R}_{i,j} = \Theta\left(\varepsilon - \|\boldsymbol{x}_i - \boldsymbol{x}_j\|\right) , \tag{5.6}$$

on a nominal scale

$$\mathbf{R}_{i,j} = \delta\left(\boldsymbol{x}_i - \boldsymbol{x}_j\right) , \tag{5.7}$$

and on an ordinal scale

$$\mathbf{R}_{i,j} = \delta\left(\pi(\boldsymbol{u}_i) - \pi(\boldsymbol{u}_j)\right) , \tag{5.8}$$

where $\Theta(x) = 1$ for $x \geq 0$ and $\Theta(x) = 0$ else, $\delta(x) = 1$ if $x = 0$ and $\delta(x) = 0$ otherwise, and $i, j = 1, \ldots, N$.

A recurrence plot (RP) is the graphical representation of a recurrence matrix [8, 20]. The RP is obtained by plotting the recurrence matrix and using different colors for its binary entries, e.g., plotting a black dot at the coordinates (i, j), where $\mathbf{R}_{i,j} = 1$, and a white dot, where $\mathbf{R}_{i,j} = 0$. Both axes of the RP are time axes. Since $\mathbf{R}_{i,i} \equiv 1$ for $i = 1 \ldots N$ by definition, the RP has always a black main diagonal line. Furthermore, the RP is symmetric with respect to the main diagonal, i.e., $\mathbf{R}_{i,j} = \mathbf{R}_{j,i}$.

RPs yield important insights into the time evolution of phase space trajectories, because typical patterns in RPs are linked to a specific behavior of the system. One important structural element are diagonal lines $\mathbf{R}_{i+k,j+k} = 1$ for $k = 0 \ldots l - 1$, where l is the length of the diagonal line. On metric scale a diagonal occurs when a segment of the trajectory runs almost in parallel to another segment (i.e., through an ε-tube around the other segment) for l time units (cf. Fig. 5.4). The length of this diagonal line is determined by the duration of such similar local evolution of the trajectory segments. The direction of these diagonal structures is parallel to the main diagonal. Since the definition of the Rényi entropy of second order K_2 is based on how long trajectories evolve within an ε-tube, it is possible to estimate K_2 by means of the distribution of diagonal lines in the RP [20, 37]. On an ordinal scale we also obtain diagonal structures when two different segments of the trajectory have the same sequence of order patterns (cf. Fig. 5.5). In particular, we will show how these diagonal lines are linked to the phase of an oscillator and will derive a measure to quantify phase synchronization.

5.3.2 Bivariate Recurrence Plots

There are two approaches to extend RPs to the analysis of bivariate data $(\boldsymbol{x}_i, \boldsymbol{y}_i)$. In the first approach, the *common auto-recurrences* are registered. This is essentially the same procedure as going from metric recurrence to

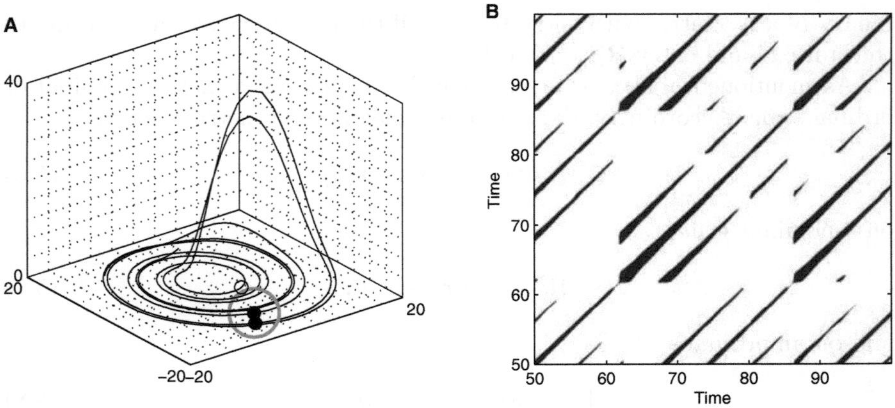

Fig. 5.4 (**A**) Segment of the phase space trajectory of the Rössler system (5.14), with $a = 0.15$, $b = 0.20$, $c = 10$, using the components (x, y, z). (**B**) The corresponding recurrence plot based on metric scale. A phase space vector at j which falls into the neighborhood (gray circle in (A)) of a given phase space vector at i is considered to be a recurrence point (black point on the trajectory in (A)). This is marked by a black point in the RP at the position (i, j). A phase space vector outside the neighborhood (empty circle in (A)) leads to a white point in the RP. The radius of the neighborhood for the RP is $\varepsilon = 5$; L_2-norm is used

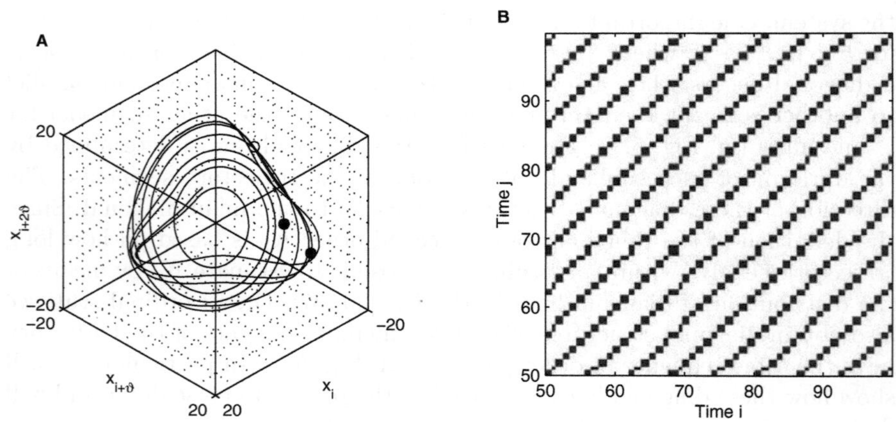

Fig. 5.5 (**A**) Segment of the phase space trajectory of the Rössler system (5.14) with $a = 0.15$, $b = 0.20$, $c = 10$, by using its time-delayed first component $(x_i, x_{i+\vartheta}, x_{i+2\vartheta})$ with $\vartheta = 2$ and (**B**) its corresponding recurrence plot based on ordinal scale. Phase space vectors in the same region are considered as recurrent points (black points on the trajectory in (A)), where a phase space vector in a different region (empty circle in (A)) is not a recurrence point

vector recurrence in Sect. 5.2, and it can be done for two qualitatively different systems: The vectors $\boldsymbol{x}_i$ and $\boldsymbol{y}_i$ can have different dimension, and represent different physical quantities.

In the other approach, we compute the recurrence of the states of one system to the other, i.e., the distances between the different systems in phase space. This requires a certain degree of similarity of the systems for $\boldsymbol{x}$ and $\boldsymbol{y}$ although the ordinal approach also allows to compare physical different systems. However, lagged dependencies can be better visualized, and the two time series can have different length. As we will show here, depending on the situation, one approach might be more appropriate than the other.

5.3.3 Joint Recurrence Plots

The first possibility to compare $\boldsymbol{x}, \boldsymbol{y}$ is to consider the recurrences of their trajectories in their respective phase spaces separately and regard the times at which both of them recur simultaneously, i.e., when a *joint recurrence* occurs [20, 27]. A joint recurrence plot (JRP) is defined as pointwise product of the two RPs of the two considered systems

$$\mathbf{JR}_{i,j}^{\boldsymbol{x},\boldsymbol{y}} = \mathbf{R}_{i,j}^{\boldsymbol{x}} \cdot \mathbf{R}_{i,j}^{\boldsymbol{y}}\,, \quad i,j = 1,\ldots,N\,. \tag{5.9}$$

In this approach, a recurrence takes place if the first system at time j described by the vector $\boldsymbol{x}_j$ returns to the neighborhood of a former point $\boldsymbol{x}_i$, and *simultaneously* the second system $\boldsymbol{y}_j$ returns at the same time j to the neighborhood of a formerly visited point $\boldsymbol{y}_i$.

Actually, joint recurrence is just the vector recurrence of the bivariate series $(\boldsymbol{x}_i, \boldsymbol{y}_i)_{i=1,\ldots,N}$. The dimensions of the vectors $\boldsymbol{x}$ and $\boldsymbol{y}$ can differ, and we can consider different norms and different thresholds ε for each system, so that the recurrence conditions can be adapted to each system separately, respecting the corresponding natural measure. Mathematically, this just means taking the norm $\|(\boldsymbol{x},\boldsymbol{y})\| = \max\{\|\boldsymbol{x}\|_1/\varepsilon_1, \|\boldsymbol{y}\|_2/\varepsilon_2\}$ on the product space of the two phase spaces.

We mention that a product representation similar to (5.9) holds for the transition from ordinary recurrence to the recurrence of m successive states: $\mathbf{R}_{i,j}^{\boldsymbol{x}} = \prod_{k=0}^{m-1} \mathbf{R}_{i+k,j+k}^{x}$, which simplifies plots for recurrence of strength m.

A delayed version of the joint recurrence matrix can be introduced by

$$\mathbf{JR}_{i,j}^{\boldsymbol{x},\boldsymbol{y}} = \mathbf{R}_{i,j}^{\boldsymbol{x}} \cdot \mathbf{R}_{i+\tau,j+\tau}^{\boldsymbol{y}}\,, \qquad i,j = 1,\ldots,N-\tau\,, \tag{5.10}$$

which is useful for the analysis of interacting delayed systems (e.g., for lag synchronization) [29, 33], and for systems with feedback.

5.3.4 Cross-Recurrence Plots

A cross-recurrence plot (CRP) visualizes dependencies between two different systems by looking at recurrences from one system to the other [19, 20]. Using a metric scale, it is defined as

$$\mathbf{CR}_{i,j}^{\boldsymbol{x},\boldsymbol{y}} = \Theta\left(\varepsilon - \|\boldsymbol{x}_i - \boldsymbol{y}_j\|\right) , \tag{5.11}$$

for the nominal scale it is

$$\mathbf{CR}_{i,j}^{\boldsymbol{x},\boldsymbol{y}} = \delta\left(\boldsymbol{x}_i - \boldsymbol{y}_j\right) , \tag{5.12}$$

and for the ordinal scale

$$\mathbf{CR}_{i,j}^{\boldsymbol{u},\boldsymbol{v}} = \delta\left(\pi(\boldsymbol{u}_i) - \pi(\boldsymbol{v}_j)\right) , \tag{5.13}$$

with $i = 1, \ldots, N,\ j = 1, \ldots, M$. The length of the trajectories of $\boldsymbol{x}$ and $\boldsymbol{y}$, or $\boldsymbol{u}$ and $\boldsymbol{v}$, respectively, need not be identical, so that $\mathbf{CR}$ need not be a square matrix. However, in the metric case both systems must be represented in the same phase space, otherwise we cannot measure distances between states of both systems. Therefore, the data under consideration should be from very similar processes and, actually, should represent the same observable.

On ordinal or nominal scale, this is not necessary. Nevertheless, we have to take the same embedding dimension for the delay vectors to define order patterns of the same length, or meaningful related decompositions with equal number of sets when symbolization is used to obtain a nominal series.

Since the values of the main diagonal $\mathbf{CR}_{i,i}$ for $i = 1 \ldots N$ are not necessarily one, there is usually no black main diagonal. The lines which are diagonally oriented are here of major interest, too. They represent segments on both trajectories, which run parallel for some time. The distribution and length of these lines are obviously related to the interaction between the dynamics of both systems. A measure based on the lengths of such lines can be used to find nonlinear interrelations between both systems (Sect. 5.4).

5.3.5 Comparison Between CRPs and JRPs

In order to illustrate the difference between CRPs and JRPs, we consider the trajectory of the Rössler system [30]

$$\begin{aligned} \dot{x} &= -y - z, \\ \dot{y} &= x + a\,y, \\ \dot{z} &= b + z\,(x - c)\,, \end{aligned} \tag{5.14}$$

in three different situations: the original trajectory (Fig. 5.6A), the trajectory rotated around the z-axis (Fig. 5.6B) and the trajectory under the time scale transformation $\tilde{t} = t^2$, which gives the same picture as the first one.

Let us consider the RPs of these three trajectories. The RP of the original trajectory is identical to the RP of the rotated one (Fig. 5.7A), but the RP of the stretched/compressed trajectory is different from the RP of the original trajectory (Fig. 5.7B): it contains bowed lines, as the recurrent structures are shifted and stretched in time with respect to the original RP.

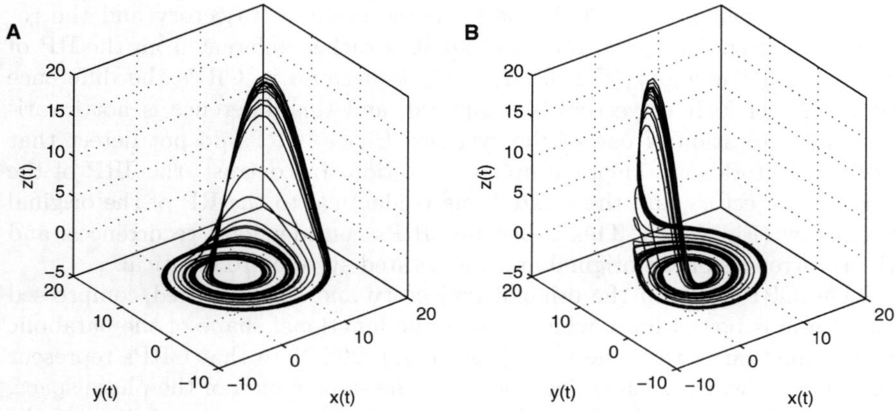

Fig. 5.6 Phase space trajectories of the Rössler system (5.14) with $a = 0.15$, $b = 0.2$ and $c = 10$: (**A**) original system, (**B**) rotated around the z-axis by $\frac{3}{5}\pi$

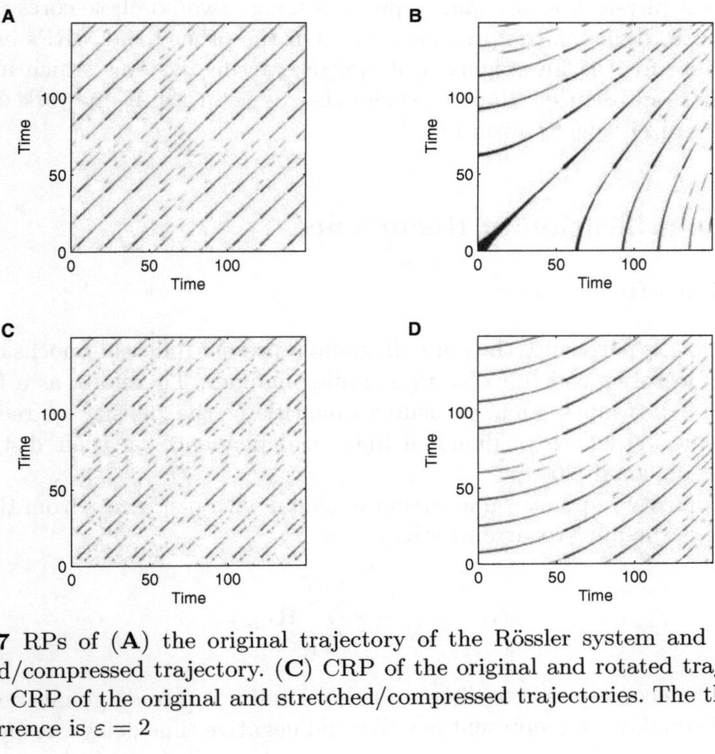

Fig. 5.7 RPs of (**A**) the original trajectory of the Rössler system and (**B**) the stretched/compressed trajectory. (**C**) CRP of the original and rotated trajectories and (**D**) CRP of the original and stretched/compressed trajectories. The threshold for recurrence is $\varepsilon = 2$

Now we calculate the CRP between the original trajectory and the rotated one (Fig. 5.7C) and observe that it is rather different from the RP of the original trajectory (Fig. 5.7A). This is because in CRPs the difference between each pair of vectors is computed, and this difference is not invariant under rotation of one of the systems. Hence, CRPs do not detect that both trajectories are identical up to a rotation. In contrast, the JRP of the original trajectory and the rotated one is identical to the RP of the original trajectory (Fig. 5.7A). This is because JRPs consider joint recurrences, and the recurrences of the original and the rotated system are identical.

The CRP between the original trajectory and the stretched/compressed one contains bowed lines, which reveals the functional shape of the parabolic transformation of the time scale (Fig. 5.7D) [20]. Note that CRPs represent the times at which both trajectories visit the same region of the phase space. On the other hand, the JRP of these trajectories – the intersection of the black sets in Fig. 5.7A and B – is almost empty, except for the main diagonal, because the recurrence structure of both systems is so different. There are almost no joint recurrences. Therefore, JRPs are not built to detect a time transformation applied to the trajectory, even though the shape of the phase space trajectories is identical.

To conclude, we can state that CRPs are more appropriate to investigate relationships between the parts of the same system which have been subjected to different physical or mechanical processes, e.g., two borehole cores in a lake subjected to different compression rates. On the other hand, JRPs are more appropriate for the investigation of two interacting systems which influence each other, and hence, adapt to each other, e.g., in the framework of phase and generalized synchronization.

5.4 Quantification of Recurrence

5.4.1 Auto-Recurrence

The diagonals parallel to the main diagonal represent different epochs of phase space trajectories which evolve in a similar manner. Therefore, as a first approach we introduce some measures quantifying the density of recurrence points and the length of diagonal lines in dependence on their distance to the main diagonal [20].

The density of points on a certain diagonal with distance τ from the main diagonal is the auto-recurrence rate

$$RR_\tau = \frac{1}{N-\tau} \sum_{i=1}^{N-\tau} \mathbf{R}_{i,i+\tau} \,, \tag{5.15}$$

where $\tau > 0$ corresponds to diagonals above and $\tau < 0$ to diagonals below the main diagonal, which represent positive and negative time delays, respectively.

The auto-recurrence rate can be considered as a *non-linear version of the auto-correlation function*, since it also describes higher order correlations between the points of the trajectory in dependence on τ [26]. It can be interpreted as an estimate of the probability $p(\tau)$ that a state recurs to its ε-neighborhood after τ time steps.

Similar to auto-correlation, auto-recurrence fulfills $RR_0 = 1$ and is symmetric: $RR_\tau = RR_{-\tau}$. However, RR_τ is always between 0 and 1. The reference line, which corresponds to the zero line for correlation functions, is given by the *average recurrence rate*

$$RR = \frac{1}{N^2} \sum_{i,j=1}^{N} \mathbf{R}_{i,j} \,. \tag{5.16}$$

It is clear that RR and hence RR_τ heavily depend on the threshold ε which therefore must be adapted carefully to the problem at hand.

The *ordinal average recurrence rate* can be exactly determined:

$$RR = \sum_{\pi} p_\pi^2 \,, \tag{5.17}$$

where $p_\pi = n_\pi / N$ is the relative and n_π the absolute frequency of the order patterns π in the time series. To explain this formula, we note that $\mathbf{R}_{i,j} = 1$ if $\pi_i = \pi_j$. For every order pattern π, the number of pairs (i,j) with $\pi_i = \pi_j = \pi$ equals n_π^2. Thus the number of entries $\mathbf{R}_{i,j} = 1$ in the matrix is $\sum n_\pi^2$ where the sum runs over all possible order patterns π. This fact together with (5.16) implies (5.17).

Let us take order patterns of length three as an example, and assume they all appear with equal probability. Then one sixth of the entries of the matrix $\mathbf{R}$ are 1. In Sect. 5.6, we use (5.17) to express coupling in the multivariate case where we cannot work with τ.

For $l \geq 1$, let $P(l)$ denote the number of (maximal) diagonal line segments of length $= l$ in $\mathbf{R}$. Since they represent l successive recurrences, we can introduce two measures of repeated recurrence, or strength of recurrence:

$$\mathrm{DET} = \frac{\sum_{l=l_{\min}}^{N-1} l\, P(l)}{\sum_{i \neq j} \mathbf{R}_{i,j}} \,, \qquad L = \frac{\sum_{l=l_{\min}}^{N-1} l\, P(l)}{\sum_{l=l_{\min}}^{N-1} P(l)} \,. \tag{5.18}$$

DET is the *fraction of recurrence points on lines of length* $\geq l_{\min}$, where $l_{\min}$ is a parameter ≥ 2, and is called determinism since it increases with the predictability of the system. L is the *average length of a diagonal line* of length $\geq l_{\min}$. Rules for choosing $l_{\min}$ can be found in [20].

If the time series is long enough, these two parameters can also be studied as functions of τ. So let $P_\tau(l)$ denote the number of diagonal lines of exact length l on the diagonal $\mathbf{RR}_{i,i+\tau}$, and

$$\mathrm{DET}_\tau = \frac{\sum_{l=l_{\min}}^{N-\tau} l\, P_\tau(l)}{\sum_{l=1}^{N-\tau} l\, P_\tau(l)} \,, \qquad L_\tau = \frac{\sum_{l=l_{\min}}^{N-\tau} l\, P_\tau(l)}{\sum_{l=l_{\min}}^{N-\tau} P_\tau(l)} \,. \tag{5.19}$$

5.4.2 Cross-Recurrence

The diagonal-wise determination of the recurrence measures is useful for the study of interrelations and synchronization. For the study of interrelations we can use CRPs. Long diagonal structures in CRPs reveal a similar time evolution of the trajectories of the two processes under study. An increasing similarity between the processes causes an increase of the recurrence point density along the main diagonal $\mathbf{CR}_{i,i}$ $(i = 1 \ldots N)$. When the processes become identical, the main diagonal appears, and the CRP becomes an RP. Thus, the occurrence of diagonal lines in CRPs can be used in order to benchmark the similarity between the considered processes. Using this approach it is possible to assess the similarity in the dynamics of two different systems in dependence on a certain time delay [19].

The cross-recurrence rate of a CRP

$$RR_\tau = RR_\tau^{\boldsymbol{x},\boldsymbol{y}} = \frac{1}{N-\tau} \sum_{i=1}^{N-\tau} \mathbf{CR}_{i,i+\tau} \,, \qquad (5.20)$$

reveals the probability of the occurrence of similar states in both systems with a certain delay τ. The average recurrence rate $RR = RR^{\boldsymbol{x},\boldsymbol{y}}$ is determined as in (5.16). It depends not only on ε, but also indicates whether trajectories of the two systems often visit the same phase space regions.

Stochastic and strongly fluctuating processes generate only short diagonals, whereas deterministic processes often admit longer diagonals. If two deterministic processes have the same or similar time evolution, i.e., parts of the phase space trajectories visit the same phase space regions for certain times, the amount of longer diagonals increases and the amount of shorter diagonals decreases. The measures DET_τ and L_τ of a CRP describe the similar time evolution of the systems' states.

As cross-correlation, cross-recurrence is not symmetric in τ. It is possible to define indices of symmetry and asymmetry (for a small range $0 \leq \tau \ll N$), as

$$Q(\tau) = \frac{RR_\tau + RR_{-\tau}}{2} \,, \quad \text{and} \quad q(\tau) = \frac{RR_\tau - RR_{-\tau}}{2} \,. \qquad (5.21)$$

By means of these indices it is possible to quantify interrelations between two systems and determine which system leads the other one (this is similar to the approach for the detection of event synchronization proposed in [25]).

Summarizing, we can state that high values of RR_τ indicate a high probability of occurrence of the same state in both systems, and high values of DET_τ and L_τ indicate a long time span, in which both systems visit the same region of phase space. The consideration of an additional CRP

$$\mathbf{CR}^{-}_{i,j} = \Theta\left(\varepsilon - \|\boldsymbol{x}_i + \boldsymbol{y}_j\|\right) \qquad (5.22)$$

with a negative signed second trajectory $-\boldsymbol{y}_j$ allows distinguishing correlations and anti-correlations between the considered trajectories [19].

5.5 Synchronization

5.5.1 Phase Synchronization on Metric Scale

The concept of recurrence can be used to detect indirectly phase synchronization (PS) in a wide class of chaotic systems and also systems corrupted by noise, where other methods are not so appropriate [26]. The distances between diagonal lines in an RP reflect the characteristic time scales of the system. In contrast to periodic dynamics, for a chaotic oscillator the diagonal lines are interrupted due to the divergence of nearby trajectories. Furthermore, the distances between the diagonal lines are not constant, i.e., we find a distribution of distances, reflecting the different time scales present in the chaotic system.

If two oscillators are in PS, the distances between diagonal lines in their respective RPs coincide, because their phases, and hence their time scales adapt to each other. However, the amplitudes of oscillators, which are only PS but not in general or complete synchronization, are in general uncorrelated. Therefore, their RPs are not identical. However, if the probability that the first oscillator recurs after τ time steps is high, then the probability that the second oscillator recurs after the same time interval will be also high, and vice versa. Therefore, looking at the probability $p(\tau)$ that the system recurs to the ε-neighborhood of a former point $\boldsymbol{x}_i$ of the trajectory after τ time steps and comparing $p(\tau)$ for both systems allows detecting and quantifying PS properly. As mentioned above, $p(\tau)$ can be estimated as recurrence rate (5.15), $\hat{p}(\tau) = RR_\tau$. Studying the coincidence of the positions of the maxima of RR_τ for two coupled systems $\boldsymbol{x}$ and $\boldsymbol{y}$, PS can be identified. More precisely, the correlation coefficient between $RR^{\boldsymbol{x}}_\tau$ and $RR^{\boldsymbol{y}}_\tau$

$$\mathrm{CPR} = \langle \widetilde{RR}^{\boldsymbol{x}}_\tau \cdot \widetilde{RR}^{\boldsymbol{y}}_\tau \rangle \,, \tag{5.23}$$

can be used to quantify PS. Here $\widetilde{RR}^{\boldsymbol{x}}_\tau$ denotes $RR^{\boldsymbol{x}}_\tau$ normalized to zero mean and standard deviation one. If both systems are in PS, the probability of recurrence will be maximal at the same time and CPR ≈ 1. On the other hand, if the systems are not in PS, the maxima of the probability of recurrence will not occur simultaneously. Then we observe a drift and hence expect low values of CPR.

5.5.2 General Synchronization on Metric Scale

It is also possible to detect generalized synchronization (GS) by means of RPs [26]. Let us consider the average probability of recurrence over time for systems $\boldsymbol{x}$ and $\boldsymbol{y}$, i.e., the recurrence rate, $RR^{\boldsymbol{x}}$ and $RR^{\boldsymbol{y}}$, determined by (5.16). The average probability of joint recurrence over time is then given by $RR^{\boldsymbol{x},\boldsymbol{y}}$, which is the recurrence rate of the JRP of the systems $\boldsymbol{x}$ and $\boldsymbol{y}$ [27]. If both systems are independent, the average probability of the joint recurrence will be $RR^{\boldsymbol{x},\boldsymbol{y}} = RR^{\boldsymbol{x}} RR^{\boldsymbol{y}}$. On the other hand, if both systems are in GS, we expect approximately the same recurrences, and hence $RR^{\boldsymbol{x},\boldsymbol{y}} \approx RR^{\boldsymbol{x}} = RR^{\boldsymbol{y}}$. For the computation of the recurrence matrices in the case of essentially different systems that undergo GS, it is more appropriate to use a fixed amount of nearest neighbors N_{n} for each column in the matrix than using a fixed threshold, which corresponds to the original definition of RPs by Eckmann et al. [8]. $RR^{\boldsymbol{x}}$ and $RR^{\boldsymbol{y}}$ are then equal and fixed by N_{n}, because of $RR^{\boldsymbol{x}} = RR^{\boldsymbol{y}} = N_{\mathrm{n}}/N$. Now we call $RR = N_{\mathrm{n}}/N$ and define the coefficient

$$S = \frac{RR^{\boldsymbol{x},\boldsymbol{y}}}{RR}$$

as an index for GS that varies from RR (independent) to 1 (GS). Furthermore, in order to be able to detect also lag synchronization (LS) [29], a time lag is included by using the time delayed JRP (5.10),

$$S(\tau) = \frac{\frac{1}{N^2}\sum_{i,j}^{N} \mathbf{JR}_{i,j}^{\boldsymbol{x},\boldsymbol{y}}(\tau)}{RR}. \tag{5.24}$$

Then, we introduce an index for GS based on the average *joint probability of recurrence* JPR by choosing the maximum value of $S(\tau)$ and normalizing it,

$$\mathrm{JPR} = \max_{\tau} \frac{S(\tau) - RR}{1 - RR}. \tag{5.25}$$

The index JPR ranges from 0 to 1. The parameter RR has to be fixed to compute JPR, but it can be shown that the JPR index does not depend crucially on the choice of RR [26].

5.5.3 Phase Synchronization on Ordinal Scale

As mentioned before, there exists a connection between the order patterns and the phase of a signal. This connection is illustrated in Fig. 5.5, which suggests a representation of the oscillatory behavior of the Rössler system by order patterns. In this section we show how the order patterns of dimension $m = 3$ and the common phase definitions are mathematically related.

Following [12] we introduce a new cylindrical coordinate system (r, ϕ, z) in terms of the time-delayed coordinates $(u_i, u_{i+\vartheta}, u_{i+2\vartheta})$. The z-coordinate corresponds to the main diagonal, and r and ϕ span a plane perpendicular to the main diagonal. The radius r describes the distance to the main diagonal and ϕ the angle. Hence, the order pattern is completely determined by ϕ. On the other hand, the order patterns can be considered as a discretization of ϕ. It has been shown in [12] that ϕ can be written in terms of time-delayed coordinates

$$\tan\phi_i = \sqrt{3}\,\frac{u_{i+2\vartheta} - u_i}{u_{i+2\vartheta} - 2\,u_{i+\vartheta} + u_i} \approx 2\sqrt{3}\,\frac{\dot{u}_{i+1}}{\ddot{u}_{i+1}}. \tag{5.26}$$

Several concepts have been introduced to define a phase for chaotic oscillators [24, 28]. Nevertheless, these approaches are in general restricted to narrow-band signals. For this reason, alternative methods based on the curvature of a phase-space-trajectory have been proposed [17, 22], where the phase is defined as $\phi = \arctan \dot{x}/\dot{y}$. In a similar sense, a phase can be defined as $\phi' = \arctan \dot{x}/\ddot{x}$, which coincides with relation (5.26), up to a constant factor.

To derive a measure for phase synchronization, we analyze the coincidence of phases of two oscillators by means of the recurrence rate RR_τ of order patterns. This yields a distribution of phase differences as a function of the time-lag τ. Following the idea of [36], we introduce a coupling index by means of the Shannon entropy

$$\rho_\pi = 1 - \frac{-\sum_{\tau=\tau_{\min}}^{\tau_{\max}} rr_\tau \log rr_\tau}{\log(\tau_{\max} - \tau_{\min})}, \tag{5.27}$$

where rr_τ is the normalized distribution $rr_\tau = RR_\tau / \sum_\tau RR_\tau$. This index ranges from 0 to 1, where 0 indicates that both systems are independent from each other. The actual maximum depends on $[\tau_{\min}, \tau_{\max}]$ if there are several maxima with distance of a mean recurrence time. Due to a close relationship between the order patterns and the phase, we expect that ρ_π is sensitive to phase synchronization.

This connection to phase indicates a main difference between recurrence plots on metric and ordinal scale. In case of a phase-coherent but chaotic oscillator such as the Rössler system the trajectory returns irregularly to itself. A metric recurrence plot as Fig. 5.4 has only short diagonals. But due to a high coherence of the phase the recurrence time is narrow-banded, and the recurrence rate shows sharp equidistant peaks (cf. Fig. 5.9). Considering recurrence on ordinal scale, only the phase is taken into account. Hence we observe long lines in the recurrence plot (Fig. 5.5), while the distances of peaks in the recurrence rate coincide with that of the metric case.

5.6 Prototypical Examples

5.6.1 Finding Nonlinear Interrelations Using Cross-Recurrence

This example shows the ability of CRPs to find nonlinear interrelations between two processes, which cannot be detected by means of linear tests [19]. We consider linear correlated noise (auto-regressive process of order 1, see for example [32]) which is nonlinearly coupled with the x-component of the Lorenz system (for standard parameters $\sigma = 10$, $r = 28$, $b = 8/3$ and a time resolution of $\Delta t = 0.01$ [2, 18]):

$$y_i = 0.86\, y_{i-1} + 0.500\, \xi_i + \kappa\, x_i^2 , \tag{5.28}$$

where ξ is Gaussian white noise and x_i ($x(t) \to x_i$, $t = i\,\Delta t$) is normalized with respect to the standard deviation. The data length is 8,000 points and the coupling κ is realized without any lag.

As expected, due to the nonlinear relationship, the cross-correlation analysis between x and y does not reveal any significant linear correlation between these data series (Fig. 5.8A). However, the mutual information as a well-established measure to detect nonlinear dependencies [15] shows a strong dependence between x and y at a delay of 0.05 (Fig. 5.8B). The CRP based τ-recurrence rate RR_τ and τ-average diagonal length L_τ exhibit maxima at a lag of about 0.05 for RR^+/L^+ and RR^-_τ/L^-_τ and additionally at 0.45 and -0.32 for RR^-_τ/L^-_τ (Fig. 5.8C, D). The maxima around 0.05 for the + and − measures are a strong indication of a nonlinear relationship between the data. The delay of approximately 0.05 stems from the auto-correlation of y and approximately corresponds to its correlation time $\Delta t/\ln 0.86 = 0.066$. The maxima at 0.45 and -0.32 correspond to half of the mean period of the Lorenz system. Since the result is rather independent of the sign of the second data, the found interrelation is of the kind of an even function. Five Hundred realizations of the AR model have been used in order to estimate the distributions of the measures. The 2σ margins of these distributions can be used to assess the significance of the results.

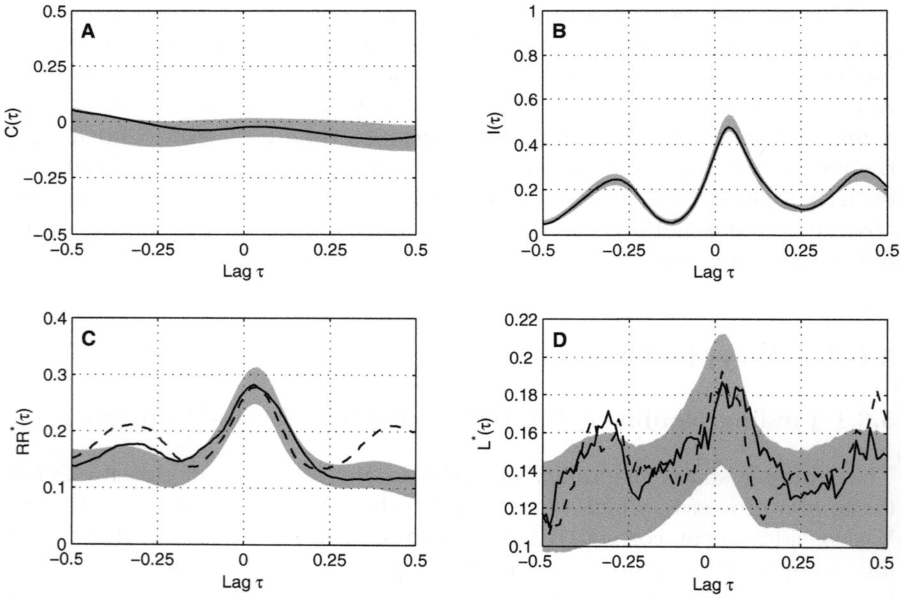

Fig. 5.8 (**A**) Cross-correlation $C(\tau)$, (**B**) mutual information $I(\tau)$, (**C**) τ-recurrence rate RR_τ for the model given in (5.28). (**D**) τ-average line length L_τ for the forced auto-regressive process and the forcing function; the curves represent the measures for one realization as functions of the delay τ for a coupling $\kappa = 0.2$. In (C) and (D) the solid lines show positive relation; the dashed lines show negative relation. The gray bands mark the 2σ margin of the distributions of the measures gained from 500 realizations. The lag τ and the average line length L_τ have units of time [19]

Due to the rapid fluctuations of y, the number of long diagonal lines in the CRP decreases. Therefore, measures based on these diagonal structures, especially DET_τ, do not perform well on such heavily fluctuating data. However, we can infer that the measures RR_τ, as well as L_τ (though less significant for rapidly fluctuating data), are suitable for finding a nonlinear relationship between the considered data series x and y, where the linear analysis is not able to detect such a relation. In contrast to mutual information, this technique is applicable to rather short and non-stationary data.

5.6.2 Synchronization in Rössler Oscillators: Metric Scale

In order to exemplify this method, we consider two mutually coupled Rössler systems

$$\begin{aligned} \dot{x}_1 &= -(1+\nu)x_2 - x_3\,, \\ \dot{x}_2 &= (1+\nu)x_1 + a\,x_2 + \mu(y_2 - x_2)\,, \\ \dot{x}_3 &= b + x_3\,(x_1 - c)\,, \end{aligned} \tag{5.29}$$

$$\begin{aligned} \dot{y}_1 &= -(1-\nu)y_2 - y_3\,, \\ \dot{y}_2 &= (1-\nu)y_1 + a\,y_2 + \mu(x_2 - y_2)\,, \\ \dot{y}_3 &= b + y_3\,(y_1 - c)\,. \end{aligned} \tag{5.30}$$

in the phase coherent regime ($a = 0.16$, $b = 0.1$, $c = 8.5$), similar to the example of Fig. 5.4. According to [22], for $\nu = 0.02$ and $\mu = 0.05$ both systems are in PS. We observe that the local maxima of $RR_\tau^{\boldsymbol{x}}$ and $RR_\tau^{\boldsymbol{y}}$ occur at $\tau = n\,T$, where T is the mean period of both Rössler systems and n is an integer (Fig. 5.9A).

Note that the heights of the local maxima are in general different for both systems if they are only in PS (and not in stronger kinds of synchronization, such as generalized or complete synchronization [24]). But the positions of the local maxima of RR_τ coincide, and the correlation coefficient is CPR$= 0.998$. For $\mu = 0.02$ the systems are not in PS and the positions of the maxima of RR_τ do not coincide anymore (Fig. 5.9B), clearly indicating that the frequencies are not locked. In this case, the correlation coefficient is CPR$= 0.115$.

It is important to emphasize that this method is highly efficient even for non-phase coherent oscillators, such as two mutually coupled Rössler systems in the rather complicated funnel regime (5.29) and (5.30), for $a = 0.2925$, $b = 0.1$, $c = 8.5$, $\nu = 0.02$ (Fig. 5.10). We analyze again two different coupling strengths: $\mu = 0.2$ and $\mu = 0.05$. The peaks in RR_τ (Fig. 5.11) are not as well-pronounced and regular as in the coherent regime, reflecting the different time scales that play a relevant role and the broad band power spectrum of these systems. However, for $\mu = 0.2$ the positions of the local maxima coincide for both oscillators (Fig. 5.11A), indicating PS, whereas for $\mu = 0.05$ the positions

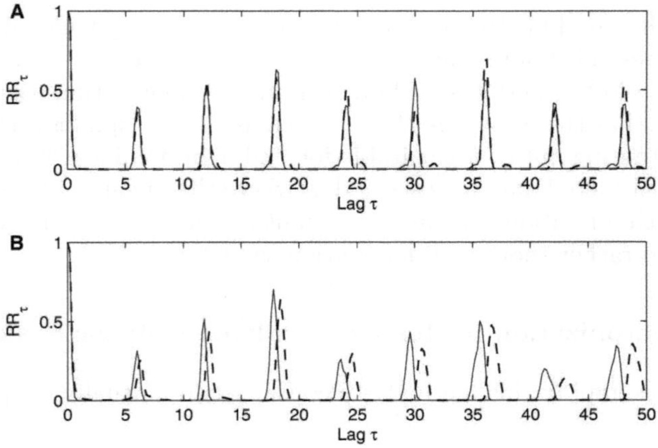

Fig. 5.9 Recurrence probability RR_τ for two mutually coupled Rössler systems (5.29) and (5.30), for $a = 0.16$, $b = 0.1$, $c = 8.5$, in (**A**) phase synchronized and (**B**) non-phase synchronized regime. Solid line: oscillator $\boldsymbol{x}$, dashed line: oscillator $\boldsymbol{y}$

of the local maxima do not coincide anymore (Fig. 5.11B), indicating non-PS. These results are in accordance with [22].

In the PS case of this latter example, the correlation coefficient is CPR = 0.988, and in the non-PS case, CPR = 0.145. Note that the positions of the first peaks in RR_τ coincide (Fig. 5.11B), although the oscillators are not in PS. This is due to the small frequency mismatch ($2\nu = 0.04$). However, by means of the index CPR we can distinguish rather well between both regimes.

Furthermore, the index CPR is able to detect PS even in time series which are strongly corrupted by noise [26]. Additionally, CPR indicates clearly the onset of PS. In [26], the results obtained for CPR in dependence on the

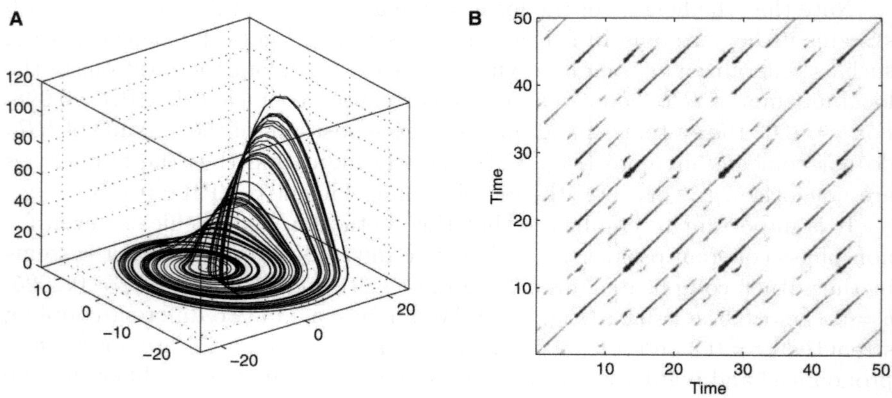

Fig. 5.10 (**A**) Trajectory and (**B**) Recurrence plot for a Rössler system in the funnel regime (5.14) for $a = 0.2925$, $b = 0.1$, $c = 8.5$. Compare with Fig. 5.4

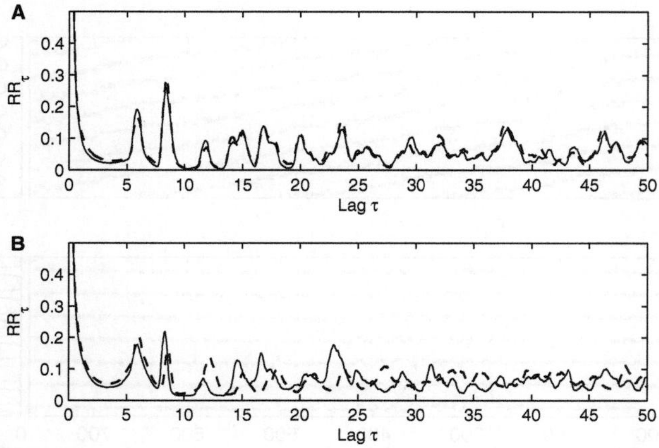

Fig. 5.11 Recurrence probability RR_τ for two mutually coupled Rössler systems in funnel regime (5.29) and (5.30), for $a = 0.2925$, $b = 0.1$, $c = 8.5$. (**A**) $\mu = 0.2$ (PS) and (**B**) $\mu = 0.05$ (non-PS). Solid line: oscillator $\boldsymbol{x}$, dashed line: oscillator $\boldsymbol{y}$

coupling strength were compared with the Lyapunov exponents, as they theoretically indicate the onset of PS (in the phase-coherent case). The results obtained with CPR coincide with the ones obtained by means of the Lyapunov exponents.

The results obtained with CPR are very robust with respect to the choice of the threshold ε. Simulations show that the outcomes are almost independent of the choice of ε corresponding to a percentage of black points in the RP between 1% and 90%, even for non-coherent oscillators. The patterns obtained in the RP, of course, depend on the choice of ε. But choosing ε for both interacting oscillators in such a way that the percentage of black points in both RPs is the same, the relationship between their respective recurrence structures does not change for a broad range of values of ε.

5.6.3 Synchronization in Rössler Oscillators: Ordinal Scale

The locking of phases in case of synchronized oscillators is also reflected by order patterns, which then become synchronized, too. The order patterns represent the phase, which allows an instantaneous study of phase interaction during the onset of phase synchronization of oscillators. A direct comparison of states has the main advantage to study synchronization behavior instantaneously. In order to study a longer range in time and to focus on a small range of the time-lag we choose a slightly different representation of the recurrence plot. In the following we consider the recurrence plot as a function of time i and time-lag $\tau = i - j$, where diagonal lines become horizontal lines.

Figure 5.12 shows cross-recurrence plots on ordinal scale of two mutually coupled Rössler systems. The parameters are the same as before. In the

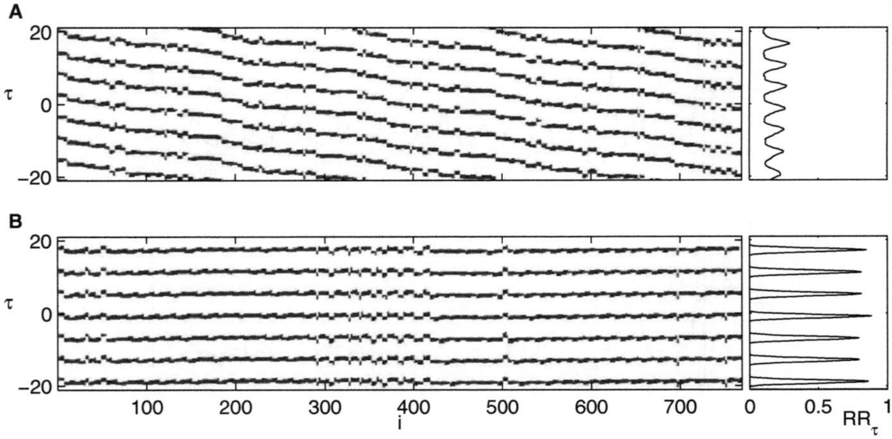

Fig. 5.12 Cross-recurrence plot on ordinal scale and corresponding recurrence rate RR_τ for two mutually coupled Rössler systems in (**A**) non-phase synchronized and (**B**) phase synchronized regime. Embedding parameters are $m = 3$ and $\vartheta = 2$

non-phase synchronized regime (Fig. 5.12A, $\mu = 0.02$) both oscillators diverge due to detuning and consequently we observe drifting lines. The corresponding recurrence rate RR_τ shows no significant values. In case of phase synchronization both oscillators pass the regions of order patterns simultaneously, which is reflected in long horizontal lines (Fig. 5.12B, $\mu = 0.05$). The recurrence rate shows distinct peaks with a distance of the mean recurrence time. With metric CRPs we do also observe the transition to PS, but the lines are

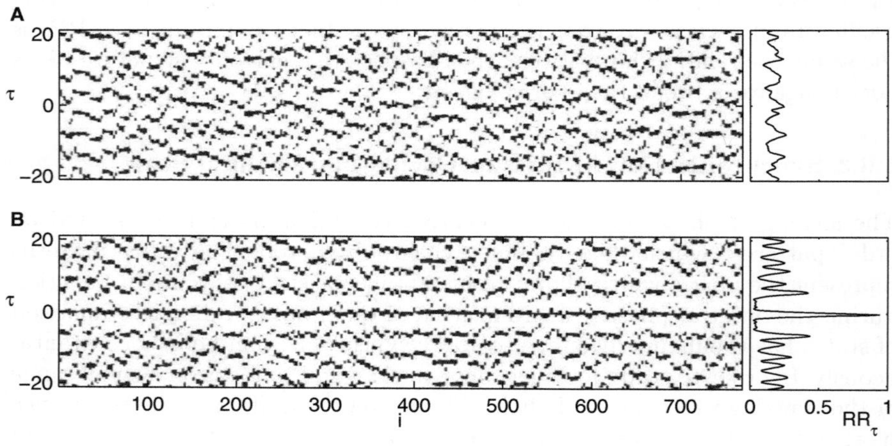

Fig. 5.13 Cross-recurrence plot on ordinal scale and corresponding recurrence rate RR_τ for two mutually coupled Rössler systems in funnel regime in (**A**) non-phase synchronized and (**B**) phase synchronized regime. Embedding parameters are $m = 3$ and $\vartheta = 2$

interrupted because amplitudes are not equal, as they would be in complete synchronization.

In the non-phase-coherent funnel regime (Fig. 5.10) the distribution of recurrence times is broad, what is already reflected in the recurrence rate on metric scale (Fig. 5.11). In case of no phase synchronization ($\mu = 0.05$) we observe only very short horizontal lines in the cross-recurrence plot on ordinal scale (Fig. 5.13A), and the recurrence rate shows no significant values. In case of phase synchronization (Fig. 5.13B, $\mu = 0.2$) the plot clearly shows a long line at $\tau \approx 0$. But in contrast to the phase-coherent regime, there are no other distinct lines and the recurrence rate exhibits only a single dominant peak.

5.7 Application to EEG Data

5.7.1 Synchronization Analysis During an Epileptic Seizure

The following application to EEG data illustrates the advantages of an analysis on ordinal scale in contrast to the metric scale. Scalp EEG data are susceptible to many artifacts which cause offset and amplitude fluctuations. We study the phenomenon of synchronization of neuronal groups during an epileptic seizure, where the specific type of phase synchronization has already been discussed to detect seizure activity [21].

We consider EEG signals from a 14-year old child with epileptic disorder, which were provided by H. Lauffer from the Department of Pediatric Medicine of the University of Greifswald. The 19-channel EEG data (international 10-20 system) were sampled with 256 Hz and band-pass filtered (0.3–70 Hz). On two representative channels the seizure onset is shown in Fig. 5.14A, indicated by a gray bar.

The data during the seizure are clearly dominated by oscillations in the alpha range (≈8–13 Hz) which are locked, indicating synchronization. This yields to a high coupling index ρ_π of the order patterns (Fig 5.14B). Before the seizure there are no dominant oscillations, and the coupling index ρ_π is clearly smaller. Although the EEG data are corrupted by artifacts, the coupling index gives reliable results to reveal the seizure period. The cross-correlation function, however, is strongly disturbed (Fig. 5.14C), and special pre-processing of the data would be inevitable for its use.

5.7.2 Multivariate Coupling in Event-Related Potentials

Our second example will show that quantization of coupling makes sense also for data which do not oscillate. Event-related potentials are multivariate time series which show the reaction of the human brain on a certain stimulus [14]. They are non-stationary, but since the experiment is repeated many times, we have a whole sample of such time series. The traditional way is to take the average time series and look for its maxima and minima which are then

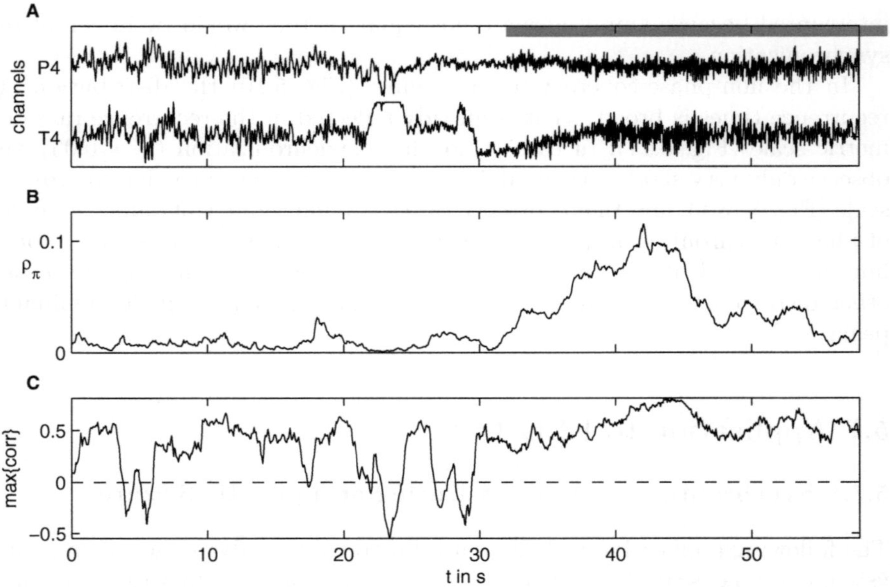

Fig. 5.14 (**A**) Coupling of two EEG channels during onset of epileptic seizure (gray bar). (**B**) Coupling index ρ_π derived from ordinal recurrence plot with embedding parameters $m = 3$ and $\vartheta = 27\,\text{ms}$. (**C**) Maximum value of cross-correlation function. Both coupling indices are determined in a sliding window of length 2 s. Considered time-lag between both channels $-0.2\,s \ldots 0.2\,s$

compared with peaks of typical reaction curves, called N100 or P300 since they appear 100 or 300 ms after the stimulus and are negative or positive.

It is not clear whether these most obvious peaks represent the most interesting brain activity, but there is little fine structure in an average curve of hundreds of time series. Today's challenge is the analysis of single trials, and the huge problem is that beside the reaction under study, there is a lot of other brain activity, measurement errors etc.

The question we address here is whether certain reactions can be characterized not only by the size of amplitudes, but also *by the coupling of different channels,* with no regard to the size of values. This is recent research of C. Bandt and D. Samaga with A. Hamm and M. Weymar from the Department of Psychology in Greifswald who performed a simple oddball experiment just for single-trial analysis. Eight male students were presented, in intervals of about two seconds, 150 equal shapes on a screen. Twenty three were red, and 127 yellow, and the red (oddball) patterns had to be counted. Raw EEG data with sample frequency 500 Hz were taken from 128 channels (Electrical Geodesics, Inc.). Preprocessing involved subtraction of the average of all channels, selection of 54 important channels in the parietal and occipital region, and a decent projection method to remove the 50 Hz mains hum.

Theory predicts that the rare oddball patterns, on which the attention is directed, are characterized by a clear P300 peak in parietal channels – in the average curve. This was true and is shown for the first four subjects and the average of nine parietal channels in Fig. 5.15A. Note that there are big differences between individuals, which casts doubt on the common practice to take "grand averages" over many persons. For instance, the first person has a distinctive N100, and the last one has a clear P200.

The question was whether order patterns in the time series will also detect the oddball effect. We took rank numbers with respect to the previous 40 values (80 ms), $r_i = \#\{k|1 \leq k \leq 40, x_{i-k} < x_i\}$ [3]. The resulting average curves were worse than amplitudes in characterizing oddball trials (Fig. 5.15B). This is because those rank numbers represent the *local* structure of the curve: even a huge P300 value in one trial, which influences the average in Fig. 5.15A, can only have rank number 40, which means it is just larger than the last 40

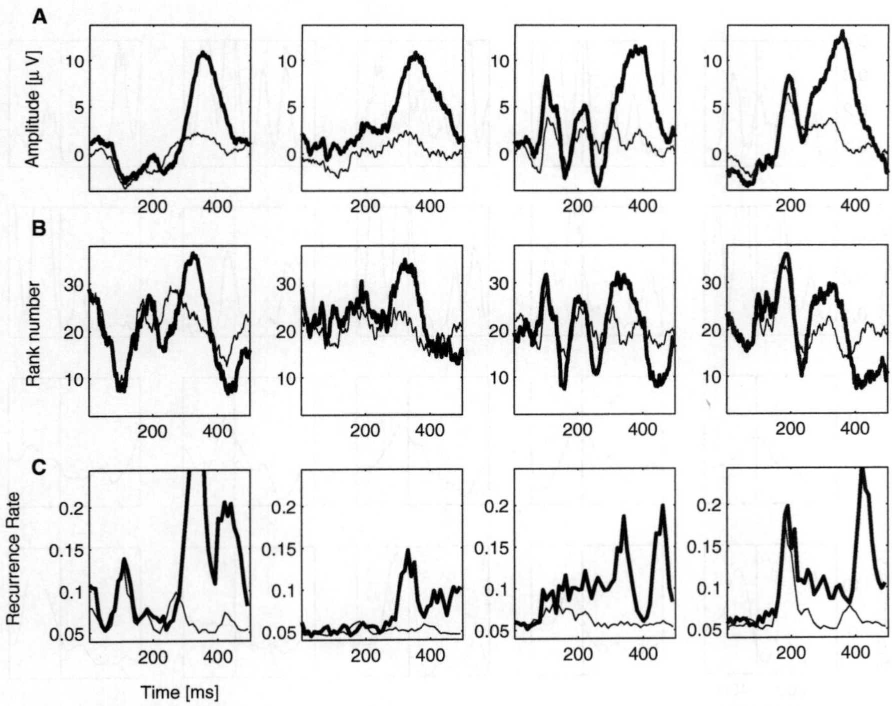

Fig. 5.15 Reaction of four persons (columns) in a visual oddball experiment. (**A**) Mean amplitude of 9 parietal channels and all ordinary (thin) resp. oddball (thick) trials clearly shows P300. (**B**) Corresponding means of rank numbers (40 values backwards) show the effect less clearly. (**C**) The recurrence rate of rank numbers for all nine channels measures the multivariate coupling of the channels. P300 comes out very clearly, but a bit different from (A). Note also N100 for the first person and P200 for the fourth

values. Other features, however, as N100 for the first person, seem to come out more properly with rank numbers than with amplitudes.

Now we took order patterns of length $m = 4$, with lag $\vartheta = 22$ ms, for these 9 channels in the parietal region, and all red (resp. yellow) trials, and took the *recurrence rate of the resulting distribution of order patterns,* as defined in (5.17). To have a better statistics for the 23 red trials, we also took disjoint windows of 5 successive time points (10 ms). (When we estimated $4! = 24$ probabilities p_π from 9×23 order patterns for each time point, the curve became rather noisy, so we took always 5 time points together.) RR is high when many patterns coincide, so it measures the coupling of the channels. The result in Fig. 5.15C shows that this coupling really distinguishes the oddball trials, at least in the overall statistics. The same was found for permutation entropy [4], and also for the rank numbers instead of order patterns.

Now let us go to *single trials,* taking only the first person. Figure 5.16 shows the first 10 oddball trials. The recurrence rates were taken over the 9 parietal

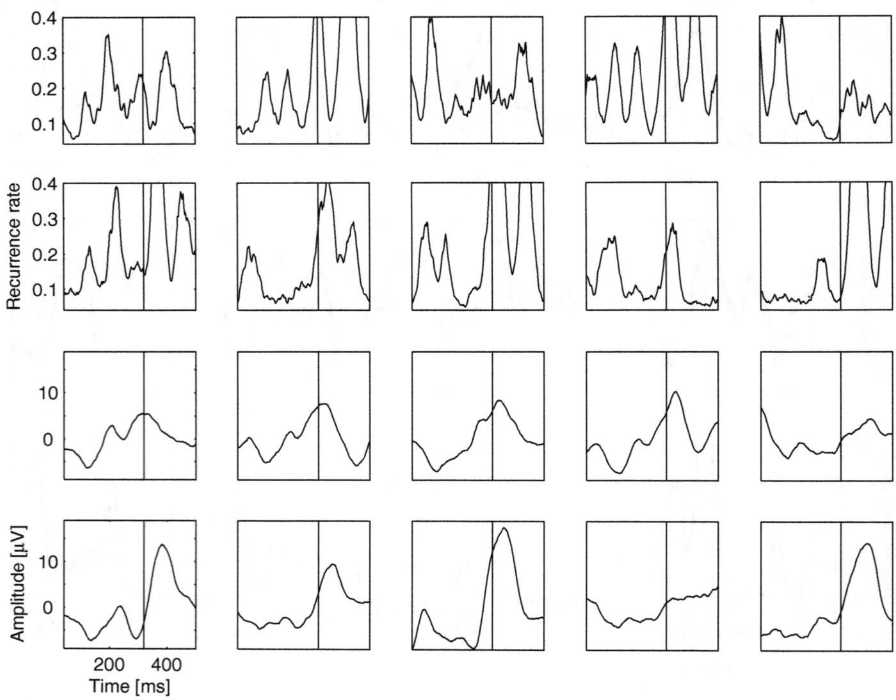

Fig. 5.16 Analysis of P300 for the first 10 oddball trials of the first person. The two upper rows show recurrence rates of order patterns, over 9 channels and time windows of 40 ms, measuring the coupling of the channels. The two lower rows show average amplitude, also taken over 9 channels and 40 ms windows. Time from 40 to 500 ms, vertical lines at 300 ms. As in Fig. 5.15, P300 usually has a single peak in amplitude and a double peak in coupling

channels and over sliding windows of 20 points. For one time point in a single trial, we have only 9 order patterns, one for each channel. With the sliding window, we had $9 \times 20 = 180$ order patterns to estimate $4! = 24$ probabilities p_π which resulted in surprisingly smooth curves. For a fair comparison, the amplitudes in the lower part of Fig. 5.16 were also averaged over the 9 channels and sliding windows of width 20. The P300 can be seen in most cases, with a single peak in amplitude and a twin peak in the coupling. This connection deserves further attention.

As was indicated in Fig. 5.15, the trials also show N100 – small amplitude and coupling around 100 ms. This will be our main concern when we now study the first 10 ordinary trials. There is a peak in coupling around 100 ms in 9 of 10 trials of Fig. 5.17. In the corresponding pictures of amplitude, it is less obvious that there is a minimum. At this point recurrence rates work better. Comparing Figs. 5.16 and 5.17, we see that the reaction at 100 ms comes later in oddball trials.

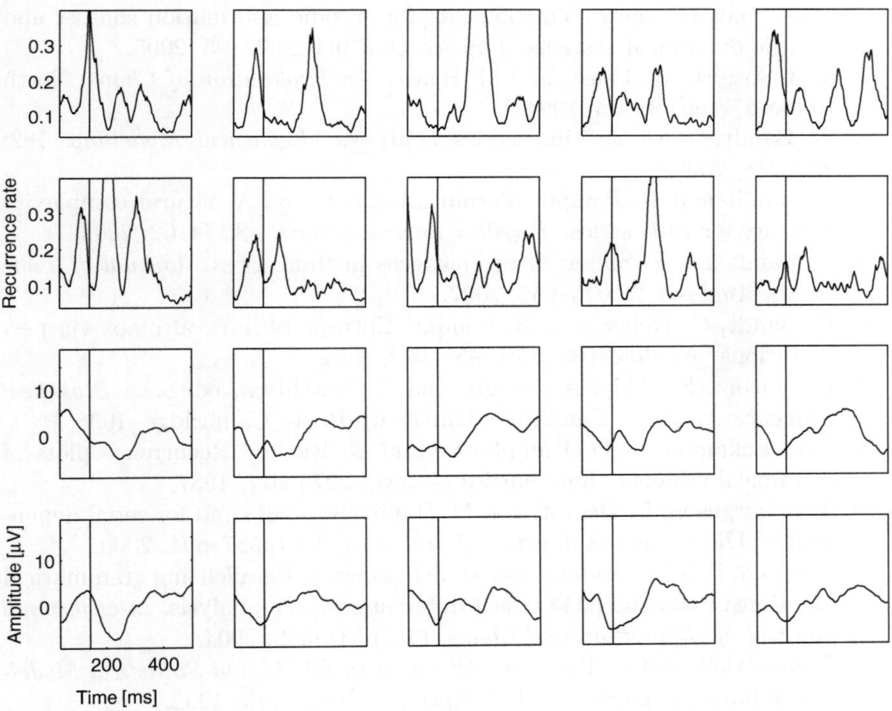

Fig. 5.17 Single-trial analysis of N100 for the first 10 ordinary trials of the first person. The two upper rows show recurrence rates of order patterns over 9 channels and time windows of 40 ms, measuring the coupling of the channels. The two lower rows show average amplitude for the corresponding trials. Time from 40 to 500 ms, vertical lines at 130 ms. To detect N100, coupling performed better than amplitude

Ordinal trials also show weak P300 peaks, which is possible from a physiological point of view. In fact, we had to accumulate the information from many channels to obtain rules for every individual which correctly classify oddball and ordinary trials in 90% of all cases. Other peaks in Fig. 5.17 which irregularly appear may have to do with further brain activity. On the whole, this example indicates that coupling concepts can be very useful for single-trial analysis.

Acknowledgement

This study was made possible by support from the DFG priority programme SPP 1114.

References

[1] J. M. Amigó, M. B. Kennel, and L. Kocarev. The permutation entropy rate equals the metric entropy rate for ergodic information sources and ergodic dynamical systems. *Physica D*, 210(1–2):77–95, 2005.

[2] J. H. Argyris, G. Faust, and M. Haase. *An Exploration of Chaos*. North Holland, Amsterdam, 1994.

[3] C. Bandt. Ordinal time series analysis. *Ecological Modelling*, 182: 229–238, 2005.

[4] C. Bandt and B. Pompe. Permutation entropy: A natural complexity measure for time series. *Physical Review Letters*, 88:174102, 2002.

[5] C. Bandt and F. Shiha. Order patterns in time series. *Journal of Time Series Analysis*, 28:646–665, 2007.

[6] C. Bandt, G. Keller, and B. Pompe. Entropy of interval maps via permutations. *Nonlinearity*, 15:1595–1602, 2002.

[7] R. Durbin, S. Eddy, A. Krogh, and G. Mitchison, editors. *Biological Sequence Analysis*. Cambridge University Press, Cambridge, 1998.

[8] J.-P. Eckmann, S. O. Kamphorst, and D. Ruelle. Recurrence plots of dynamical systems. *Europhysics Letters*, 4:973–977, 1987.

[9] T. S. Ferguson, C. Genest, and M. Hallin. Kendall's tau for serial dependence. *The Canadian Journal of Statistics*, 28(3):587–604, 2000.

[10] S. Frisch, P. beim Graben, and M. Schlesewsky. Parallelizing grammatical functions: P600 and p345 reflect different cost of reanalysis. *International Journal of Bifurcation and Chaos*, 14(2):531–549, 2004.

[11] A. Goettlein and H. Pruscha. *Advances in GLIM and Statistical Modeling*, volume 78, pages 113–118. Springer, New York, 1992.

[12] A. Groth. Visualization of coupling in time series by order recurrence plots. *Physical Review E*, 72:046220, 2005.

[13] M. Hallin and J. Jurečkova. Optimal tests for autoregressive models based on autoregression rank scores. *The Annals of Statistics*, 27(4):1385–1414, 1999.

[14] T. C. Handy, editor. *Event-Related Potentials.* MIT Press, Cambridge, Mass., 2005.
[15] H. Kantz and T. Schreiber. *Nonlinear Time Series Analysis.* Cambridge University Press, Cambridge, second edition, reprint edition, 2005.
[16] M. G. Kendall and J. D. Gibbons. *Rank Correlation Methods.* Oxford University Press, New York, 5th edition, 1990.
[17] I. Z. Kiss, Q. Lv, and J. L. Hudson. Synchronization of non-phase coherent chaotic electrochemical oscillations. *Physical Review E*, 71:035201, 2005.
[18] E. N. Lorenz. Deterministic nonperiodic flow. *Journal of the Atmospheric Sciences*, 20:120–141, 1963.
[19] N. Marwan and J. Kurths. Nonlinear analysis of bivariate data with cross recurrence plots. *Physics Letters A*, 302(5–6):299–307, 2002.
[20] N. Marwan, M.C. Romano, M. Thiel, and J. Kurths. Recurrence plots for the analysis of complex systems. *Physics Reports*, 438(5–6):237–329, 2007.
[21] F. Mormann, K. Lehnertz, P. David, and C. E. Elger. Mean phase coherence as a measure for phase synchronization and its application to the eeg of epilepsy patients. *Physica D*, 144:358–369, 2000.
[22] G. Osipov, B. Hu, C. Zhou, M. Ivanchenko, and J. Kurths. Three types of transition to phase synchronization in coupled chaotic oscillators. *Physical Review Letters*, 91(2):024101, 2003.
[23] K. Petersen. *Ergodic Theory.* Cambridge University Press, Cambridge, 1983.
[24] A. Pikovsky, M. Rosenblum, and J. Kurths. *Synchronization – A Universal Concept in Nonlinear Sciences.* Cambridge University Press, Cambridge, 2003.
[25] R. Quian Quiroga, T. Kreuz, and P. Grassberger. Event synchronization: A simple and fast method to measure synchronicity and time delay patterns. *Physical Review* E66:041904, 2002.
[26] M. C. Romano, M. Thiel, J. Kurths, I. Z. Kiss, and J. L. Hudson. Detection of synchronization for non-phase-coherent and non-stationary data. *Europhysics Letters*, 71(3):466–472, 2005.
[27] M.C. Romano, M. Thiel, J. Kurths, and W. von Bloh. Multivariate recurrence plots. *Physics Letters A*, 330(3-4):214–223, 2004.
[28] M. Rosenblum, A. Pikovsky, and K. Kurths. Phase synchronization of chaotic oscillators. *Physical Review Letters*, 76(11):1804–1807, 1996.
[29] M. G. Rosenblum, A. S. Pikovsky, and J. Kurths. From phase to lag synchronization in coupled chaotic oscillators. *Physical Review Letters*, 78(22):4193–4196, 1997.
[30] O. E. Rössler. An equation for continuous chaos. *Physics Letters A*, 57(5):397–398, 1976.
[31] T. Sauer, J. A. Yorke, and M. Casdagli. Embedology. *Journal of Statistical Physics*, 65(3–4):579–616, 1991.

[32] R. H. Shumway and D.S. Stoffer. *Time Series Analysis and Its Applications*. Springer, New York, 2006.
[33] O. V. Sosnovtseva, A. G. Balanov, T. E. Vadivasova, V. V. Astakhov, and E. Mosekilde. Loss of lag synchronization in coupld chaotic systems. *Physical Review E*, 60(6):6560–6565, 1999.
[34] S. S. Stevens. On the theory of scales of measurement. *Science*, 103: 677–680, 1946.
[35] F. Takens. Detecting strange attractors in turbulence. *Lecture Notes in Mathematics*, volume 898, pages 366–381. Springer, Berlin, 1981.
[36] P. Tass, M. Rosenblum, J.Weule, J. Kurths, A. Pikovsky, J. Volkmann, A. Schnitzler, and H.-J. Freund. Detection of n:m phase locking from noisy data: Application to magnetoencephalography. *Physical Review Letters*, 81(15):3291–3294, 1998.
[37] M. Thiel, M.C. Romano, P.L. Read, and J.Kurths. Estimation of dynamical invariants without embedding by recurrence plots. *Chaos*, 14(2):234–243, 2004.
[38] J. P. Zbilut and Charles L. Webber, Jr. Embeddings and delays as derived from quantification of recurrence plots. *Physics Letters A*, 171(3–4): 199–203, 1992.

6

Structural Adaptive Smoothing Procedures

Jürgen Franke[1], Rainer Dahlhaus[2], Jörg Polzehl[3], Vladimir Spokoiny[3], Gabriele Steidl[4], Joachim Weickert[5], Anatoly Berdychevski[3], Stephan Didas[5], Siana Halim[6], Pavel Mrázek[7], Suhasini Subba Rao[8], and Joseph Tadjuidje[2]

[1] Department of Mathematics, University of Kaiserslautern, Kaiserslautern, Germany
{franke,tadjuidj}@mathematik.uni-kl.de

[2] Institute for Applied Mathematics, University of Heidelberg, Heidelberg, Germany
dahlhaus@statlab.uni-heidelberg.de

[3] Weierstrass Institute for Applied Analysis and Stochastics, Berlin, Germany
{spokoiny,berdichevski,polzehl}@wias-berlin.de

[4] Faculty of Mathematics and Computer Science, University of Mannheim, Mannheim, Germany
steidl@math.uni-mannheim.de

[5] Department of Mathematics, Saarland University, Saarbrücken, Germany
{weickert,didas}@mia.uni-saarland.de

[6] Industrial Engineering Department, Petra Christian University, Surabaya, Indonesia
halim@peter.petra.ac.id

[7] Upek R& D s.r.o., Prague, Czech Republic
pavel.mrazek@upek.com

[8] Department of Statistics, Texas A& M University, College Station, TX, USA
suhasini@stat.tamu.edu

6.1 Local Smoothing in Signal and Image Analysis

An important problem in image and signal analysis is denoising. Given data y_j at locations $x_j, j = 1, \ldots, N$, in space or time, the goal is to recover the original image or signal $m_j, j = 1, \ldots, N$, from the noisy observations $y_j, j = 1, \ldots, N$. Denoising is a special case of a function estimation problem: If $m_j = m(x_j)$ for some function $m(x)$, we may model the data y_j as real-valued random variables Y_j satisfying the regression relation

$$Y_j = m(x_j) + \varepsilon_j\,, \quad j = 1, \ldots, N\,, \tag{6.1}$$

where the additive noise $\varepsilon_j, j = 1, \ldots, N$, is independent, identically distributed (i.i.d.) with mean $\mathbb{E}\,\varepsilon_j = 0$. The original denoising problem is solved by finding an estimate $\widehat{m}(x)$ of the regression function $m(x)$ on some subset containing all the x_j. More generally, we may allow the function arguments to be random variables $X_j \in \mathbb{R}^d$ themselves ending up with a regression model with stochastic design

$$Y_j = m(X_j) + \varepsilon_j\,, \quad j = 1, \ldots, N\,, \tag{6.2}$$

where X_j, Y_j are identically distributed, and $\mathbb{E}\{\varepsilon_j \,|\, X_j\} = 0$. In this case, the function $m(x)$ to be estimated is the conditional expectation of Y_j given $X_j = x$.

In signal and image denoising, the design variables x_j are typically deterministic and equidistant, i.e., if we standardize the observation times and the pixels to the unit interval resp. to the unit square, we have $x_j = j/n$, $j = 0, \ldots, n$, with $N = n + 1$, and $x_{i,j} = (i/n, j/n)$, $i, j = 0, \ldots, n$, with $N = (n+1)^2$ in the regression (6.1). If, e.g., the data Y_j are part of a stationary time series instead (6.2), becomes a nonlinear autoregressive model of order p if we choose $X_j = (Y_{j-1}, \ldots, Y_{j-p})$. Then, $m(X_j)$ becomes the best predictor in the mean-square sense of the next Y_j given the last p observations.

If in (6.1) or (6.2), $m(x) \equiv m_0$ does not depend on x, we just have $m_0 = \mathbb{E}Y_j$. It is consistently estimated by its sample mean $\widehat{m}_0 = \sum_j Y_j/N$ which is a least-squares estimate, i.e., it solves

$$\sum_{j=1}^{N} (Y_j - u)^2 = \min_{u \in \mathbb{R}}!$$

In general, $m(x)$ will not be constant, but it frequently may be approximated locally by a constant, i.e., $m(z) \approx m(x)$ in a neighborhood $|z - x| \leq h$ of x, where $|z|$ denotes the Euclidean norm of z. Then, $m(x)$ may be estimated by a local least-squares approach, i.e., by the solution $\widehat{m}(x)$ of

$$\sum_{j=1}^{N} w_j(x)(Y_j - u)^2 = \min_{u \in \mathbb{R}}! \tag{6.3}$$

where the localizing weights $w_j(x) \approx 0$ for $|X_j - x| > h$. Setting the derivative with respect to u to 0, we get

$$\widehat{m}(x) = \frac{\sum_{j=1}^{N} w_j(x) Y_j}{\sum_{j=1}^{N} w_j(x)}\,, \tag{6.4}$$

which essentially is a local average of the data Y_j from (6.2) corresponding to $X_j \approx x$. The local least-squares approach may be generalized by approximating $m(x)$ locally not by a constant but by a polynomial of low degree resulting in local polynomial smoothers [19]. Also, the quadratic distance in (6.3) may be replaced by the absolute value, and the solution of the corresponding minimization problem is a median smoother. With more general loss functions $\rho(Y_j - u)$, we analogously get M-smoothers [28].

A popular choice are weights $w_j(x) = K\Big(|x - X_j|/h\Big)$ generated by some kernel function $K : \mathbb{R}^d \to \mathbb{R}$ satisfying at least

$$\int K(z)\,\mathrm{d}z = 1\,, \quad \int zK(z)\mathrm{d}z = 0\,, \quad V_K = \int |z|^2 K(z)\,\mathrm{d}z < \infty\,. \qquad (6.5)$$

With this choice of weights, the estimate (6.4) becomes the familiar Nadaraya-Watson kernel estimate

$$\widehat{m}(x,h) = \frac{\frac{1}{Nh^d}\sum_{j=1}^N K(\frac{|x-X_j|}{h})Y_j}{\widehat{p}(x,h)}, \quad \widehat{p}(x,h) = \frac{1}{Nh^d}\sum_{j=1}^N K\Big(\frac{|x-X_j|}{h}\Big). \qquad (6.6)$$

$\widehat{p}(x,h)$ is the Rosenblatt-Parzen estimate of the probability density of the X_j. Frequently, K is also assumed to be nonnegative and bounded. Then $S_K = \int K^2(z)\,\mathrm{d}z < \infty$. If, in particular, K has compact support, say $K(z) = 0$ for $|z| > 1$, then $w_j(x) = 0$ for $|x - X_j| > h$.

In the case of (6.1), where the x_j are deterministic and equidistant, $\widehat{p}(x,h)$ converges quite fast to 1, such that the estimate of $m(x)$ may be simplified to the Priestley-Chao kernel estimate

$$\widehat{m}(x,h) = \frac{1}{Nh^d}\sum_{j=1}^N K\Big(\frac{|x-X_j|}{h}\Big)Y_j\,. \qquad (6.7)$$

The performance of the estimates (6.6), (6.7) crucially depends on the choice of the *bandwidth* h. For model (6.1) with dimension $d = 1$ and the Priestley-Chao estimate we have, e.g.,

$$\mathrm{var}\ \widehat{m}(x,h) = \frac{S_K\sigma_\varepsilon^2}{Nh} + o\Big(\frac{1}{Nh}\Big) \qquad (6.8)$$

$$\mathrm{bias}\ \widehat{m}(x,h) = \mathrm{E}\ \widehat{m}(x,h) - m(x) = \frac{V_K m''(x)}{2}h^2 + o(h^2)\,, \qquad (6.9)$$

where we assume $\sigma_\varepsilon^2 = \mathrm{var}\ \varepsilon_j < \infty$ and $m \in C^2$. Therefore, $\widehat{m}(x,h) \to m(x)$ in mean-square if $N \to \infty, h \to 0, Nh \to \infty$, but the optimal bandwidth minimizing the mean-squared error depends not only on the sample size N but also on the local smoothness of $m(x)$ measured by its curvature $(m''(x))^2$. In general, the optimal degree of smoothing, specified for kernel weights by the choice of h, depends on the location x. It should be large where m is smooth, and small where m has rapidly changing values. The common theme of this chapter is the search for procedures adapting automatically to the structure of the data generating process in specifying the degree of smoothing.

In Sect. 6.2, the problem of choosing an asymptotically optimal local bandwidth $h = h(x)$ is discussed for the image denoising problem where correlation in the noise is allowed for.

Section 6.3 considers a more general dependence between Y_j and X_j. Here, the conditional density $p(y,\theta)$ of Y_j given $X_j = x$ is specified by some unknown parameter θ depending on x, and, correspondingly, local least-squares is replaced by a local maximum likelihood approach. Regression models with additive noise like (6.2) are covered by choosing $p(y,\theta) = p(y-\theta)$, i.e., $\theta = m(x)$ is just a location parameter.

Local smoothing procedures are not the only methods for denoising signals and images. Another large class is based on the idea of regularization. Section 6.4 presents a unified framework which combines local smoothing and regularisation and which allows for a better understanding of the relations between different denoising methods. In Sect. 6.5, some first steps towards an asymptotic theory for that type of nonparametric function estimates are discussed.

Section 6.6 considers another type of structural adaptation in the context of time series analysis. Here, the functions to be estimated sometimes are not constant over time, but are changing slowly compared to the observation rate. These only locally stationary models still allow for applying familiar smoothing methods. The generally applicable approach is illustrated with an example from financial time series analysis where noise is no longer additive but multiplicative, i.e., the autoregressive model which we get from (6.2) with $X_j = (Y_{j-1}, \ldots, Y_{j-p})$ is replaced by an ARCH model $Y_j = \sigma(X_j)\varepsilon_j$.

6.2 Fully Adaptive Local Smoothing of Images with Correlated Noise

We consider the problem of denoising an image by local smoothing where the bandwidth parameters are chosen from the data to adapt to the spatial structure of the original image. We assume that the noisy image is generated by a regression model (6.1) where we do not number the data consecutively but by the coordinates of the respective pixels spread evenly over the interior of the unit square $[0,1]^2$:

$$Y_{ij} = m(x_{ij}) + \varepsilon_{ij} \quad \text{with pixels } x_{ij} = \frac{1}{n}\left(i - \frac{1}{2}, j - \frac{1}{2}\right),\; i,j = 1, \ldots, n.$$

The noise variables ε_{ij} are not necessarily i.i.d., but, more generally, are part of a strictly stationary random field on the integer lattice with $\mathbb{E}\,\varepsilon_{ij} = 0$, $\operatorname{var}\varepsilon_{ij} = c(0,0) < \infty$. Let $c(k,l) = \operatorname{cov}(\varepsilon_{i+k,j+l}, \varepsilon_{ij}), -\infty < k,l < \infty$, denote the corresponding autocovariances and

$$f(0,0) = \sum_{k,l=-\infty}^{\infty} c(k,l) \tag{6.10}$$

the spectral density at (0,0), assuming that it exists. For our smoothing algorithm, $f(0,0)$ is all we have to know about the stochastic structure of the noise. For estimating the original image $m(x), x \in [0,1]^2$ we consider a local average

$$\widehat{m}(x,H) = \sum_{i,j=1}^{n} \int_{A_{ij}} K_H(x-u)du\, Y_{ij} \tag{6.11}$$

where A_{ij}, $i,j = 1,\dots n$, is a partition of the unit square into n^2 subsquares with midpoint x_{ij} and side length $1/n$ each, and where

$$K_H(z) = \frac{1}{\det H} K(H^{-1}z) = \frac{1}{h_1 h_2} K(H^{-1}z),\; z \in \mathbb{R}^2 .$$

Equation (6.11) is a Gasser–Müller type kernel estimate [25] which is closely related to the Priestley-Chao estimate (6.7). However, we do not restrict out considerations to an isotropic kernel $K(z), z \in \mathbb{R}^2$, depending only on $|z|$, but we allow for a full spatial adaption. K may be an arbitrary kernel function satisfying certain symmetry and regularity conditions. In particular, we assume (6.5) and that K has a compact support, is nonnegative, Lipschitz continuous and symmetric in both directions, i.e., $K(-u_1,u_2) = K(u_1,u_2) = K(u_1,-u_2)$ for all u_1,u_2. $K(u_1,u_2) = K(u_2,u_1)$ for all u_1,u_2. Even more important, the one bandwidth parameter h is now replaced by a 2×2 bandwidth matrix $H = C^T H_d C$ with a diagonal matrix H_d and a rotation matrix C:

$$H_d = \begin{pmatrix} h_1 & 0 \\ 0 & h_2 \end{pmatrix}, \quad C = \begin{pmatrix} \cos\alpha & -\sin\alpha \\ \sin\alpha & \cos\alpha \end{pmatrix}.$$

So we have 3 free parameters $0 < h_1, h_2, 0 \le \alpha < 2\pi$. If we fix $\alpha = 0$, then $h_1 = h_1(x), h_2 = h_2(x)$ determine the degree of smoothing at location x along the two coordinate axes. If, e.g., in a neighborhood of x, the image is roughly constant along the first axis, but changes rapidly in the orthogonal direction, h_1 should be large and h_2 small. In general, however, image features like steep slopes, edges or ridges will not necessarily be parallel to the coordinate axes. Therefore, it may be helpful, to rotate the direction of maximal local smoothing until it is parallel to such a feature. Here, the selection of an arbitrary $\alpha = \alpha(x)$ provides the necessary flexibility.

In the following, we assume that $m(x)$ is twice continuously differentiable on the unit square with derivatives

$$m^{(\alpha,\beta)}(x) = \frac{\partial^\alpha}{\partial x_1^\alpha} \frac{\partial^\beta}{\partial x_2^\beta}\, m(x)\,,\;\; \alpha,\beta \ge 0\,.$$

We need this smoothness condition only locally, such that the denoising algorithm practically works also in the presence of an edge, and the asymptotic theory applies everywhere except for a thin stripe around the edge with width shrinking to 0 for $n \to \infty$. Analogously to (6.8) and (6.9), we have the variance and bias expansions

$$\operatorname{var} \widehat{m}(x, H) = \frac{f(0,0) S_K}{n^2 h_1 h_2} + o\left(\frac{1}{n^2 h_1 h_2}\right) \tag{6.12}$$

$$\operatorname{bias} \widehat{m}(x, H) = \frac{1}{2} V_K \operatorname{tr}\left(H \; \nabla^2 m(x) \; H\right) + o(|h|^2), \tag{6.13}$$

where $\nabla^2 m(x)$ denotes the Hessian of m at x, and $V_K = \int u_1^2 K(u)\, du = \int u_2^2 K(u)\, du$, $S_K = \int K^2(u)\, du$ as in Sect. 6.1.

We are interested in a bandwidth matrix which results in a small mean-squared error $\mathbb{E}|\widehat{m}(x,H) - m(x)|^2 = \operatorname{var} \widehat{m}(x,H) + \{\operatorname{bias} \widehat{m}(x,H)\}^2$. Equations (6.12) and (6.13) immediately provide an asymptotically valid approximation for that error where only $f(0,0)$ and $\nabla^2 m(x)$ are unknown. The basic idea of the plug-in method, going back to [2], is to replace the unknown quantities by estimates and, then, to minimize with respect to H. In the following, we apply this approach to the problem of image denoising.

6.2.1 Estimating the Noise Variability

For correlated noise, $f(0,0)$ replaces var ε_{ij} as a measure of the total variability of the noise. As discussed in [31] for the one-dimensional signal denoising problem, it suffices to consider a rather simple estimate of the spectral density for the purpose of adaptive smoothing. First, we estimate the unobservable noise by asymmetric differences. To simplify notation, we use the abbreviations $\mathbf{i} = (i_1, i_2)$, $\mathbf{j} = (j_1, j_2)$ for arbitrary integers i_1, i_2, j_1, j_2, $\mathbf{0} = (0,0)$, and for some $1 \ll M \ll n$ we, in particular, set

$$\mathbf{m}^+ = (M+1, M+1), \;\; \mathbf{m}^- = (M+1, M-1)$$
$$\text{and} \quad \Delta_\mathbf{j} = \frac{2(M+1)}{2(M+1) + |j_1| + |j_2|}.$$

We define

$$\begin{aligned}
\widetilde{\varepsilon}_{\mathbf{i},\mathbf{j}} &= Y_\mathbf{i} - \Delta_\mathbf{j}\, Y_{\mathbf{i}-\mathbf{j}} - (1-\Delta_\mathbf{j})\, Y_{\mathbf{i}+\mathbf{m}^+}, && j_1 \geq 0,\ j_2 > 0, \\
\widetilde{\varepsilon}_{\mathbf{i},\mathbf{j}} &= Y_\mathbf{i} - \Delta_\mathbf{j}\, Y_{\mathbf{i}-\mathbf{j}} - (1-\Delta_\mathbf{j})\, Y_{\mathbf{i}+\mathbf{m}^-}, && j_1 > 0,\ j_2 \leq 0, \\
\widetilde{\varepsilon}_{\mathbf{i},0} &= Y_\mathbf{i} - \frac{1}{2}(Y_{\mathbf{i}-\mathbf{m}^+} + Y_{\mathbf{i}+\mathbf{m}^+}) && \\
\widetilde{\varepsilon}_{\mathbf{i},\mathbf{j}} &= Y_\mathbf{i} - \Delta_\mathbf{j}\, Y_{\mathbf{i}-\mathbf{j}} - (1-\Delta_\mathbf{j})\, Y_{\mathbf{i}-\mathbf{m}^+}, && j_1 \leq 0,\ j_2 < 0, \\
\widetilde{\varepsilon}_{\mathbf{i},\mathbf{j}} &= Y_\mathbf{i} - \Delta_\mathbf{j}\, Y_{\mathbf{i}-\mathbf{j}} - (1-\Delta_\mathbf{j})\, Y_{\mathbf{i}-\mathbf{m}^-}, && j_1 < 0,\ j_2 \geq 0.
\end{aligned}$$

For given $\mathbf{j}$, let $\mathbf{I}_\mathbf{j}$ be the set of all $\mathbf{i}$ for which $\widetilde{\varepsilon}_{\mathbf{i},\mathbf{j}}$ can be calculated from the available sample $\{Y_\mathbf{i},\ 1 \leq i_1, i_2 \leq n\}$, e.g., for $j_1 \geq 0,\ j_2 > 0$, $\mathbf{I}_\mathbf{j}$ consists of all $\mathbf{i} = (i_1, i_2)$ with $1 \leq i_1, i_2, i_1 + M + 1, i_2 + M + 1 \leq n$. Let $N_\mathbf{j}$ denote the number of elements in $\mathbf{I}_\mathbf{j}$. Then, consider the following estimates of the autocovariances $c(\mathbf{j})$, $|j_1|, |j_2| \leq M$:

$$\widehat{c}(0) = \frac{2}{3}\frac{1}{N_0}\sum_{\mathbf{i}\in\mathbf{I_0}} \widetilde{\varepsilon}_{\mathbf{i},0}^2$$

$$\widehat{c}(\mathbf{j}) = -\frac{1}{2\Delta_{\mathbf{j}}}\frac{1}{N_{\mathbf{j}}}\sum_{\mathbf{i}\in\mathbf{I_j}} \widetilde{\varepsilon}_{\mathbf{i},\mathbf{j}}^2 + \Gamma_{\mathbf{j}}\widehat{c}(0)\,,\ |j_1|,|j_2| \le M,\ \mathbf{j}\neq\mathbf{0}\,, \tag{6.14}$$

$$\text{with } \Gamma_{\mathbf{j}} = \frac{1}{2\Delta_{\mathbf{j}}}\{1+\Delta_{\mathbf{j}}^2 + (1-\Delta_{\mathbf{j}})^2\}\,.$$

Then we get an estimate of $f(0,0)$ by replacing the autocovariances in (6.10) by their estimates and truncating the sum

$$\widehat{f}_M(0,0) = \sum_{|j_1|,|j_2|\le M} \widehat{c}(\mathbf{j})\,.$$

Under appropriate assumptions $\widehat{f}_M(0,0)$ is a consistent estimate of $f(0,0)$ for $n, M \to \infty$ such that $M^2/n \to 0$ (compare Proposition 2 of [22]). Moreover, for appropriately chosen M, the effect of replacing $f(0,0)$ by its estimate may be neglected asymptotically for the purpose of the algorithm discussed in the next section.

6.2.2 Adaptive Choice of Smoothing Parameters

The main idea of adaptive smoothing by the plug-in method is minimizing the mean-squared error mse $\widehat{m}(x,H) = \text{var}\ \widehat{m}(x,H) + \{\text{bias}\ \widehat{m}(x,H)\}^2$ with respect to H where variance and bias are given by (6.12), (6.13) and where the unknown quantities are replaced by estimates. An estimate for $f(0,0)$ has been given in the last subsection. For the second derivatives in $\nabla^2 m(x)$ we use again Gasser–Müller estimates

$$\widehat{m}^{(\alpha,\beta)}(x,B) = \sum_{i,j=1}^{n}\int_{A_{ij}} K_B^{(\alpha,\beta)}(x-u)\mathrm{d}u\ Y_{ij}$$

where $K_B^{(\alpha,\beta)}$ denotes the corresponding second derivative of the kernel $K_B(u)$. The resulting approximation for the mean-squared error is

$$\text{amse}\ \widehat{m}(x,H) = \frac{1}{n^2 h_1 h_2}\widehat{f}_M(0,0)S_K + \frac{1}{4}V_K^2\left\{\text{tr}\left(H\ \nabla^2\widehat{m}(x,B)\ H\right)\right\}^2. \tag{6.15}$$

Of course, we now have the problem of choosing the bandwidth matrix B for the estimate of second derivatives. Based on the original approach of [2], we apply an iterative scheme described in detail in [22]. In this algorithm, minimizing (6.15) locally at each pixel $x = x_{kl}$ from the beginning would lead to instabilities. Therefore, in the first steps of the iteration, a good global bandwidth matrix H, which does not depend on x, is determined by minimizing an analogous approximation of the mean-integrated squared error mise $\widehat{m}(\cdot,H) = \int \mathbb{E}\big(\widehat{m}(x,H) - m(x)\big)^2\mathrm{d}x$:

$$\text{amise } \widehat{m}(\cdot, H) = \frac{1}{n^2 h_1 h_2} \widehat{f}_M(0,0) S_K + \frac{1}{4} V_K^2 \int \left\{ \text{tr}\left(H \, \nabla^2 \widehat{m}(x, B) \, H\right) \right\}^2 \mathrm{d}x \,, \qquad (6.16)$$

where the integrals run over the interior of the unit square excluding a stripe of width $\max(h_1, h_2)$ around the boundary to avoid the common boundary effects of local smoothers. After a good global bandwidth H^{glo} is determined by iteratively minimizing (6.16) with respect to H with B also changing in the course of the iteration, it serves as the starting point for the second iteration minimizing now (6.15) and resulting in a local bandwidth selection $H^{\text{loc}} = H^{\text{loc}}(x)$ at every pixel x which is not too close to the boundary of the unit square.

This algorithm leads to an asymptotically optimal choice of bandwidth, i.e., H^{loc} will minimize mse $\widehat{m}(x, H)$ for $n \to \infty$. Moreover, H^{loc} has asymptotically a Gaussian distribution (compare [22]).

6.2.3 A Practical Illustration

We illustrate the performance of the kernel smoother with adaptively chosen local bandwidth by an example from a larger case study. For a wide spectrum of different test images (compare www.math.uni-bremen.de/zetem/DFG-Schwerpunkt/), the denoising methods described in this section prove to be quite competitive to standard methods (compare Chap. 2 and the Appendix of [29]).

Figure 6.1 shows the original image without noise. For the calculations the gray values of this image are scaled to the interval [0,1]. The noisy images are generated from independent Gaussian noise with standard deviation 0.25, 0.5 and 1 respectively, by the following process:

1. Scale the gray values to the interval [0.2, 0.8],
2. add the Gaussian noise,
3. truncate the resulting values back to the interval [0,1], i.e., values below 0 or above 1 are clipped to 0 resp. 1.

Fig. 6.1 Original image without noise

The resulting noise, which is common in image processing applications, is no longer identically distributed as, for values of the original image close to 0 or 1, truncation is more likely to happen. Moreover, if the image is, e.g., close to 1 at a pixel then it frequently is close to 1 in a whole neighborhood which introduces kind of a positive pseudo-correlation in the noise.

We compare the kernel smoother with adaptive bandwidth matrix with four common denoising methods: a simple median filter (MedF), a simple local average with constant weights (AveF), a Gaussian Filter (GauF) and wavelet denoising (Wave) using the Wavelet Toolbox of MATLAB 6 (coif wavelet, soft thresholding). As kind of a benchmark, we also look at the raw data (RawD), i.e., the noisy image before smoothing. We apply three different versions of the plug-in kernel smoother. The first two do not allow for rotation, i.e., $\alpha = 0$ and $H = H_d$ is diagonal, and only 2 bandwidth parameters h_1, h_2 have to be chosen. (AdB2I) assumes independent noise, whereas (AdB2C) allows for correlation. (AdB3C) takes noise correlation into account and admits fully adaptive local smoothing with 3 bandwidth parameters h_1, h_2, α. Figure 6.2 displays some results.

Fig. 6.2 From top left to bottom right: the noisy image, smoother with 2 bandwidth parameters assuming uncorrelated noise, smoother with 2 bandwidth parameters allowing for correlated noise, smoother with 3 bandwidth parameters allowing for correlated noise

Table 6.1 L_1 and L_2-denoising errors

Methods:	RawD	MedF	AveF	GauF	Wave	AdB2I	AdB2I	AdB3C
$100 \times L_1$-norm								
σ=0.25	17.53	6.85	7.08	11.55	6.22	1.90	1.78	1.85
σ=0.5	28.61	9.25	10.69	19.06	9.89	2.71	2.49	2.55
σ=1.0	38.10	13.93	14.91	19.06	14.25	4.87	3.54	3.63
$100 \times L_2$-norm								
σ=0.25	4.91	0.94	0.95	2.17	0.93	0.11	0.09	0.09
σ=0.5	12.31	1.91	1.87	5.64	1.63	0.21	0.14	0.14
σ=1.0	20.22	3.60	3.41	5.64	3.17	0.72	0.26	0.27

Table 6.1 shows the L_1-norm $\sum_{i,j} \left|m(x_{ij}) - \widehat{m}(x_{ij})\right|$ and the L_2-norm $\sum_{i,j} \left|m(x_{ij}) - \widehat{m}(x_{ij})\right|^2$ of the denoising errors. For other measures of performance, we get a similar picture (compare Appendix of [29]). Our method performs much better than the other smoothing procedures which, however, is not surprising as they do not include a local adaptation of the degree of smoothing. However, bandwidth selection by the plug-in method turns out to be competitive too, if compared with other local procedures like the AWS denoising [49] which is based on the approach of the next section.

Finally, we remark that for low signal to noise ratio, it makes not much difference if we allow for correlation in the noise. For $\sigma = 1$, however, where truncation of the original Gaussian noise becomes frequent, the smoothers taking a possible correlation into account perform better (compare the discussion of the effect of truncation above). For this particular example, the additional third bandwidth parameter α does not lead to an improvement. This may be due to the low resolution of the image used in the application.

6.3 Structural Adaptive Smoothing Procedures Using the Propagation-Separation-Approach

Regression is commonly used to describe and analyze the relation between explanatory input variables X and one or multiple responses Y. In many applications such relations are too complicated to be modeled by a parametric regression function. Classical nonparametric regression, see Sect. 6.1 and, e.g., [19, 35, 61, 73] and varying coefficient models, see, e.g., [5, 6, 30], allow for a more flexible form. We focus on methods that allow to efficiently handle discontinuities and spatial inhomogeneity of the regression function in such models.

Let us assume that we have a random sample $Z_1, \ldots, Z_n$ of the form $Z_i = (X_i, Y_i)$. Every X_i is a vector of explanatory variables which determines the distribution of an observed response Y_i. Let the X_i's be valued in the finite dimensional Euclidean space $\mathcal{X} = \mathbb{R}^d$ and the Y_i's belong to $\mathcal{Y} \subseteq \mathbb{R}^q$.

The explanatory variables X_i may, e.g., quantify some experimental conditions, coordinates within an image or a time.

The response Y_i in these cases identifies the observed outcome of the experiment, the gray value or color at the given location or the value of a time series, respectively.

We assume that the distribution of each Y_i is determined by a finite dimensional parameter $\theta = \theta(X_i)$ which may depend on the value X_i of the explanatory variable. In the following we formally introduce our approach within different settings.

6.3.1 One Parameter Exponential Families

Let $\mathcal{P} = (P_\theta, \theta \in \Theta)$ be a family of probability measures on $\mathcal{Y}$ where Θ is a subset of the real line $\mathbb{R}^1$. We assume that this family is dominated by a measure P and denote $p(y,\theta) = \mathrm{d}P_\theta/\mathrm{d}P(y)$. We suppose that each Y_i is, conditionally on $X_i = x$, distributed with density $p(\cdot, \theta(x))$. The density is parameterized by some unknown function $\theta(x)$ on $\mathcal{X}$ which we aim to estimate. A global parametric structure simply means that the parameter θ does not depend on the location, that is, the distribution of every "observation" Y_i coincides with P_θ for some $\theta \in \Theta$ and all i. This assumption reduces the original problem to an estimation problem in a well established parametric model. Here, the maximum likelihood estimate $\widetilde{\theta} = \widetilde{\theta}(Y_1, \ldots, Y_n)$ of θ which is defined by maximization of the log-likelihood $L(\theta) = \sum_{i=1}^n \log p(Y_i, \theta)$ is root-n consistent and asymptotically efficient under rather general conditions.

Such a global parametric assumption is typically too restrictive. The classical nonparametric approach is based on the idea of localization: for every point x, the parametric assumption is only fulfilled locally in a vicinity of x. We therefore use a local model concentrated in some neighborhood of the point x. The most general way to describe a local model is based on weights. Let, for a fixed x, a nonnegative weight $w_i = w_i(x) \le 1$ be assigned to the observations Y_i at X_i, $i = 1, \ldots, n$. When estimating the local parameter $\theta(x)$, every observation Y_i is used with the weight $w_i(x)$. This leads to the local (weighted) maximum likelihood estimate

$$\widetilde{\theta}(x) = \arg \sup_{\theta \in \Theta} L(W(x), \theta) \tag{6.17}$$

with

$$L(W(x), \theta) = \sum_{i=1}^n w_i(x) \log p(Y_i, \theta). \tag{6.18}$$

Note that this definition is a special case of a more general local linear (polynomial) likelihood modeling when the underlying function θ is modeled linearly (polynomially) in x, see, e.g., [18]. However, our approach focuses on the choice of localizing weights in a data-driven way rather than on the method of local approximation of the function θ.

A common choice is to define weights as $w_i(x) = K_{\mathrm{loc}}(\boldsymbol{l}_i)$ with $\boldsymbol{l}_i = |x - X_i|/h$ where h is a bandwidth, $|x - X_i|$ is the Euclidean distance between x and the design point X_i and K_{loc} is a *location kernel*. This approach is intrinsically based on the assumption that the function θ is smooth. It leads to a local approximation of $\theta(x)$ within a ball of some small radius h centered in the point x, see, e.g., [4, 6, 18, 30, 69].

An alternative approach is *localization by a window*. This simply restricts the model to a subset (window) $U = U(x)$ of the design space which depends on x, that is, $w_i(x) = I_{X_i \in U(x)}$. Observations Y_i with X_i outside the region $U(x)$ are not used when estimating the value $\theta(x)$. This kind of localization arises, e.g., in the regression tree approach, in change point estimation, see, e.g., [43, 64], and in image denoising, see [51, 56] among many others.

In our procedure we allow for arbitrary configurations of weights $w_i(x)$. The weights are computed in an iterative way from the data. In what follows we identify the set $W(x) = \{w_1(x), \ldots, w_n(x)\}$ and the local model in x described by these weights. For simplicity we will assume the case where $\theta(x)$ describes the conditional expectation $m(x) = \mathbb{E}\{Y/X = x\}$ and the local estimate is obtained explicitly as

$$\widetilde{\theta}(x) = \sum_i w_i(x) Y_i / \sum_i w_i(x) ,$$

compare Sect. 6.1. In particular, for kernel weights, $\widetilde{\theta}(x)$ is the Nadaraya–Watson estimate (6.6). The quality of estimation heavily depends on the localizing scheme we selected. We illustrate this issue by considering kernel weights $w_i(x) = K_{\mathrm{loc}}(|x - X_i|/h)$ where the kernel K_{loc} is supported on $[0, 1]$. Then the positive weights $w_i(x)$ are concentrated within the ball of radius h at the point x. A small bandwidth h leads to a very strong localization. In particular, if the bandwidth h is smaller than the distance from x to the nearest neighbor, then the resulting estimate coincides with the observation at x. The larger bandwidth we select, the more noise reduction can be achieved. However, the choice of a large bandwidth may lead to a bias if the local parametric assumption of a homogeneous structure is not fulfilled in the selected neighborhood.

The classical approach to solving this problem is selection of a local bandwidth h that may vary with the point x. See Sect. 6.2 and, e.g., [18] for more details.

We employ a related but more general approach. We consider a family of localizing models, one per design point X_i, and denote them as $W_i = W(X_i) = \{w_{i1}, \ldots, w_{in}\}$. Every W_i is built in an iterative data-driven way, and its support may vary from point to point.

6.3.2 Structural Adaptation

Let us assume that for each design point X_i the regression function θ can be well approximated by a constant within a local vicinity $U(X_i)$ containing X_i. This serves as our structural assumption.

Our estimation problem can now be viewed as consisting of two parts. For a given weighting scheme $W(X_i) = \{w_{i1}, \dots, w_{in}\}$ we can estimate the function θ in the design point X_i by (6.17). Initially, a trivial weighting scheme that satisfies our structural assumption is given by $U(X_i) = \{X_i\}$ and $w_{ij} = I_{X_j \in U(X_i)}$.

In order to efficiently estimate the function θ in a design point X_i, we need to describe a local model, i.e., to assign weights $W(X_i) = \{w_{i1}, \dots, w_{in}\}$. If we knew the neighborhood $U(X_i)$ by an oracle, we would define local weights as $w_{ij} = w_j(X_i) = I_{X_j \in U(X_i)}$ and use these weights to estimate $\theta(X_i)$. Since θ and therefore $U(X_i)$ are unknown the assignments will have to depend on the information on θ that we can extract from the observed data. If we have good estimates $\widehat{\theta}_j = \widehat{\theta}(X_j)$ of $\theta(X_j)$ we can use this information to infer on the set $U(X_i)$ by testing the hypothesis

$$H : \theta(X_j) = \theta(X_i) .$$

A weight w_{ij} can be assigned based on the value of a test statistic T_{ij}, assigning zero weights if $\widehat{\theta}_j$ and $\widehat{\theta}_i$ are significantly different. This provides us with a set of weights $W(X_i) = \{w_{i1}, \dots, w_{in}\}$ that determines a local model in X_i.

We utilize both steps in an iterative procedure. We start with a very local model in each point X_i given by weights

$$w_{ij}^{(0)} = K_{\mathrm{loc}}(\boldsymbol{l}_{ij}^{(0)}) \quad \text{with} \quad \boldsymbol{l}_{ij}^{(0)} = |X_i - X_j|/h^{(0)} .$$

The initial bandwidth $h^{(0)}$ is chosen very small. K_{loc} is a kernel function supported on $[-1, 1]$, i.e., weights vanish outside a ball $U_i^{(0)}$ of radius $h^{(0)}$ centered in X_i. We then iterate two steps, estimation of $\theta(x)$ and refining the local models. In the k-th iteration new weights are generated as

$$w_{ij}^{(k)} = K_{\mathrm{loc}}(\boldsymbol{l}_{ij}^{(k)}) K_{\mathrm{st}}(\boldsymbol{s}_{ij}^{(k)}) \quad \text{with} \tag{6.19}$$

$$\boldsymbol{l}_{ij}^{(k)} = |X_i - X_j|/h^{(k)} \quad \text{and} \quad \boldsymbol{s}_{ij}^{(k)} = T_{ij}^{(k)}/\lambda. \tag{6.20}$$

The kernel function K_{st} is monotone nonincreasing on the interval $[0, \infty)$. The bandwidth h is increased by a constant factor with each iteration k. The test statistic

$$T_{ij}^{(k)} = N_i^{(k)} \mathcal{K}(\widehat{\theta}_i^{(k-1)}, \widehat{\theta}_j^{(k-1)}) \tag{6.21}$$

is used to specify the penalty $\boldsymbol{s}_{ij}^{(k)}$. This term effectively measures the statistical difference of the current estimates in X_i and X_j. In (6.21) the term $\mathcal{K}(\theta, \theta')$ denotes the Kullback–Leibler distance of the probability measures P_θ and $P_{\theta'}$.

Additionally we may introduce a kind of memory in the procedure, that ensures that the quality of estimation will not be lost with iterations. This basically means that we compare a new estimate $\widetilde{\theta}_i^{(k)} = \widetilde{\theta}^{(k)}(X_i)$ with the previous estimate $\widehat{\theta}_i^{(k-1)}$ to define a memory parameter $\eta_i = K_{\mathrm{me}}(\boldsymbol{m}_i^{(k)})$ using a kernel function K_{me} and

$$\boldsymbol{m}_i^{(k)} = \tau^{-1} \sum_j K_{\mathrm{loc}}(\boldsymbol{l}_{ij}^{(k)}) \mathcal{K}(\widetilde{\theta}_i^{(k)}, \widehat{\theta}_i^{(k-1)}) .$$

This leads to an estimate

$$\widehat{\theta}_i^{(k)} = \eta_i \widetilde{\theta}_i^{(k)} + (1 - \eta_i) \widehat{\theta}_i^{(k-1)} .$$

6.3.3 Adaptive Weights Smoothing

We now formally describe the resulting algorithm.

- **Initialization:** Set the initial bandwidth $h^{(0)}$, $k = 0$ and compute, for every i the statistics

  $$N_i^{(k)} = \sum_j w_{ij}^{(k)}, \quad \text{and} \quad S_i^{(k)} = \sum_j w_{ij}^{(k)} Y_j$$

  and the estimates

  $$\widehat{\theta}_i^{(k)} = S_i^{(k)} / N_i^{(k)}$$

  using $w_{ij}^{(0)} = K_{\mathrm{loc}}(\boldsymbol{l}_{ij}^{(0)})$. Set $k = 1$ and $h^{(1)} = c_h^{(0)}$.

- **Adaptation:** For every pair i, j, compute the penalties

  $$\boldsymbol{l}_{ij}^{(k)} = |X_i - X_j| / h^{(k)} , \tag{6.22}$$
  $$\boldsymbol{s}_{ij}^{(k)} = \lambda^{-1} T_{ij}^{(k)} = \lambda^{-1} N_i^{(k-1)} \mathcal{K}(\widehat{\theta}_i^{(k-1)}, \widehat{\theta}_j^{(k-1)}) . \tag{6.23}$$

  Now compute the weights $w_{ij}^{(k)}$ as

  $$w_{ij}^{(k)} = K_{\mathrm{loc}}\big(\boldsymbol{l}_{ij}^{(k)}\big) K_{\mathrm{st}}\big(\boldsymbol{s}_{ij}^{(k)}\big)$$

  and specify the local model by $W_i^{(k)} = \{w_{i1}^{(k)}, \ldots, w_{in}^{(k)}\}$.
- **Local estimation:** Now compute new local MLE estimates $\widetilde{\theta}_i^{(k)}$ of $\theta(X_i)$ as

  $$\widetilde{\theta}_i^{(k)} = S_i^{(k)} / \widetilde{N}_i^{(k)} \qquad \text{with} \qquad \widetilde{N}_i^{(k)} = \sum_j w_{ij}^{(k)}, \quad S_i^{(k)} = \sum_j w_{ij}^{(k)} Y_j .$$

- **Adaptive control:** compute the memory parameter as $\eta_i = K_{\mathrm{me}}(\boldsymbol{m}_i^{(k)})$. Define

  $$\widehat{\theta}_i^{(k)} = \eta_i \widetilde{\theta}_i^{(k)} + (1 - \eta_i) \widehat{\theta}_i^{(k-1)} \quad \text{and}$$
  $$N_i^{(k)} = \eta_i \widetilde{N}_i^{(k)} + (1 - \eta_i) N_i^{(k-1)} .$$

- **Stopping:** Stop if $h^{(k)} \geq h_{\max}$, otherwise set $h^{(k)} = c_h h^{(k-1)}$, increase k by 1 and continue with the adaptation step.

Figure 6.3 illustrates the properties of the algorithm for an image with artificial i.i.d. Gaussian noise. Figure 6.3(d) clearly indicates the adaptivity of the procedure.

6.3.4 Choice of Parameters – Propagation Condition

The proposed procedure involves several parameters. The most important one is the scale parameter λ in the statistical penalty s_{ij}. The special case $\lambda = \infty$ simply leads to a kernel estimate with bandwidth $h_{\max}$. We propose to choose λ as the smallest value satisfying a propagation condition. This condition requires that, if the local assumption is valid globally, i.e., $\theta(x) \equiv$

Fig. 6.3 (**a**) Original image, (**b**) image with additive Gaussian noise, (**c**) smoothing result for bandwidth $h_{\max} = 4$ and (**d**) image of pixelwise sum of weights N_i

θ does not depend on x, then with high probability the final estimate for $h_{\max} = \infty$ coincides in every point with the global estimate. More formally we request that in this case for each iteration k

$$\mathbf{E}|\widehat{\theta}^{(k)}(X) - \theta(X)| < (1+\alpha)\mathbf{E}|\breve{\theta}^{(k)}(X) - \theta| \tag{6.24}$$

for a specified constant $\alpha > 0$. Here

$$\breve{\theta}^{(k)}(X_i) = \sum_j K_{\mathrm{loc}}(\boldsymbol{l}_{ij}^{(k)})Y_j / \sum_j K_{\mathrm{loc}}(\boldsymbol{l}_{ij}^{(k)})$$

denotes the nonadaptive kernel estimate employing the bandwidth $h^{(k)}$ from step k. The value λ provided by this condition does not depend on the unknown model parameter θ and can therefore be approximately found by simulations. This allows to select default values for λ depending on the specified family of the probability distribution $\mathcal{P} = (P_\theta, \theta \in \Theta)$. Default values for λ in the examples are selected for a value of $\alpha = 0.2$.

The second parameter of interest is the maximal bandwidth $h_{\max}$ which controls both numerical complexity of the algorithm and smoothness within homogeneous regions.

The scale parameter τ in the memory penalty $\boldsymbol{m}_i$ can also be chosen to meet the propagation condition (6.24). The special case $\tau = \infty$ turns off the adaptive control step.

Additionally we specify a number of parameters and kernel functions that have less influence on the resulting estimates. As a default the kernel functions are chosen as $K_{\mathrm{loc}}(x) = K_{\mathrm{me}}(x) = (1-x^2)_+$ and $K_{\mathrm{st}}(x) = I_{x \le p} + [(1-x)/(1-p)]_+ I_{x>p}$. If the design is on a grid, e.g., for images, the initial bandwidth $h^{(0)}$ is chosen as the distance between neighboring pixel. The bandwidth is increased after each iteration by a default factor $c_h = 1.25^{1/d}$.

For theoretical results on properties of the algorithm see [53].

6.3.5 Local Polynomial Smoothing

In Subsect. 6.3.2 we assumed that the expected value of Y is locally constant. This assumption is essentially used in the form of the stochastic penalty $\boldsymbol{s_{ij}}$. The effect can be viewed as a regularization in the sense that in the limit for $h_{\max} \to \infty$ the reconstructed regression function is forced to a local constant structure even if the true function is locally smooth. Such effects can be avoided if a local polynomial structural assumption is employed. Due to the increased flexibility of such models this comes at the price of a decreased sensitivity to discontinuities.

The Propagation-Separation approach from [52] assumes that within a homogeneous region containing X_i, i.e., for $X_j \in U(X_i)$, the Y_j can be modeled as

$$Y_j = \theta(X_i)^\top \Psi(X_j - X_i) + \varepsilon_j\,, \tag{6.25}$$

where the components of $\Psi(\delta)$ contain values of basis functions

$$\psi_m(\delta_1, \ldots, \delta_d) = (\delta_1)^{m_1} \ldots (\delta_d)^{m_d}$$

for integers $m_1, \ldots m_d \geq 0$, $\sum_k m_k \leq p$ and some polynomial order p. For a given local model $W(X_i)$ estimates of $\theta(X_i)$ are obtained by local Least Squares as

$$\widetilde{\theta}(X_i) = B_i^{-1} \sum_j w_{ij} \Psi(X_j - X_i) Y_j ,$$

with

$$B_i = \sum_j w_{ij} \Psi(X_j - X_i) \Psi(X_j - X_i)^\top .$$

The parameters $\theta(X_i)$ are defined with respect to a system of basis functions centered in X_i. Parameter estimates $\widehat{\theta}(X_{j,i})$ in the local model $W(X_j)$ with respect to basis functions centered at X_i can be obtained by a linear transformation from $\widehat{\theta}(X_j)$, see [52]. In iteration k a statistical penalty can now be defined as

$$\boldsymbol{s}_{\boldsymbol{ij}}^{(\boldsymbol{k})} = \frac{1}{\lambda 2\sigma^2} \big(\widehat{\theta}^{(k-1)}(X_i) - \widehat{\theta}^{(k-1)}(X_{j,i})\big)^\top B_i \big(\widehat{\theta}^{(k-1)}(X_i) - \widehat{\theta}^{(k-1)}(X_{j,i})\big) .$$

In a similar way a memory penalty is introduced as

$$\boldsymbol{m}_{\boldsymbol{ij}}^{(\boldsymbol{k})} = \frac{1}{\tau 2\sigma^2} \big(\widetilde{\theta}^{(k)}(X_i) - \widehat{\theta}^{(k-1)}(X_i)\big)^\top \widetilde{B}_i^{(k)} \big(\widetilde{\theta}^{(k)}(X_i) - \widehat{\theta}^{(k-1)}(X_i)\big) ,$$

where $\widetilde{B}_i$ is constructed like B_i employing location weights $K_{\mathrm{loc}}(l_{ij}^{(k)})$. The main parameters λ and τ are again chosen by a propagation condition requiring free propagation of weights in the specified local polynomial model. A detailed description and discussion of the resulting algorithm and corresponding theoretical results can be found in [52].

Figure 6.4 provides an illustration for local polynomial smoothing based on a piecewise smooth image (a). The local constant reconstruction (c) suffers from bias effects induced by the inappropriate structural assumption while the local quadratic reconstruction (d) benefits from local smoothness while preserving discontinuities.

6.3.6 Using Anisotropy

Applying the adaptive smoothing has a problem of local oversmoothing and loosing some local features if the noise level becomes too large. An improvement of the performance is only possible if some additional structural information is taken into account. One possible approach is based on the notion of the local image flow or local image direction. Namely, it is assumed that

Fig. 6.4 Local polynomial smoothing: (**a**) Original image, (**b**) image with additive Gaussian noise, (**c**) local constant reconstruction $h_{\text{max}} = 7$ and (**d**) local quadratic reconstruction $h_{\text{max}} = 15$

the image is either locally constant or it changes just in one particular direction. This idea is closely related to the notion of intrinsic dimension, see also subsection on adaptive anisotropic filtering in Chap. 7, of the image: within large homogeneous regions this local intrinsic dimension is equal to zero, for edges between two regions it is one, and for the majority of the points in the image the intrinsic dimension is either zero or one.

The corresponding structural assumption can be stated as follows: The image function $\theta(\cdot)$ can, locally around a fixed point x, be represented as $\theta(s) = g(\phi^\top(s - x))$ where $g(\cdot)$ is a univariate function and ϕ is a vector in $\mathbb{R}^d$. The vector ϕ may also vary with x but our structural assumption precisely means that the variability in ϕ is much smaller than in the image function. This assumption leads to anisotropic smoothing methods: the function $g(\cdot)$ is modeled as locally constant or polynomial and the function $\theta(\cdot)$ is approximated as a polynomial of $\phi^\top(s - x)$ in the anisotropic elliptic neighborhood stretched in the direction ϕ and expanded in orthogonal direction(s).

The whole procedure can be decomposed in two main steps which are iterated. One step is the estimation of the local image directions and the second step is the adaptive anisotropic smoothing using the estimated directions. The first step is based on the simple observation that the gradient of the image function is proportional to the image direction ϕ. This suggests to estimate the image direction ϕ from a neighborhood $\mathcal{V}(x)$ of the point x as the first principle direction of the collection of the estimated gradients $\widehat{\nabla\theta}(t)$, $t \in \mathcal{V}(x)$, or equivalently, as the first eigenvector of the matrix

$$D(x) = \sum_{t \in \mathcal{V}(x)} \widehat{\nabla\theta}(t)\widehat{\nabla\theta}(t)^\top .$$

The corresponding eigenvalue can be used to determine the local variability (coherence) of the image function in the estimated direction.

In our implementation, the neighborhoods $\mathcal{V}_i = \mathcal{V}(X_i)$ are taken in the form of the isotropic balls of a rather large bandwidth. The gradient estimates $\widehat{\nabla\theta}(t)$ are computed using usual numerical differences from the 3×3 squares around the point t from the currently estimated image values $\widehat{\theta}_j$.

The corresponding eigenvalue measures the local coherence (variability) of the image in the direction ϕ.

In the second step, the image is estimated again using the anisotropic elliptic neighborhoods. This basically means that we differently localize in the direction ϕ and in the orthogonal directions, leading to the location penalty $\boldsymbol{l}_{ij}$ in the form

$$\boldsymbol{l}_{ij} = \frac{\left|\phi^{\top}(X_i - X_j)\right|^2}{h_1^2} + \frac{\left|(I - \Pi_{\phi})(X_i - X_j)\right|^2}{h_2^2},$$

where Π_{ϕ} means the projector on the direction ϕ. Such penalty defines an elliptic neighborhood around X_i of the points X_j with $\boldsymbol{l}_{ij} \leq 1$. The axis h_1, h_2 are taken to provide that the volume of this neighborhood is equal to the volume of the ball with the radius $h^{(k)}$ at the iteration k of the algorithm. The degree of anisotropy defined by the ratio h_1/h_2 is selected due to the estimated coherence in the central point X_i. Figure 6.5 illustrates the behavior of the algorithm.

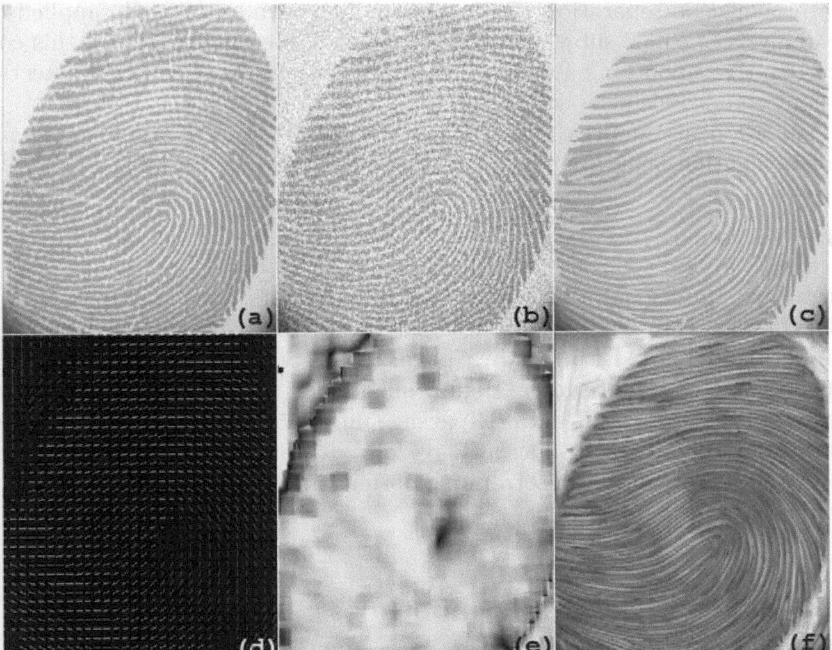

Fig. 6.5 Anisotropic AWS: (**a**) Original image, (**b**) image with Gaussian noise, (**c**) reconstruction with maximal anisotropy 25 and $h_{\max} = 12$, (**d**) estimates anisotropy direction, (**e**) coherence and (**f**) sum of weights N_i

6.3.7 Digital Images

Digital images may carry much more complicated noise. In a CCD sensor a Bayer mask is used to obtain color information. Within a square of four pixels one is red filtered, one blue and two green. The measurement in each voxel in principle consists of a random photon count plus thermal noise leading in the ideal case to a linear dependence between noise variance and signal intensity. To obtain a color image in full resolution, the missing color values are filled in by interpolation (demosaicing). This results in spatial correlation within the image. Other sources of spatial correlation are noise reduction filters and preprocessing steps applied within the camera. Correlation between color channels is induced by rotation into the RGB color space.

The algorithm described in [54] handles the effects of correlation by an adjustment of the statistical penalty $\boldsymbol{s}_{ij}$ and incorporates a model for the variance-intensity dependence.

6.3.8 Applications in MRI

The propagation-separation (PS) approach has been successfully applied for the analysis of single subject fMRI experiments, see [50, 68, 72]. In this context adaptive smoothing allows for an improved sensitivity of signal detection without compromising spatial resolution.

A method for adaptive smoothing of diffusion weighted MR images using the Propagation-Separation approach has been proposed in [67].

6.3.9 Software

A reference implementation for the adaptive weights procedure described in Subsect. 6.3.3 is available as a package (aws) of the R-Project for Statistical Computing [57] from http://www.r-project.org/. Software for specific applications, e.g., adaptive filtering of digital images (R-package adimpro, [54]), the analysis of functional MR data (R-package fmri, [55]) and adaptive smoothing of diffusion weighted MR images (R-package dti) are available from the same site. Structural adaptive anisotropic smoothing is implemented in a standalone demo program.

6.4 Nonlocal Data and Smoothness Terms for Image Smoothing

Image smoothing with the goal of simplification or denoising is an important part of many applications in image processing and computer vision. This chapter reviews a unifying variational framework for image smoothing based on an energy function with *nonlocal data and smoothness terms (NDS)*. In [39] this function has first been presented, and it was shown that depending on its parameters its minimization can lead to a large variety of existing algorithms for image smoothing. Several numerical methods to minimize the NDS energy function have been investigated in [14] with their theoretical and practical properties. In [47] the question of how to choose sensible parameters was discussed.

This section is organized as follows: In Subsect. 6.4.1 we explain the NDS model and relate it to several well-known filtering techniques. To deepen the understanding of the model and its parts, the question of how to choose the parameters is discussed in Subsect. 6.4.3. Several numerical methods to minimize the energy function are presented and compared in the following Subsect. 6.4.4. The section is concluded with Subsect. 6.4.6.

6.4.1 The Variational NDS Model

First we are going to present the NDS model and take a look at special cases which coincide with classical filtering methods. The model shown here is working with images on a discrete domain.

Let $n \in \mathbb{N}$ be the number of pixels in our images and $\Omega = \{1, \dots, n\}$ the corresponding index set. Let $y \in \mathbb{R}^n$ be the given image. In the following, $u \in \mathbb{R}^n$ stands for a filtered version of y. The energy function E of the NDS filter presented in [39] can be decomposed into two parts: the *data* and the *smoothness term.* The data term can be written as

$$E_D(u) = \sum_{i,j \in \Omega} \Psi_D \left(|u_i - y_j|^2 \right) w_D \left(|x_i - x_j|^2 \right) , \tag{6.26}$$

where $\Psi_D : [0, \infty) \longrightarrow [0, \infty)$ is an increasing function which plays the role of a penalizer for the difference between u and the initial image y, the so-called *tonal weight function.* This weight function is important for the behavior of the filtering method with respect to image edges and the robustness against outliers. Six possibilities for such penalizing functions are displayed in Table 6.2. The upper three functions are convex while the lower three ones are nonconvex. The convex ones have the advantage that the existence and uniqueness of a minimizer can be proven directly with this property. On the other hand, the nonconvex functions yield edge enhancement and high robustness with respect to noise and outliers which can significantly improve the reconstruction results. We notice that the data term not only compares the gray values of u and y at the pixel x_i, but it also takes a nonlocal neighborhood into account. This neighborhood is defined with the help of the *spatial weight function* $w_D : [0, \infty) \longrightarrow [0, \infty)$ depending on the Euclidean distance between the pixels x_i and x_j. Table 6.3 shows the two most important types of spatial windows, the so-called *hard* and *soft window.* By choosing special cases of w_D,

Table 6.2 Possible choices for tonal weights Ψ. Source: [14]

$\Psi(s^2)$		$\Psi'(s^2)$	Known in the context of
s^2	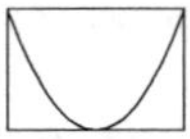	1	Tikhonov regularization [70]
$2\lambda^2\left(\sqrt{1+\frac{s^2}{\lambda^2}}-1\right)$	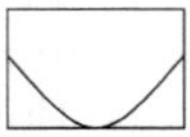	$\left(1+\frac{s^2}{\lambda^2}\right)^{-\frac{1}{2}}$	Nonlinear regularization, Charbonnier et al. [7]
$2\left(\sqrt{s^2+\varepsilon^2}-\varepsilon\right)$	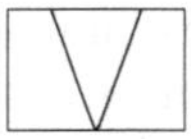	$(s^2+\varepsilon^2)^{-\frac{1}{2}}$	Regularized total variation [60]
$\lambda^2\log\left(1+\frac{s^2}{\lambda^2}\right)$	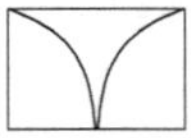	$\left(1+\frac{s^2}{\lambda^2}\right)^{-1}$	Nonlinear diffusion, Perona and Malik [46]
$\lambda^2\left(1-\exp\left(-\frac{s^2}{\lambda^2}\right)\right)$	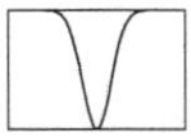	$\exp\left(-\frac{s^2}{\lambda^2}\right)$	Nonlinear diffusion, Perona and Malik [46]
$\min(s^2,\lambda^2)$	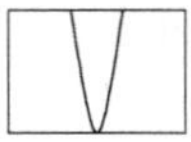	$\begin{cases}1 & \lvert s\rvert<\lambda\\ 0 & \text{else}\end{cases}$	Segmentation, Mumford and Shah [44]

one can determine the amount of locality of the filter from one pixel up to the whole image domain.

The second ingredient of the NDS function is the smoothness term

$$E_S(u) \;=\; \sum_{i,j\in\Omega} \Psi_S\left(|u_i-u_j|^2\right) w_S\left(|x_i-x_j|^2\right) \tag{6.27}$$

which differs from the data term by the fact that not the difference between u and y is calculated as argument of the penalizer, but the difference between gray values of u inside a larger neighborhood. Since E_S does not depend on the initial image y, each constant image $u \equiv c \in \mathbb{R}$ would always yield a minimum: Thus we are rather interested in intermediate solutions during the minimization process.

The complete NDS energy function is then the convex combination of these two parts:

Table 6.3 Possible choices for spatial weights w. Source: [14]

$w(s^2)$			Known in the context of
$\begin{cases}1 & \lvert s\rvert<\lambda\\ 0 & \text{else}\end{cases}$	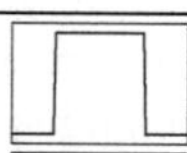	Hard window	Locally orderless images, Koenderink and van Doorn [32]
$\exp\left(-\frac{s^2}{\lambda^2}\right)$	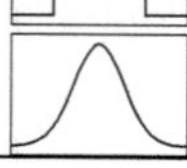	Soft window	Chu et al. [9]

$$E(u) = \alpha\, E_D(u) + (1-\alpha)\, E_S(u) \tag{6.28}$$
$$= \alpha \sum_{i,j\in\Omega} \Psi_D\left(|u_i - y_j|^2\right) w_D\left(|x_i - x_j|^2\right)$$
$$+ (1-\alpha) \sum_{i,j\in\Omega} \Psi_S\left(|u_i - u_j|^2\right) w_S\left(|x_i - x_j|^2\right) \tag{6.29}$$

for $0 \leq \alpha \leq 1$. This function combines two ways of smoothing: Besides the smoothness term, also choosing a window with larger size in the data term introduces some averaging over this neighborhood. The presence of the initial image y in the data term makes nonflat minima of E possible.

6.4.2 Included Classical Methods

Figure 6.6 gives a first overview over the methods which can be expressed as minimization of the energy function (6.29).

In the following, we are going to take a closer look at the special cases mentioned there:

- **M-estimators:** The estimation of an unknown value $y \in \mathbb{R}$ with the help of n noisy measurements can be written as minimizing the energy function

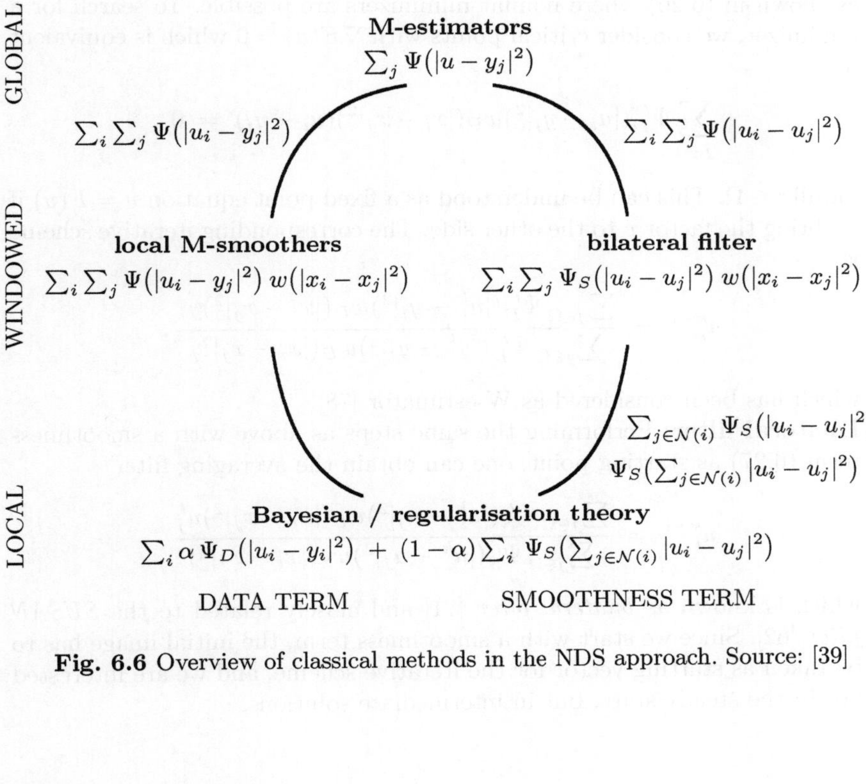

Fig. 6.6 Overview of classical methods in the NDS approach. Source: [39]

$$E(u) \;=\; \sum_{j=1}^{n} \Psi(|u - y_j|^2)\,. \tag{6.30}$$

This can be understood as a very reduced form of a data term where we only have one unknown.

- **Histogram Operations:** Extending the function E in (6.30) from one unknown to n unknown pixel values yields a data term (6.26) with global spatial window

$$E(u) \;=\; \sum_{i,j\in\Omega} \Psi(|u_i - y_j|^2)\,. \tag{6.31}$$

Since the function treats the gray value u_i independent of its position x_i, and since for each pixel all gray values y_j of the initial image are considered, the result only depends on the histogram of the image y. Note that this is also the reason why the global minimum is a flat image in this context. We are thus interested in intermediate states of iterative minimization processes.

- **Local M-smoothers / W-estimators:** Restricting the neighborhood size leads to the data term

$$E_D(u) \;=\; \sum_{i,j\in\Omega} \Psi_D(|u_i - y_j|^2) w_D(|x_i - x_j|^2)$$

as shown in (6.26) where nonflat minimizers are possible. To search for a minimizer, we consider critical points with $\nabla E(u) = 0$ which is equivalent to

$$\sum_{j\in\Omega} \Psi_D'(|u_i - y_j|^2) w_D(|x_i - x_j|^2)(u_i - y_j) \;=\; 0$$

for all $i \in \Omega$. This can be understood as a fixed point equation $u = F(u)$ if we bring the factor u to the other side. The corresponding iterative scheme is

$$u_i^{k+1} \;=\; \frac{\sum_{j\in\Omega} \Psi_D'(|u_i^k - y_j|^2) w_D(|x_i - x_j|^2) y_j}{\sum_{j\in\Omega} \Psi_D'(|u_i^k - y_j|^2) w_D(|x_i - x_j|^2)}\,,$$

which has been considered as W-estimator [78].

- **Bilateral filter:** Performing the same steps as above with a smoothness term (6.27) as starting point, one can obtain the averaging filter

$$u_i^{k+1} \;=\; \frac{\sum_{j\in\Omega} \Psi_S'(|u_i^k - u_j|^2) w_S(|x_i - x_j|^2) u_j^k}{\sum_{j\in\Omega} \Psi_S'(|u_i^k - u_j|^2) w_S(|x_i - x_j|^2)}$$

which is known as *bilateral filter* [71] and closely related to the *SUSAN filter* [62]. Since we start with a smoothness term, the initial image has to be taken as starting vector for the iterative scheme, and we are interested not in the steady state, but in intermediate solutions.

- **Bayesian / regularization approaches:** Classical regularization approaches can be expressed with the NDS function by using very small local neighborhoods. Typically, one will use only the central pixel in the data term, and only the four direct neighbors $\mathcal{N}(i)$ of the pixel i in the smoothness term. This results in the energy function

$$E(u) = \alpha \sum_{i\in\Omega} \Psi_D(|u_i - y_i|^2) + (1-\alpha) \sum_{i\in\Omega} \sum_{j\in\mathcal{N}(i)} \Psi_S(|u_j - u_i|^2),$$

which is some kind of discrete anisotropic regularization function since the differences to all four neighbors are penalized independently. Exchanging the sum and the penalizer in the smoothness term would yield the isotropic variant with an approximation of $|\nabla u|^2$ inside the penalizer.

6.4.3 Choice of Parameters

As we have seen in the last section, the NDS model is a very general model which is capable of yielding various kinds of filtering results depending on the choice of parameters. To obtain denoising results with good visual quality, one has to determine an appropriate set of parameters and weights depending on the noise and the properties of the desired result. For example, nonconvex tonal weights tend to yield images which can be seen as compositions of regions with constant gray value, while quadratic weights usually blur the image edges. In this context the question arises if there is some redundancy in the set of parameters. We will display one experiment concerning this question here – further experiments and remarks addressing the problem of choosing the parameters can be found in [47]. In our experiment we compare the influence of the neighborhood size in the data term with the weight α between data and smoothness term. To this end we consider a pure data term

$$E_D(u) = \sum_{i,j\in\Omega} (u_i - y_j)^2 w_D(|x_i - x_j|^2)$$

with a quadratic tonal penalizer and a disc-shaped hard window function w_D with radius r_D. On the other hand, we have a function with a local data term and a smoothness term that involves only the direct neighbors:

$$E_C(u) = \alpha \sum_{i\in\Omega} (u_i - y_i)^2 + (1-\alpha) \sum_{i\in\Omega} \sum_{j\in\mathcal{N}(i)} (u_i - u_j)^2 .$$

The only parameter α determines how smooth the result is in this case.

We are looking for a quantification of the difference between the results obtained by minimizing these two functions. It is especially interesting how the parameters α and r_D are connected. Figure 6.7 shows that the filtering results obtained by the two approaches are hardly distinguishable when the parameters are suitably chosen. There is also a graph relating the size of the window r_D, and the value for α such that the difference is minimized in Fig. 6.8.

Fig. 6.7 Example of the trade-off between different parameters. **Left**: Original image, 256 × 256 pixels, with additive Gaussian noise, standard deviation $\sigma = 50$. **Middle**: Denoising with E_D, radius $r_D = 5$. **Right**: Denoising with E_C, $\alpha = 0.424$. Source: [47]

6.4.4 Numerical Minimization Methods

This section gives a short overview over several possibilities to minimize the NDS energy function (6.29). There are two classes of methods described here: The first two methods are fixed point iterations, and the last two are based on Newton's method:

- **Jacobi method:** As already sketched in Sect. 6.4.1, we start with the search for critical points with $\nabla E(u) = 0$ here. With the abbreviations $d_{i,j}^k := \Psi'_D(|u_i^k - y_j|^2) w_D(|x_i - x_j|^2)$ and $s_{i,j}^k := \Psi'_S(|u_i^k - u_j^k|^2) w_S(|x_i - x_j|^2)$ we obtain the fixed point iteration scheme

$$u_i^{k+1} = \frac{\alpha \sum_{j\in\Omega} d_{i,j}^k y_j + 2(1-\alpha)\sum_{j\in\Omega} s_{i,j}^k u_j^k}{\alpha \sum_{j\in\Omega} d_{i,j}^k + 2(1-\alpha)\sum_{j\in\Omega} s_{i,j}^k} =: F(u^k)$$

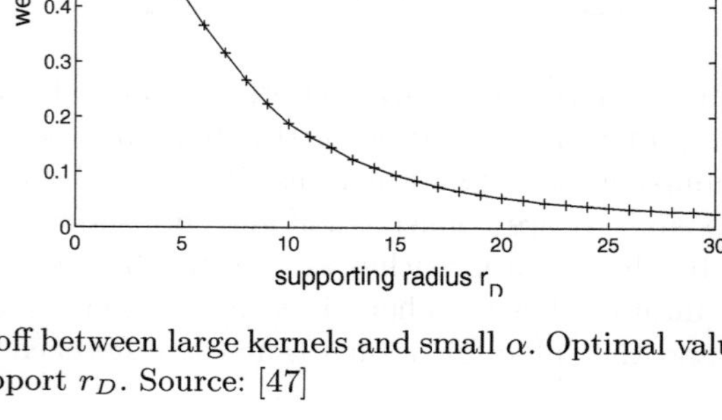

Fig. 6.8 Trade-off between large kernels and small α. Optimal value of α depending on the radial support r_D. Source: [47]

for all $i \in \Omega, k \in \mathbb{N}$ starting with $u_i^0 := y_i$ for all $i \in \Omega$. Since Ψ_D and Ψ_S are assumed to be nonnegative, it can easily be seen in this formulation that the scheme satisfies a maximum–minimum principle. With Brouwer's fixed point theorem [79, p. 51] and the continuity of the scheme function F, this implies the existence of a fixed point.

- **Gauß–Seidel method:** The difference of the Gauß–Seidel method to the Jacobi method described above is that variables of the old and the new iteration level are used in a local fixed point scheme. We set $x^0 := u_i$ and perform $m \in \mathbb{N}$ steps of a local fixed point iteration in the i-th component

$$x^{l+1} = F_i(u_1^{k+1}, \ldots, u_{i-1}^{k+1}, x^l, u_{i+1}^k, \ldots, u_n^k), \quad l = 0, \ldots, m-1$$

and set $u_i^{k+1} := x^m$ afterwards. Here F_i denotes the i-th component of the vector-valued function F defined above.

- **Newton's method:** To find a point where the gradient vanishes, one can also use Newton's method for ∇E:

$$u^{k+1} = u^k - H(E, u^k)^{-1} \nabla E(u^k).$$

Note that this method is only applicable if the tonal weight functions Ψ_D and Ψ_S are convex, because this implies that the Hessian $H(E, u^k)$ is positive definite and thus invertible. Practical experiments have shown that it makes sense to combine Newton's method with a line-search strategy to steer the length of the steps adaptively. This results in the scheme

$$u^{k+1} = u^k - \sigma^k H(E, u^k)^{-1} \nabla E(u^k),$$

where $\sigma^k := \frac{1}{2^l}$ is chosen maximal such that the energy is decreasing: $E(u^{k+1}) < E(u^k)$.

- **Gauß–Seidel Newton method:** The Gauß–Seidel Newton method is a Gauss–Seidel method with a local Newton's method in each component instead of the local fixed point scheme. Here we have also used an adaptive step size, but we used a local energy where all pixels except the evolving one are fixed for the step size criterion.

Experiments have shown that there is no unique method which is preferable over all others in any case. The methods based on the fixed point scheme have the clear advantage that a maximum–minimum principle holds even for nonconvex tonal weights where the other two methods are not even applicable. On the other hand, Newton's method can be faster for special choices of the parameters. Further details about the numerical implementation of NDS can be found in [14]. The combination of the presented methods with a coarse-to-fine strategy is a possible extension which could be helpful to further improve the running time.

6.4.5 Diffusion Filters and Wavelet Shrinkage

As discussed in Sect. 6 of [39], the NDS approach also covers nonlinear diffusion filtering. The latter, on the other hand, is related to wavelet shrinkage which is the basis for another popular class of smoothing procedures. Here, we give a brief overview of our results concerning the relations between these two methods.

For the case of space-discrete 1 D diffusion, we proved that Haar wavelet shrinkage on a single scale is equivalent to a single step of TV diffusion or TV regularization of two-pixel pairs [65]. For the general case of N-pixel signals, this led us to a numerical scheme for TV diffusion, more precisely, translationally invariant Haar wavelet shrinkage on a single scale can be used as an absolutely stable explicit discretization of TV diffusion. In addition, we proved that space-discrete TV diffusion and TV regularization are identical in 1 D, and that they are also equivalent to a dynamical system called SIDEs when a specific force function is chosen [3, 66].

Both considerations were then extended to 2 D images, where an analytical solution for 2×2 pixel images served as a building block for a wavelet-inspired numerical scheme for diffusion with singular diffusivities which we called "locally analytic scheme" (LAS) [75, 76]. Replacing space-discrete diffusion by a fully discrete one with an explicit time discretization, we obtained a general relation between the shrinkage function of a shift-invariant Haar wavelet shrinkage on a single scale and the diffusivity of a nonlinear diffusion filter [40, 41]. This allowed us to study novel, diffusion-inspired shrinkage functions with competitive performance and to suggest new shrinkage rules for 2 D images with better rotation invariance [38]. We proposed coupled shrinkage rules for color images where a desynchronization of the color channels is avoided [38]. Moreover, the technique was extended to anisotropic diffusion where a "locally semi-analytic scheme" (LSAS) was developed [75].

By using wavelets with a higher number of vanishing moments, equivalences to higher-order diffusion-like PDEs were discovered in [74].

A new explicit scheme for nonlinear diffusion which directly incorporates ideas from multiscale Haar wavelet shrinkage was proposed in [48]. Numerical experiments concerning multiscale Haar wavelet shrinkage and nonlinear explicit diffusion schemes are contained in [42]. In [15] it was shown that 1D multiscale continuous wavelet shrinkage can be linked to novel integrodifferential equations. They differ from nonlinear diffusion filtering and corresponding regularization methods by the fact that they involve smoothed derivative operators and perform a weighted averaging over all scales.

6.4.6 Conclusions

We have given an overview about a very general optimization framework for image denoising and simplification with nonlocal data and smoothness terms

(NDS). It has been demonstrated that this framework allows to find a transition between several classical models in one single energy function. The price for this generality is that the function depends on several parameters. Nevertheless, experiments have shown that there is some redundancy in this parameter set, and thus one can reduce the complexity of the choice of optimal parameters. We also took a look at numerical methods for minimizing the NDS energy function. Some of the corresponding schemes turn out to be maximum–minimum stable while others can reduce the running time significantly.

6.5 Nonlocal Data and Smoothness Terms: Some Asymptotic Theory

In this section we present some asymptotic theory for the general class of smoothers introduced in Sect. 6.4. We consider a particularly simple situation where the function Ψ_D measuring the distance between the smoothed signal or image u and the original observations y and the function Ψ_S specifying the regularity of u both are quadratic:

$$\Psi_D(s^2) = s^2 = \Psi_S(s^2) \,. \tag{6.32}$$

For this particular choice, we get an explicit formula for the minimizer $\widehat{u}$ of the NDS function $E(u)$. We start with a simple local average $\widehat{\mu} = (\widehat{\mu}_1, \dots, \widehat{\mu}_N)$ of the noisy image given by (6.4) where the standardized weights are chosen from (6.26) according to $w_j(x) = w_D(|x - x_j|^2) / \sum_{i=1}^N w_D(|x - x_i|^2)$, i.e.,

$$\widehat{\mu}_k = \frac{\sum_{j=1}^N w_D(|x_k - x_j|^2) y_j}{\sum_{i=1}^N w_D(|x_k - x_i|^2)} \,. \tag{6.33}$$

Let W_D, W_S be the $N \times N$ matrices of weights before standardization

$$W_{D,kl} = w_D(|x_k - x_l|^2) \,, \quad W_{S,kl} = w_S(|x_k - x_l|^2) \,, \quad k, l = 1, \dots, N \,,$$

and let V_D, V_S denote the diagonal matrices with entries

$$V_{D,kk} = \sum_{i=1}^N w_D(|x_k - x_i|^2) \,, \quad V_{S,kk} = \sum_{i=1}^N w_S(|x_k - x_i|^2) \,, \quad k = 1, \dots, N \,.$$

In particular, we have $\widehat{\mu} = V_D^{-1} W_D y$. By setting the partial derivatives of $E(u)$ with respect to $u_1, \dots, u_N$ to 0, we get, using the abbreviation $\lambda = 2(1-\alpha)/\alpha$ and denoting the $N \times N$ identity matrix by I:

Proposition 6.1 *If $I + \lambda V_D^{-1}(V_S - W_S)$ is invertible,*

$$\widehat{u} \;=\; \left\{I + \lambda V_D^{-1}(V_S - W_S)\right\}^{-1} \widehat{\mu} \;=\; \left\{V_D + \lambda(V_S - W_S)\right\}^{-1} W_D \, y \,.$$

Therefore, the NDS estimate $\widehat{u}$ is the result of applying a second smoothing operation to the local average $\widehat{\mu}$ where the weights now are specified mainly by the regularization term (6.27) via W_S, as V_D, V_S only depend on the sums of weights and will typically be close to multiples of I.

For kernel weights $w_D(|x|^2) = h^{-d}K(|x|/h)$, $\widehat{\mu}$ is just the familiar Nadaraya-Watson estimate (6.6).

6.5.1 Consistency and Asymptotic Normality of NDS Smoothers

If we assume a regression model (6.1), the local averages $\widehat{\mu}_k$ are decomposed into

$$\widehat{\mu}_k = \sum_{j=1}^{N} w_j(x_k)m(x_j) + \sum_{j=1}^{N} w_j(x_k)\varepsilon_j\,,\ k = 1,\dots,N\,.$$

The first term is a local average of function values $m(x_j)$ in a neighborhood of x_k, and, for large N, it will be close to $m(x_k)$ if m is smooth enough and if the width of the neighborhood converges to 0 for $N \to \infty$. The second term is a weighted average of i.i.d. random variables ε_j, and, under appropriate assumptions on the weights, it will converge to the mean $\mathbb{E}\varepsilon_j = 0$ for $N \to \infty$. Moreover, by an appropriate central limit theorem, it will have an symptotic normal distribution after rescaling.

As the NDS smoother $\widehat{u}$ coincides with the local average $\widehat{\mu}$ up to a deterministic matrix factor, it is plausible that both estimates share the same type of asymptotics. To make this heuristic argument precise with a minimum amount of notation, we restrict our discussion here to the signal denoising problem, where $d = 1$, and, additionally to assuming (6.32), we choose spatial kernel weights

$$w_D(x^2) = \frac{1}{h}K\big(\frac{x}{h}\big), \quad w_S(x^2) = \frac{1}{g}L\big(\frac{x}{g}\big)$$

for both parts of the NDS function. $K(z), L(z)$ are kernel functions satisfying (6.5), having compact support $[-1, +1]$ and which additionally are Lipschitz continuous, symmetric around 0 and nonincreasing in $[0, 1]$.

Theorem 6.2 *Let the data satisfy the regression model (6.1) with* $\sigma_\varepsilon^2 = \operatorname{var} \varepsilon_j < \infty$*. Let* $m(x)$ *be twice continuously differentiable, and* $m''(x)$ *be Hoelder continuous on [0,1]. With* V_K, S_K *as in Sect. 6.1 and known rational functions* $1 \le r_b(\lambda) \to 1$, $1 \ge r_v(\lambda) \to 1$ *for* $\lambda \to 0$*, let*

$$a_{k,N} = \left(r_b(\lambda)\,\frac{h^2}{2}m''(x_k)V_K\right)^2 + r_v^2(\lambda)\,\frac{\sigma_\varepsilon^2 S_K}{Nh}\,.$$

Then, for $N \to \infty$, $h \to 0, g \to 0, \lambda \to 0$, such that

$$\frac{g}{h} \to 0\,, \quad Nh^3 \left(\frac{g}{h}\right)^4 \to \infty\,, \quad \frac{\lambda^2}{(h^2\, g)} \to 0\,,$$

we have

$$\frac{\text{mse}\ \widehat{u}_k}{a_{k,N}} = \frac{\mathbb{E}\left(\widehat{u}_k - m(x_k)\right)^2}{a_{k,N}} \to 1 \quad \textit{for } N \to \infty.$$

For the proof, compare Theorem 2.1 of [13]. As $a_{k,N} \to 0$, this result implies consistency, i.e., $\widehat{u}_k - m(x_k) \to 0$ in mean-square for $N \to \infty$, and it specifies the rate of convergence. The following result, compare Theorem 3.1 of [13], shows that the NDS estimate has asymptotically a Gaussian distribution.

Theorem 6.3 *Under the assumptions of Theorem 6.2, we have*

$$\frac{1}{r_v(\lambda)}\sqrt{Nh}\left(\widehat{u}_k - \mathbb{E}\ \widehat{u}_k\right) \to \mathcal{N}(0, \sigma_\varepsilon^2 S_K)$$

in distribution for $N \to \infty$, and

$$\text{bias}\ \widehat{u}_k = \mathbb{E}\ \widehat{u}_k - m(x_k) = r_b(\lambda)\ \frac{h^2}{2} m''(x_k) V_K + R_{k,N}$$

for $N \to \infty$, where the remainder term $R_{k,N}$ is of smaller order than the first term.

6.5.2 Outlook

The asymptotic theory of the previous subsection forms the basis for solving a large variety of practical image processing problems. Theorem 6.2 is the starting point for deriving plug-in methods which lead to a data-adaptive choice of the smoothing parameters h, α and g like in Sect. 6.2.

Theorem 6.3 may be used for constructing tests which allow for checking if two different noisy images show the same pattern or if, in particular, one of them shows defects, similar to [23] where only local averages are considered. If asymptotic normality does not provide a reasonable approximation, which frequently happens for complicated test statistics and small or medium sample sizes, then resampling methods like the bootstrap may be used instead, compare [21] for kernel smoothers. For a corresponding theoretical basis, uniform consistency is needed which also holds under appropriate regularity conditions, i.e., we have $\sup\{|\widehat{u}_k - m(x_k)|, \delta \le x_k \le 1-\delta\} \to 0$ for $n \to \infty$ for arbitrary $\delta > 0$.

Finally, the results of Subsect. 6.5.1 may be extended to other loss functions. The advantage of an explicit formula for the NDS smoother which is guaranteed by the quadratic loss (6.32) is lost, but instead standard asymptotic approximations may be used to get a similar theory for M-smoothers, median smoothers, etc. with additional regularization.

6.6 Statistical Inference for Time Varying ARCH-Models

6.6.1 Time Varying ARCH-Models

To model volatility in time series [17], introduced the ARCH model where the conditional variance is stochastic and dependent on past observations. More precisely, the process $\{X_t\}$ is called an ARCH(p) process if it satisfies the representation

$$X_t = Z_t \sigma_t \qquad \sigma^2 = a_0 + \sum_{j=1}^{p} a_j X_{t-j}^2$$

where $\{Z_t\}$ are independent identically distributed (i.i.d.) random variables with $\mathbb{E}(Z_t) = 0$ and $\mathbb{E}(Z_t^2) = 1$. We observe from the definition of the ARCH model that $\{X_t\}$ is uncorrelated, but the conditional variance of X_t is determined by the previous p observations.

The ARCH model and several of its related models have gained widespread recognition because they model quite well the volatility in financial markets over relatively short periods of time (cf. [1, 45], and for a recent review [27]). However, underlying all these models is the assumption of stationarity. Now given the changing pace of the world's economy, modeling financial returns over long intervals using stationary time series models may be inappropriate. It is quite plausible that structural changes in financial time series may occur causing the time series over long intervals to deviate significantly from stationarity. It is, therefore, plausible that by relaxing the assumption of stationarity in an adequate way, we may obtain a better fit. In this direction, [16] have proposed the simple nonlinear model $X_t = \mu + \sigma(t) Z_t$, where Z_t are i.i.d. random variables and $\sigma(\cdot)$ is a smooth function, which they estimate using a nonparametric regression method. Essentially, though it is not mentioned, the authors are treating $\sigma(t)$ as if it were of the form $\sigma(t) = \widetilde{\sigma}(t/N)$, with N being the sample size. Through this rescaling device it is possible to obtain a framework for a meaningful asymptotic theory. Feng [20] has also studied time inhomogeneous stochastic volatility, by introducing a multiplicative seasonal and trend component into the GARCH model.

To model the financial returns over long periods of time [11] generalized the class of ARCH(∞) models (cf. [26] and [59]) to models with time-varying parameters. In order to obtain a framework for a meaningful asymptotic theory, they rescale the parameter functions as in nonparametric regression and for (linear) locally stationary processes to the unit interval, that is they assume

$$X_{t,N} = \sigma_{t,N} Z_t$$

$$\text{where} \quad \sigma_{t,N}^2 = a_0\left(\frac{t}{N}\right) + \sum_{j=1}^{\infty} a_j\left(\frac{t}{N}\right) X_{t-j,N}^2 \quad \text{for } t = 1 \ldots, N\,, \tag{6.34}$$

where Z_t are i.i.d. random variables with $\mathbb{E}Z_t = 0$, $\mathbb{E}Z_t^2 = 1$. The sequence of stochastic processes $\{X_{t,N} : t = 1, \ldots, N\}$ which satisfy (6.34) is called a time-varying ARCH (tvARCH) process. As shown below the tvARCH process can locally be approximated by stationary ARCH processes, i.e., it is a locally stationary process.

We mention that the rescaling technique is mainly introduced for obtaining a meaningful asymptotic theory and by this device it is possible to obtain adequate approximations for the non-rescaled case. In particular, the rescaling does not effect the estimation procedure. Furthermore, classical ARCH-models are included as a special case (if the parameters are constant in time). The same rescaling device has been used for example in nonparametric time series by [58] and by [10] in his definition of local stationarity.

Reference [11] prove under certain regularity conditions that $\{X_{t,N}^2\}$ defined in (6.34) has an almost surely well-defined unique solution in the set of all causal solutions. The solution has the form of a time-varying Volterra series expansion.

We now consider a *stationary* process which locally approximates the tvARCH-process in some neighborhood of a fixed point t_0 (or in rescaled time u_0). For each given $u_0 \in (0,1]$ the stochastic process $\{X_t(u_0)\}$ is the stationary ARCH process associated with the tvARCH(∞) process at time point u_0 if it satisfies

$$X_t(u_0) = \sigma_t(u_0) Z_t \,,$$

$$\text{where} \quad \sigma_t(u_0)^2 = a_0(u_0) + \sum_{j=1}^{\infty} a_j(u_0) X_{t-j}(u_0)^2 \,. \qquad (6.35)$$

Comparing (6.35) with (6.34), it seems clear if t/N is close to u_0, then $X_{t,N}^2$ and $X_t(u_0)^2$ should be close, and the degree of the approximation should depend both on the rescaling factor N and the deviation $|t/N - u_0|$. To make this precise, [11] show that there exists a stationary, ergodic, positive process $\{U_t\}$ independent of u_0 with finite mean and a constant C independent of t and N such that

$$|X_{t,N}^2 - X_t(u_0)^2| \leq C\Big(|\frac{t}{N} - u_0| + \frac{1}{N}\Big) U_t \quad \text{almost sure}. \qquad (6.36)$$

As a consequence of (6.36) we have

$$X_{t,N}^2 = X_t(u_0)^2 + O_p\Big(|\frac{t}{N} - u_0| + \frac{1}{N}\Big) .$$

Therefore, we can locally approximate the tvARCH process by a stationary process. The above approximation can be refined by using the so called "derivative processes". Using the derivative process, [11] show that the following expansion holds:

$$X_{t,N}^2 = X_t(u_0)^2 + \left(\frac{t}{N} - u_0\right)\frac{\partial X_t(u)^2}{\partial u}\rfloor_{u=u_0}$$
$$+\frac{1}{2}\left(\frac{t}{N} - u_0\right)^2 \frac{\partial^2 X_t(u)^2}{\partial u^2}\rfloor_{u=u_0} + O_p\left((\frac{t}{N} - u_0)^3 + \frac{1}{N}\right). \quad (6.37)$$

The nice feature of this result is that it gives a Taylor expansion of the non-stationary process $X_{t,N}^2$ around $X_t(u_0)^2$ in terms of stationary processes. This is a particularly interesting result since it allows for using well known results for stationary processes (such as the ergodic theorem) in describing properties of $X_{t,N}$. The result is of high importance for deriving the asymptotic properties of the tvARCH-process.

6.6.2 The Segment Quasi-Likelihood Estimate

In this section we consider a kernel type estimator of the parameters $\boldsymbol{a}(u_0)$ of a tvARCH(p) model given the sample $\{X_{t,N} : t = 1, \ldots, N\}$. Let $t_0 \in \mathbb{N}$ be such that $|u_0 - t_0/N| < 1/N$. The estimator considered in this section is the minimizer of the weighted conditional log-likelihood

$$L_{t_0,N}(\boldsymbol{\alpha}) := \sum_{k=p+1}^{N} \frac{1}{bN} K\left(\frac{t_0 - k}{bN}\right) \ell_{k,N}(\boldsymbol{\alpha}), \quad (6.38)$$

where

$$\ell_{k,N}(\boldsymbol{\alpha}) = \frac{1}{2}\left(\log w_{k,N}(\boldsymbol{\alpha}) + \frac{X_{k,N}^2}{w_{k,N}(\boldsymbol{\alpha})}\right) \quad (6.39)$$

$$\text{with } w_{k,N}(\boldsymbol{\alpha}) = \alpha_0 + \sum_{j=1}^{p} \alpha_j X_{k-j,N}^2,$$

and $K : [-1/2, 1/2] \to \mathbb{R}$ is a kernel function of bounded variation with $\int_{-1/2}^{1/2} K(x)\mathrm{d}x = 1$ and $\int_{-1/2}^{1/2} x\, K(x)\mathrm{d}x = 0$. That is we consider

$$\widehat{\boldsymbol{a}}_{t_0,N} = \arg\min_{\boldsymbol{\alpha}\in\Omega} L_{t_0,N}(\boldsymbol{\alpha}). \quad (6.40)$$

Obviously $\ell_{t,N}(\boldsymbol{\alpha})$ is the conditional log-likelihood of $X_{t,N}$ given $X_{t-1,N}, \ldots,$ $X_{t-p,N}$ and the parameters $\boldsymbol{\alpha} = (\alpha_0, \ldots, \alpha_p)^T$ provided the Z_t are normally distributed. However, all the results discussed below also hold if the Z_t are not normally distributed. For this reason and the fact that the conditional likelihood is not the full likelihood $\mathcal{L}_{t_0,N}(\boldsymbol{\alpha})$ is called a quasi-likelihood.

In the derivation of the asymptotic properties of this estimator the local approximation of $X_{t,N}^2$ by the stationary process $X_t(u_0)^2$ plays a major role. Therefore, similar to the above, we define the stationary approximation weighted log-likelihood

$$\widetilde{L}_N(u_0, \boldsymbol{\alpha}) := \sum_{k=p+1}^{N} \frac{1}{bN} K\left(\frac{t_0 - k}{bN}\right) \widetilde{\ell}_k(u_0, \boldsymbol{\alpha}), \tag{6.41}$$

where $|u_0 - t_0/N| < 1/N$ and

$$\widetilde{\ell}_t(u_0, \boldsymbol{\alpha}) = \frac{1}{2}\left(\log \widetilde{w}_t(u_0, \boldsymbol{\alpha}) + \frac{\widetilde{X}_t(u_0)^2}{\widetilde{w}_t(u_0, \boldsymbol{\alpha})}\right)$$

$$\text{with } \widetilde{w}_t(u_0, \boldsymbol{\alpha}) = \alpha_0 + \sum_{j=1}^{p} \alpha_j \widetilde{X}_{t-j}(u_0)^2 .$$

It can be shown that both, $L_{t_0,N}(\boldsymbol{\alpha})$ and $\widetilde{L}_N(u_0, \boldsymbol{\alpha})$, converge to

$$L(u_0, \boldsymbol{\alpha}) := \mathbb{E}(\widetilde{\ell}_0(u_0, \boldsymbol{\alpha})) \tag{6.42}$$

as $N \to \infty$, $b \to 0$, $bN \to \infty$ and $|u_0 - t_0/N| < 1/N$. It is easy to show that $L(u_0, \boldsymbol{\alpha})$ is minimized by $\boldsymbol{\alpha} = (\boldsymbol{a}_0(u_0), \ldots, \boldsymbol{a}_p(u_0))$. Furthermore, let

$$\begin{aligned} B_{t_0,N}(\boldsymbol{\alpha}) &:= L_{t_0,N}(\boldsymbol{\alpha}) - \widetilde{L}_N(u_0, \boldsymbol{\alpha}) \\ &= \sum_{k=p+1}^{N} \frac{1}{bN} K(\frac{t_0 - k}{bN})(\ell_{k,N}(\boldsymbol{\alpha}) - \widetilde{\ell}_k(u_0, \boldsymbol{\alpha})) . \end{aligned} \tag{6.43}$$

Since $\widetilde{L}_N(u_0, \boldsymbol{\alpha})$ is the likelihood of the stationary approximation $X_t(u_0)$, $B_{t_0,N}(\boldsymbol{\alpha})$ is a bias caused by the deviation from stationarity. Let

$$\Sigma(u_0) = \frac{1}{2}\, \mathbb{E}\left\{\frac{\nabla \widetilde{w}_0(u_0, \boldsymbol{a}_{u_0}) \nabla \widetilde{w}_0(u_0, \boldsymbol{a}_{u_0})^T}{\widetilde{w}_0(u_0, \boldsymbol{a}_{u_0})^2}\right\} . \tag{6.44}$$

Since $X_k(u_0)/\widetilde{w}_k(u_0, \boldsymbol{a}_0) = Z_k^2$ and Z_k^2 is independent of $\widetilde{w}_k(u_0, \boldsymbol{a}_0)$ we have

$$\mathbb{E}\left(\nabla^2 \widetilde{\ell}_0(u_0, \boldsymbol{a}_{u_0})\right) = -\Sigma(u_0)$$

and

$$\mathbb{E}\left(\nabla \widetilde{\ell}_0(u_0, \boldsymbol{a}_{u_0}) \nabla \widetilde{\ell}_0(u_0, \boldsymbol{a}_{u_0})^T\right) = \frac{\mathrm{var}(Z_0^2)}{2} \Sigma(u_0) .$$

The following theorem gives the asymptotic distribution of $\widehat{\boldsymbol{a}}_{t_0,N}$. It may for example be used to construct approximate confidence intervals for $\widehat{\boldsymbol{a}}_{t_0,N}$. Details on the assumptions may be found in [11].

Theorem 6.4 *Let* $S_K = \int_{-1/2}^{1/2} K(x)^2 \mathrm{d}x$ *and* $V_K = \int_{-1/2}^{1/2} K(x) x^2 \mathrm{d}x$. *Under suitable regularity assumptions we have for* $|u_0 - t_0/N| < 1/N$

(i) if $b^3 \ll N^{-1}$ *, then* $\sqrt{bN} B_{t_0,N}(\boldsymbol{a}_{u_0}) \xrightarrow{\mathcal{P}} 0$ *and*

$$\sqrt{bN}(\widehat{\boldsymbol{a}}_{t_0,N} - \boldsymbol{a}_{u_0}) \xrightarrow{\mathcal{D}} \mathcal{N}(0, S_K \frac{\mathrm{var}(Z_0^2)}{2} \Sigma(u_0)^{-1});$$

(ii)if $b^{13} << N^{-1}$ *,, then*

$$\sqrt{bN}\Sigma(u_0)^{-1}\nabla B_{t_0,N}(\boldsymbol{a}_{u_0}) = \sqrt{bN} b^2 \mu(u_0) + o_p(1)$$

and

$$\sqrt{bN}(\widehat{\boldsymbol{a}}_{t_0,N} - \boldsymbol{a}_{u_0}) + \sqrt{bN} b^2 \mu(u_0) \xrightarrow{\mathcal{D}} \mathcal{N}\left(0, S_K \frac{\mathrm{var}(Z_0^2)}{2} \Sigma(u_0)^{-1}\right) \quad (6.45)$$

where

$$\mu(u_0) = \frac{1}{2} V_K \Sigma(u_0)^{-1} \frac{\partial^2 \nabla L(u, \boldsymbol{a}_{u_0})}{\partial u^2} \rfloor_{u=u_0} . \quad (6.46)$$

$S_K = \int_{-1/2}^{1/2} K(x)^2 \mathrm{d}x$ *and* $V_K = \int_{-1/2}^{1/2} K(x) x^2 \mathrm{d}x$ *.*

We mention that for Z_t normally distributed $\mathrm{var}(Z_0^2) = 2$ holds.

We recall the structure of this result: The asymptotic Gaussian distribution is the same as for the stationary approximation. In addition we have a bias term which comes from the deviation of the true process from the stationary approximation on the segment. In particular this bias term is zero if the true process is stationary. A simple example is given below. By estimating and minimizing the mean squared error (i.e., by balancing the variance and the bias due to non-stationarity on the segment), we may find an estimator for the optimal segment length.

Example 6.5 We consider the tvARCH(0) process

$$X_{t,N} = \sigma_{t,N} Z_t , \quad \sigma_{t,N}^2 = a_0\left(\frac{t}{N}\right) ,$$

which [16] have also studied. In this case $\frac{\partial X_t(u)^2}{\partial u} = a_0'(u) Z_t^2$, and we have

$$\frac{\partial^2 \nabla L(u, \boldsymbol{a}_{u_0})}{\partial u^2} \rfloor_{u=u_0} = -\frac{1}{2} \frac{a_0''(u_0)}{a_0(u_0)^2} \quad \text{and} \quad \Sigma(u_0) = \frac{1}{2 a_0(u_0)^2}$$

that is

$$\mu(u_0) = -\frac{1}{2} V_K a_0''(u_0) .$$

This example illustrates well how the bias is linked to the non-stationarity of the process – if the process were stationary, the derivatives of $a_0(\cdot)$ would be zero causing the bias also to be zero. Conversely, sudden variations in $a_0(\cdot)$

about the time point u_0 would be reflected in $a_0''(u_0)$ and manifest as a large $\mu(u_0)$. Straightforward minimization of the asymptotic variance

$$(bN)^{-1} S_K \frac{\mathrm{var}(Z_0^2)}{2} \Sigma(u_0)^{-1}$$

and probabilistic bias

$$b^2 \frac{1}{2} V_K \Sigma(u_0)^{-1} \frac{\partial^2 \nabla L(u, \boldsymbol{a}_{u_0})}{\partial u^2} \rfloor_{u=u_0}$$

leads to the optimal bandwidth which in this case (and for Gaussian Z_t) takes the form

$$b_{opt} = \left(\frac{2S_K}{V_K^2}\right)^{1/5} N^{-1/5} \left[\frac{a_0(u_0)}{a_0''(u_0)}\right]^{2/5},$$

leading to a large bandwidth if $a_0''(u_0)$ is small and vice versa. Thus the optimal theoretical choice of the bandwidth (of the segment length) depends on the degree of stationarity of the process.

6.6.3 Recursive Estimation

We now present an "online" method, which uses the estimate of the parameters at time point $(t-1)$ and the observation at time point t to estimate the parameter at time point t. There exists a huge literature on recursive algorithms - mainly in the context of linear systems (cf. [34, 63] or in the context of neural networks (cf. [8, 77]). For a general overview see also [33]. Reference [12] has considered the following online recursive algorithm for tvARCH models:

$$\widehat{\underline{a}}_{t,N} = \widehat{\underline{a}}_{t-1,N} + \lambda\{X_{t,N}^2 - \widehat{\underline{a}}_{t-1,N}^T \mathcal{X}_{t-1,N}\} \frac{\mathcal{X}_{t-1,N}}{|\mathcal{X}_{t-1,N}|_1^2}, \tag{6.47}$$

$t = p+1, \ldots, N$, with $\mathcal{X}_{t-1,N}^T = (1, X_{t-1,N}^2, \ldots, X_{t-p,N}^2)$, $|\mathcal{X}_{t-1,N}|_1 = 1 + \sum_{j=1}^p X_{t-j,N}^2$ and initial conditions $\widehat{\underline{a}}_{p,N} = (0, \ldots, 0)$. This algorithm is linear in the estimators, despite the nonlinearity of the tvARCH process. We call the stochastic algorithm defined in (6.47) the ARCH normalized recursive estimation (ANRE) algorithm. Let $\underline{a}(u)^T = (a_0(u), \ldots, a_p(u))$, then $\widehat{\underline{a}}_{t,N}$ is regarded as an estimator of $\underline{a}(t/N)$ or of $\underline{a}(u)$ if $|t/N - u| < 1/N$. We note that the step size λ plays in some sense a similar role to the bandwidth b used in the segment quasi-likelihood. In fact in the asymptotic considerations λ can be treated as if it were of the order $O((bN)^{-1})$. It is also worth noting that if we believe the tvARCH process were highly non-stationary, then λ should be large, in order to place greater emphasis on the current observation and "capture" the rapidly changing behavior of $\underline{a}$.

The ANRE algorithm resembles the NLMS-algorithm investigated in [37]. Rewriting (6.47), we have

$$\widehat{\underline{a}}_{t,N} = \left(I - \lambda \frac{\mathcal{X}_{t-1,N}\mathcal{X}_{t-1,N}^T}{|\mathcal{X}_{t-1,N}|_1^2} \right) \widehat{\underline{a}}_{t-1,N} + \lambda \frac{X_{t,N}^2 \mathcal{X}_{t-1,N}}{|\mathcal{X}_{t-1,N}|_1^2}. \tag{6.48}$$

We can see from (6.48) that the convergence of the ANRE algorithm relies on showing some type of exponential decay of the past. This is sometimes referred to as persistence of excitation. Persistence of excitation guarantees convergence of the algorithm. Besides the practical relevance this is used to prove the asymptotic properties of $\widehat{\underline{a}}_{t,N}$.

Reference [12] prove that the difference $\widehat{\underline{a}}_{t_0,N} - \underline{a}(u_0)$ is dominated by two terms, that is

$$\widehat{\underline{a}}_{t_0,N} - \underline{a}(u_0) = \mathcal{L}_{t_0}(u_0) + \mathcal{R}_{t_0,N}(u_0) + O_p(\delta_N), \tag{6.49}$$

where

$$\delta_N = \left(\frac{1}{(N\lambda)^{2\beta}} + \frac{\sqrt{\lambda}}{(N\lambda)^{\beta}} + \lambda + \frac{1}{N^{\beta}} \right), \tag{6.50}$$

$$\mathcal{L}_{t_0}(u_0) = \sum_{k=0}^{t_0-p-1} \lambda \{I - \lambda F(u_0)\}^k \mathcal{M}_{t_0-k}(u_0) \tag{6.51}$$

$$\mathcal{R}_{t_0,N}(u_0) = \sum_{k=0}^{t_0-p-1} \lambda \{I - \lambda F(u_0)\}^k \Big(\{\mathcal{M}_{t_0-k}(\frac{t_0-k}{N}) - \mathcal{M}_{t_0-k}(u_0)\} + F(u_0)\{\underline{a}(\frac{t_0-k}{N}) - \underline{a}(u_0)\} \Big),$$

with

$$\mathcal{M}_t(u) = (Z_t^2 - 1)\sigma_t(u)^2 \frac{\mathcal{X}_{t-1}(u)}{|\mathcal{X}_{t-1}(u)|_1^2} \text{ and } F(u) = \mathbb{E}\left(\frac{\mathcal{X}_0(u)\mathcal{X}_0(u)^T}{|\mathcal{X}_0(u)|_1^2} \right). \tag{6.52}$$

We note that $\mathcal{L}_{t_0}(u_0)$ and $\mathcal{R}_{t_0,N}(u_0)$ play two different roles. $\mathcal{L}_{t_0}(u_0)$ is the weighted sum of the stationary random variables $\{X_t(u_0)\}_t$, which locally approximate the tvARCH process $\{X_{t,N}\}_t$, whereas $\mathcal{R}_{t_0,N}(u_0)$ is the (stochastic) bias due to non-stationarity; if the tvARCH process were stationary this term would be zero. It is clear from the above that the magnitude of $\mathcal{R}_{t_0,N}(u_0)$ depends on the regularity of the time-varying parameters $\underline{a}(u)$, e.g., the Hölder class that $\underline{a}(u)$ belongs to. Reference [12] prove the following result on the asymptotic normality of this recursive estimator. Unlike in most other work in the area of recursive estimation it is assumed that the true process is a process with time-varying coefficients, i.e., a non-stationary process.

Theorem 6.6 *Let $\underline{a}(u) \in Lip(\beta)$ where $\beta \leq 1$. Under suitable regularity assumptions we have for $|u_0 - t_0/N| < 1/N$*

(i) if $\lambda \gg N^{-4\beta/(4\beta+1)}$ *and* $\lambda \gg N^{-2\beta}$*, then*

$$\lambda^{-1/2}\{\underline{\widehat{a}}_{t_0,N} - \underline{a}(u_0)\} - \lambda^{-1/2}\mathcal{R}_{t_0,N}(u_0) \xrightarrow{\mathcal{D}} \mathcal{N}(0, \Sigma(u_0)), \tag{6.53}$$

(ii) if $\lambda \gg N^{-\frac{2\beta}{2\beta+1}}$*, then*

$$\lambda^{-1/2}\{\underline{\widehat{a}}_{t_0,N} - \underline{a}(u_0)\} \xrightarrow{\mathcal{D}} \mathcal{N}(0, \Sigma(u_0)), \tag{6.54}$$

where $\lambda \to 0$ *as* $N \to \infty$ *and* $N\lambda \gg (\log N)^{1+\varepsilon}$*, for some* $\varepsilon > 0$*, with*

$$\Sigma(u) = \frac{\mu_4}{2} F(u)^{-1} \mathbb{E}\Big(\frac{\sigma_1(u)^4 \mathcal{X}_0(u)\mathcal{X}_0(u)^T}{|\mathcal{X}_0(u)|_1^4}\Big), \quad \mu_4 = \mathbb{E}(Z_0^4) - 1. \tag{6.55}$$

We now make a stronger assumption on $\underline{a}(u)$. Let $\dot{f}(u)$ denote the derivative of the vector or matrix $f(\cdot)$ with respect to u. Suppose now that $0 < \beta' \le 1$ and $\underline{a}(u) \in Lip(1+\beta')$, i.e., $\dot{\underline{a}}(u) \in Lip(\beta')$. Under this assumption it can be shown that

$$\mathbb{E}\{\underline{\widehat{a}}_{t_0,N} - \underline{a}(u_0)\} = -\frac{1}{N\lambda} F(u_0)^{-1} \dot{\underline{a}}(u_0) + O\left(\frac{1}{(N\lambda)^{1+\beta'}}\right). \tag{6.56}$$

By using this expression for the bias, [12] prove the following result.

Theorem 6.7 *Under suitable regularity assumptions we have for* $|u_0 - t_0/N| < 1/N$

$$\mathbb{E}|\underline{\widehat{a}}_{t_0,N} - \underline{a}(u_0)|^2 = \lambda\, tr\{\Sigma(u_0)\} + \frac{1}{(N\lambda)^2}|F(u_0)^{-1}\dot{\underline{a}}(u_0)|^2 \tag{6.57}$$

$$+ O\left(\frac{1}{(N\lambda)^{2+\beta'}} + \frac{\lambda^{1/2}}{(N\lambda)^{1+\beta'}} + \frac{1}{(N\lambda)^2}\right),$$

and if λ *is such that* $\lambda^{-1/2}/(N\lambda)^{1+\beta'} \to 0$*, then*

$$\lambda^{-1/2}(\underline{\widehat{a}}_{t_0,N} - \underline{a}(u_0)) + \lambda^{-1/2}\frac{1}{N\lambda} F(u_0)^{-1}\dot{\underline{a}}(u_0) \xrightarrow{\mathcal{D}} \mathcal{N}(0, \Sigma(u_0)), \tag{6.58}$$

where $\lambda \to 0$ *as* $N \to \infty$ *and* $\lambda N \gg (\log N)^{1+\varepsilon}$*, for some* $\varepsilon > 0$.

The above result can be used to achieve a bias reduction and the the optimal rate of convergence by running two ANRE algorithms with different stepsizes λ_1 and λ_2 in parallel: Let $\underline{\widehat{a}}_{t,N}(\lambda_1)$ and $\underline{\widehat{a}}_{t,N}(\lambda_2)$ be the ANRE algorithms with stepsize λ_1 and λ_2 respectively, and assume that $\lambda_1 > \lambda_2$. By using (6.56) for $i = 1, 2$, we have

$$\mathbb{E}\{\underline{\widehat{a}}_{t_0,N}(\lambda_i)\} = \underline{a}(u_0) - \frac{1}{N\lambda_i} F(u_0)^{-1}\dot{\underline{a}}(u_0) + O\left(\frac{1}{(N\lambda_i)^{1+\beta'}}\right). \tag{6.59}$$

Since $\underline{a}(u_0) - (1/N\lambda_i)F(u_0)^{-1}\dot{\underline{a}}(u_0) \approx \underline{a}\left(u_0 - (1/N\lambda_i)F(u_0)^{-1}\right)$ we heuristically estimate $\underline{a}(u_0 - (1/N\lambda_i)F(u_0)^{-1})$ instead of $\underline{a}(u_0)$ by the algorithm. By using two different λ_i, we can find a linear combination of the corresponding estimates such that we "extrapolate" the two values $\underline{a}(u_0) - (1/N\lambda_i)F(u_0)^{-1}\,\dot{\underline{a}}(u_0)$ $(i = 1,2)$ to $\underline{a}(u_0)$. Formally let $0 < w < 1$, $\lambda_2 = w\lambda_1$ and

$$\breve{\underline{a}}_{t_0,N}(w) = \frac{1}{1-w}\widehat{\underline{a}}_{t_0,N}(\lambda_1) - \frac{w}{1-w}\widehat{\underline{a}}_{t_0,N}(\lambda_2)\,.$$

If $|t_0/N - u_0| < 1/N$, then by using (6.59) we have

$$\mathbb{E}\{\breve{\underline{a}}_{t_0,N}(w)\} = \underline{a}(u_0) + O\left(\frac{1}{(N\lambda)^{1+\beta'}}\right).$$

By using Theorem 6.7, we have

$$\mathbb{E}|\breve{\underline{a}}_{t_0,N} - \underline{a}(u_0))|^2 = O\left(\lambda + \frac{1}{(N\lambda)^{2(1+\beta')}}\right),$$

and choosing $\lambda = \text{const} \times \mathrm{N}^{-(2+2\beta')/(3+2\beta')}$ gives the optimal rate. It remains the problem of choosing λ (and w). It is obvious that λ should be chosen adaptively to the degree of non-stationarity. That is λ should be large if the characteristics of the process are changing more rapidly. However, a more specific suggestion would require more investigations – both theoretically and by simulations.

Finally we mention that choosing $\lambda_2 < w\lambda_1$ will lead to an estimator of $\underline{a}(u_0 + \Delta)$ with some $\Delta > 0$ (with rate as above). This could be the basis for the prediction of volatility of time varying ARCH processes.

6.6.4 Implications for Non-Rescaled Processes

Suppose that we observe data from a (non-rescaled) time-varying ARCH process in discrete time

$$X_t = Z_t\sigma_t\,, \quad \sigma_t^2 = a_0(t) + \sum_{j=1}^{p} a_j(t)X_{t-j}^2\,, \quad t \in \mathbb{Z}\,. \tag{6.60}$$

In order to estimate $\underline{a}(t)$ we may use the segment quasi-likelihood estimator as given in (6.40) or the recursive estimator as given in (6.47). An approximation for the distribution of the estimators is given by Theorem 6.4 and Theorem 6.6 respectively, which, however, are formulated for rescaled processes.

We now demonstrate how these results can be used for the non-rescaled estimators. In particular we show why the results do not depend on the specific N used in the rescaling.

We start with the second result on recursive estimation: Theorem 6.6(ii) can be used directly since it is completely formulated without N. The matrices $F(u_0)$ and $\Sigma(u_0)$ depend on the unknown stationary approximation $\mathcal{X}_t(u_0)$

of the process at $u_0 = t_0/N$, i.e., at time t_0 in non-rescaled time. Since this approximation is unknown we may use instead the process itself in a small neighborhood of t_0, i.e., we may estimate for example $F(u_0)$ by

$$\frac{1}{m}\sum_{j=0}^{m-1} \frac{\mathcal{X}_{t_0-j}\mathcal{X}_{t_0-j}^T}{|\mathcal{X}_{t_0-j}|_1^2}$$

with m small and $\mathcal{X}_{t-1}^T = (1, X_{t-1}^2, \ldots, X_{t-p}^2)$. An estimator which fits better to the recursive algorithm is

$$[1-(1-\lambda)^{t_0-p+1}]^{-1} \sum_{j=0}^{t_0-p} \lambda(1-\lambda)^j \, \frac{\mathcal{X}_{t_0-j}\mathcal{X}_{t_0-j}^T}{|\mathcal{X}_{t_0-j}|_1^2}.$$

In the same way we can estimate $\Sigma(u_0)$ which altogether leads, e.g., to an approximate confidence interval for $\widehat{\underline{a}}_t$. In a similar way Theorem 6.6(i) can be used.

The situation is more difficult with Theorem 6.7, since here the results depend (at first sight) on N. Suppose that we have parameter functions $\widetilde{a}_j(\cdot)$ and some $N > t_0$ with $\widetilde{a}_j(t_0/N) = a_j(t_0)$ (i.e., the original function has been rescaled to the unit interval). Consider Theorem 6.7 with the functions $\widetilde{a}_j(\cdot)$. The bias in (6.56) and (6.57) contains the term

$$\frac{1}{N}\dot{\widetilde{a}}_j(u_0) \approx \frac{1}{N}\frac{\widetilde{a}_j(\frac{t_0}{N}) - \widetilde{a}_j(\frac{t_0-1}{N})}{\frac{1}{N}} = a_j(t_0) - a_j(t_0-1)$$

which again is independent of N. To avoid confusion, we mention that $1/N\,\dot{\widetilde{a}}_j(u_0)$ of course depends on N once the function $\widetilde{a}_j(\cdot)$ has been fixed (as in the asymptotic approach of this paper), but it does not depend on N when it is used to approximate the function $a_j(t)$ since then the function $\widetilde{a}_j(\cdot)$ is a different one for each N. In the spirit of the remarks above we would, e.g., use as an estimator of $1/N\,\dot{\widetilde{a}}_j(u_0)$ in (6.56) and (6.57) the expression $[1-(1-\lambda)^{t_0-p+1}]^{-1}\sum_{j=0}^{t_0-p}\lambda(1-\lambda)^j\,\big[a_j(t_0) - a_j(t_0-j)\big]$.

For the segment quasi-likelihood estimate the situation is similar: bN in Theorem 6.4 is the sample size on the segment. Therefore, Theorem 6.4(i) can immediately be applied to construct, e.g., an approximative confidence interval for the (non-rescaled) estimator. In part (ii) in addition the term

$$\sqrt{bN}\,b^2\mu(u_0) = \sqrt{bN}\,\frac{1}{2}w(2)\Sigma(u_0)^{-1}(bN)^2\frac{1}{N^2}\frac{\partial^2\nabla\widetilde{\mathcal{L}}(u,\widetilde{\boldsymbol{a}}_{u_0})}{\partial u^2}\rfloor_{u=u_0} \quad (6.61)$$

occurs. As above

$$\frac{1}{N^2}\frac{\partial^2\nabla\widetilde{\mathcal{L}}(u,\widetilde{\boldsymbol{a}}_{u_0})}{\partial u^2}\rfloor_{u=u_0} \approx \nabla\mathcal{L}(t_0,\boldsymbol{a}_{t_0}) - 2\nabla\mathcal{L}(t_0-1,\boldsymbol{a}_{t_0}) + \nabla\mathcal{L}(t_0-2,\boldsymbol{a}_{t_0})$$

which again is independent on N (here $\widetilde{\mathcal{L}}$ an $\widetilde{\boldsymbol{a}}_{u_0}$ denote the likelihood and the parameter with rescaled data while $\mathcal{L}$ an $\boldsymbol{a}_{u_0}$ are the corresponding values with non-rescaled data).

These considerations also demonstrate the need for the asymptotic approach of this paper: While it is not possible to set down a meaningful asymptotic theory for the model (6.60) and to derive, e.g., a central limit theorem for the estimator $\widehat{\underline{a}}_t$, approaching the problem using the rescaling device and the rescaled model (6.34) leads to such results. This is achieved by the "infill asymptotics" where more and more data become available of each local structure (e.g., about time u_0) as $N \to \infty$. The results can then be used also for approximations in the model (6.60) – e.g., for confidence intervals.

6.6.5 Concluding Remark

Results by cf. [36] and [24]) indicate that several stylized facts of financial log returns, for example the often discussed long range dependence of the squared log returns or the IGARCH (1,1) – effect, are in truth due to non-stationarity of the data. Furthermore, we conjecture that for example the empirical kurtosis of financial log returns is much smaller with a time-varying model than with a classical ARCH model. For this reason the results in this chapter on time varying ARCH models are of particular relevance. It is worth mentioning that [24] fitted the tvARCH model to both exchange rate data and also FTSE stock index data sets. Furthermore, forecasts of future volatility was made using the tvARCH model and often the forecasts were better than the forecasts using the benchmark stationary GARCH(1, 1) process.

References

[1] T. Bollerslev. Generalized autoregressive conditional heteroscedasticity. *Journal of Econometrics*, 31:301–327, 1986.

[2] M. Brockmann, Th. Gasser, and E. Herrmann. Local adaptive bandwidth choice for kernel regression estimates. *Journal of the American Statistical Association*, 88:1302–1309, 1993.

[3] T. Brox, M. Welk, G. Steidl, and J. Weickert. Equivalence results for TV diffusion and TV regularisation. In L. D. Griffin and M. Lillholm, editors, *Scale -Space Methods in Computer Vision*, volume 2695 of *Lecture Notes in Computer Science*, pages 86–100. Springer, Berlin-Heidelberg-New York, 2003.

[4] Z. Cai, J. Fan, and R. Li. Efficient estimation and inference for varying coefficients models. *Journal of the American Statistical Association*, 95:888–902, 2000.

[5] Z. Cai, J. Fan, and Q. Yao. Functional-coefficient regression models for nonlinear time series. *Journal of the American Statistical Association*, 95:941–956, 2000.

[6] R. J. Carroll, D. Ruppert, and A.H. Welsh. Nonparametric estimation via local estimating equation. *Journal of the American Statistical Association*, 93:214–227, 1998.

[7] P. Charbonnier, L. Blanc-Feraud, G. Aubert, and M. Barlaud. Two deterministic half-quadratic regularization algorithms for computed imaging. In *Image Processing, 1994. Proceedings of ICIP-94*, vol. 2, *IEEE International Conference*, volume 2, pages 168–172, 1994. ISBN 0-8186-6950-0
[8] X. Chen and H. White. Nonparametric learning with feedback. *Journal of Econometric Theory*, 82:190–222, 1998.
[9] C. K. Chu, I. K. Glad, F. Godtliebsen, and J. S. Marron. Edge-preserving smoothers for image processing. *Journal of the American Statistical Association*, 93(442):526–541, 1998.
[10] R. Dahlhaus. Fitting time series models to nonstationary processes. *Annals of Statistics*, 16:1–37, 1997.
[11] R. Dahlhaus and S. Subba Rao. Statistical inference of time varying arch processes. *Annals of Statistics*, 34:1074–1114, 2006.
[12] R. Dahlhaus and S. Subba Rao. A recursive onlilne algorithm for the estimation of time varying arch processes. *Bernoulli*, 13:389–422, 2007.
[13] S. Didas, J. Franke, J. Tadjuidje, and J. Weickert. Some asymptotics for local least-squares regression with regularization. Report in Wirtschaftsmathematik 107, University of Kaiserslautern, 2007.
[14] S. Didas, P. Mrázek, and J. Weickert. Energy-based image simplification with nonlocal data and smoothness terms. In *Algorithms for Approximation – Proceedings of the 5th International Conference, Chester*, pages 51–60, Springer, Berlin, Heidelberg, New York, 2007.
[15] S. Didas and J. Weickert. Integrodifferential equations for continuous multiscale wavelet shrinkage. *Inverse Problems and Imaging*, 1:29–46, 2007.
[16] H. Drees and C. Starica. A simple non-stationary model for stock returns, 2003. Preprint.
[17] R. Engle. Autoregressive conditional heteroscedasticity with estimates of the variance of the united kingdom inflation. *Econometrica*, 50:987–1006, 1982.
[18] J. Fan, M. Farmen, and I. Gijbels. Local maximum likelihood estimation and inference. *Journal of the Royal Statistical Society*, 60:591–608, 1998.
[19] J. Fan and I. Gijbels. *Local Polynomial Modelling and Its Applications*. Chapman & Hall, London, 1996.
[20] Y. Feng. Modelling different volatility components in high-frequency financial returns, 2002. Preprint.
[21] J. Franke and S. Halim. A bootstrap test for comparing images in surface inspection. Preprint 150, dfg-spp 1114, University of Bremen, 2006.
[22] J. Franke and S. Halim. Data-adaptive bandwidth selection for kernel estimates of two-dimensional regression functions with correlated errors. Report in wirtschaftsmathematik, University of Kaiserslautern, 2007.
[23] J. Franke and S. Halim. Wild bootstrap tests for signals and images, *IEEE Signal Processing Magazine*, 24(4):31–37, 2007.

[24] P. Fryzlewicz, T. Sapatinas, and S. Subba Rao. Normalised least-squares estimation in time-varying arch models, 2007. To appear in: *Annals of Statistics*
[25] Th. Gasser and H. G. Müller. Kernel estimation of regression functions. In Th. Gasser and M. Rosenblatt, editors, *Smoothing Techniques for Curve Estimation*, Springer, Berlin, Heidelberg, New York, 1979.
[26] L. Giraitis, P. Kokoskza, and R. Leipus. Stationary arch models: Dependence structure and central limit theorem. *Econometric Theory*, 16:3–22, 2000.
[27] L. Giraitis, R. Leipus, and D. Surgailis. Recent advances in arch modelling. In A. Kirman and G. Teyssiere, editors, *Long memory in Economics*, pages 3–39, Springer, Berlin, Heidelberg, New York, 2005.
[28] W. Haerdle. *Applied Nonparametric Regression.* Cambridge University Press, Cambridge, 1990.
[29] S. Halim. *Spatially adaptive detection of local disturbances in time series and stochastic processes on the integer lattice Z^2.* PhD thesis, University of Kaiserlautern, 2005.
[30] T. J. Hastie and R.J. Tibshirani. Varying-coefficient models (with discussion). *Journal of the Royal Statistical Society*, 55:757–796, 1993.
[31] E. Herrmann, Th. Gasser, and A. Kneip. Choice of bandwidth for kernel regression when residuals are correlated. *Biometrika*, 79:783–795, 1992.
[32] J. J. Koenderink and A. L. Van Doorn. The structure of locally orderless images. *International Journal of Computer Vision*, 31(2/3):159–168, 1999.
[33] H. Kushner and G. Yin. *Stochastic Approximation and Recursive Algorithms and Applications.* Springer, Berlin, Heidelberg, New York, 2003.
[34] L. Ljung and T. Söderstróm. *Theory and Practice of Recursive Identification.* MIT Press, Cambridge, MA, 1983.
[35] C. Loader. *Local regression and likelihood.* Springer, Berlin, Heidelberg, New York, 1999.
[36] T. Mikosch and C. Starica. Is it really long memory we see in financial returns? In P. Embrechts, editor, *Extremes and Integrated Risk Management*, pages 149–168, Risk Books, London, 2000.
[37] E. Moulines, P. Priouret, and F. Roueff. On recursive estimation for locally stationary time varying autoregressive processes. *Annals of Statistics*, 33:2610–2654, 2005.
[38] P. Mrázek and J. Weickert. Rotationally invariant wavelet shrinkage. In *Pattern Recognition*, volume 2781 of *Lecture Notes in Computer Science*, pages 156–163. Springer, Berlin-Heidelberg, New York, 2003.
[39] P. Mrázek, J. Weickert, and A. Bruhn. robust estimation and smoothing with spatial and tonal kernels. In R. Klette, R. Kozera, L. Noakes, and J. Weickert, editors, *Geometric Properties for Incomplete Data*, volume 31 of *Computational Imaging and Vision*, pages 335–352. Springer, Berlin-Heidelberg, New York, 2006.

[40] P. Mrázek, J. Weickert, and G. Steidl. Correspondences between wavelet shrinkage and nonlinear diffusion. In *Scale-Space Methods in Computer Vision*, volume 2695 of *Lecture Notes in Computer Science*, pages 101–116. Springer, Berlin-Heidelberg, New York, 2003.

[41] P. Mrázek, J. Weickert, and G. Steidl. Diffusion-inspired shrinkage functions and stability results for wavelet shrinkage. *International Journal of Computer Vision*, 64(2(3)):171–186, 2005.

[42] P. Mrázek, J. Weickert, and G. Steidl. On iterations and scales of nonlinear filters. In O. Drbohlav, editor, *Proceedings of the Eighth Computer Vision Winter Workshop*, pages 61–66, February 2003. Valtice, Czech Republic,. Czech Pattern Recognition Society.

[43] H. Müller. Change-points in nonparametric regression analysis. *Annals of Statistics*, 20:737–761, 1992.

[44] D. Mumford and J. Shah. Optimal approximation of piecewise smooth functions and associated variational problems. *Communications on Pure and Applied Mathematics*, 42:577–685, 1989.

[45] D. Nelson. Conditional heteroskedasity in assit returns: A new approach. *Econometrica*, 59:347–370, 1990.

[46] P. Perona and J. Malik. Scale space and edge detection using anisotropic diffusion. *IEEE Transactions on Pattern Analysis and Machine Intelligence*, 12:629–639, 1990.

[47] L. Pizarro, S. Didas, F. Bauer, and J. Weickert. Evaluating a general class of filters for image denoising. In B. K. Ersboll and K. S. Pedersen, editors, *Image Analysis*, volume 4522 of *Lecture Notes in Computer Science*, pages 601–610, Springer, Berlin, Heidelberg, New York, 2007.

[48] G. Plonka and G. Steidl. A multiscale wavelet-inspired scheme for nonlinear diffusion. *International Journal of Wavelets, Multiresolution and Information Processing*, 4:1–21, 2006.

[49] J. Polzehl and V. Spokoiny. Adaptive weights smoothing with applications to image restorations. *Journal of the Royal Statistical Society Ser. B,* 62:335–354, 2000.

[50] J. Polzehl and V. Spokoiny. Functional and dynamic magnetic resonance imaging using vector adaptive weights smoothing. *Journal of the Royal Statistical Society Ser. C,* 50:485–501, 2001.

[51] J. Polzehl and V. Spokoiny. Image denoising: pointwise adaptive approach. *Annals of Statistics*, 31:30–57, 2003.

[52] J. Polzehl and V. Spokoiny. Spatially adaptive regression estimation: Propagation-separation approach. Technical report, WIAS, Berlin, 2004. Preprint 998.

[53] J. Polzehl and V. Spokoiny. Propagation-separation approach for local likelihood estimation. *Theory and Related Fields*, 135:335–336, 2006.

[54] J. Polzehl and K. Tabelow. Adaptive smoothing of digital images: the r package *adimpro*. *Journal of Statistical Software*, 19(1), 2007.

[55] J. Polzehl and K. Tabelow. fmri: A package for analyzing fmri data, 2007. To appear in: R News.

[56] P. Qiu. Discontinuous regression surface fitting. *Annals of Statistics*, 26:2218–2245, 1998.

[57] r. *R: A language and environment for statistical computing*. R Foundation for Statistical Computing, Vienna, Austria, 2005. ISBN 3-900051-07-0 2005.

[58] P. Robinson. Nonparametric estimation of time-varying parameters. In P. Hackl, editor, *Statistical analysis and Forecasting of Economic Structural Change*, pages 253–264. Springer, Berlin, Heidelberg, New York, 1989.

[59] P. Robinson. Testing for strong serial correlation and dynamic conditional heteroskedasity in multiple regression. *Journal of Econometrics*, 47:67–78, 1991.

[60] L. I. Rudin, S. Osher, and E. Fatemi. Nonlinear total variation based noise removal algorithms. *Physica D*, 60:259–268, 1992.

[61] J. Simonoff. *Smoothing Methods in Statistics.* Springer, Berlin, Heidelberg, New York, 1996.

[62] S. M. Smith and J. M. Brady. SUSAN – A new approach to low lewel image processing. *International Journal of Computer Vision*, 23(1):43–78, 1997.

[63] V. Solo. The second order properties of a time series recursion. *Annals of Statistics*, 9:307–317, 1981.

[64] V. Spokoiny. Estimation of a function with discontinuities via local polynomial fit with an adaptive window choice. *Annals of Statistics*, 26:1356–1378, 1998.

[65] G. Steidl and J. Weickert. Relations between soft wavelet shrinkage and total variation denoising. *Lecture Notes in Computer Science*, 2449:198–205, 2002.

[66] G. Steidl, J. Weickert, T. Brox, P. Mrázek, and M. Welk. On the equivalence of soft wavelet shrinkage, total variation diffusion, total variation regularization, and SIDEs. *SIAM Journal on Numerical Analysis*, 42(2):686–713, May 2004.

[67] K. Tabelow, J. Polzehl, V. Spokoiny, and H. U. Voss. Diffusion tensor imaging: Structural adaptive smoothing. Technical report, WIAS, Berlin, 2007. Preprint 1232.

[68] K. Tabelow, J. Polzehl, H. U. Voss, and V. Spokoiny. Analyzing fMRI experiments with structural adaptive smoothing procedures. *Neuroimage*, 33:55–62, 2006.

[69] R. Tibshirani and T.J. Hastie. Local likelihood estimation. *Journal of the American Statistical Association*, 82:559–567, 1987.

[70] A. Tikhonov. Solution of incorrectly formulated problems and the regularization method. In *Soviet Mathematics Doklady*, volume 4, pages 1035–1038, 1963.

[71] C. Tomasi and R. Manduchi. Bilateral filtering for gray and color images. In *ICCV '98: Proceedings of the Sixth International Conference on Computer Vision*, pages 839–846, Washington, DC, USA, 1998. IEEE Computer Society.

[72] H. U. Voss, K. Tabelow, J. Polzehl, O. Tchernichovsky, K. K. Maul, D. Salgado-Commissariat, D. Ballon, and S. A. Helekar. Functional MRI of the zebra finch brain during song stimulation suggests a lateralized response topography. *PNAS*, 104:10667–10672, 2007.

[73] M.P. Wand and M.C. Jones. *Kernel Smoothing.* Chapman & Hall, London, 1995.

[74] J. Weickert, G. Steidl, P. Mrazek, M. Welk, and T. Brox. Diffusion filters and wavelets: What can they learn from each other? In *Handbook of Mathematical Models of Computer Vision*, pages 3–16. Springer, Berlin Heidelberg New York, 2005.

[75] M. Welk, G. Steidl, and J. Weickert. Locally analytic schemes: a link between diffusion filtering and wavelet shrinkage, 2007. To appear in: Applied and Computational Harmonic Analysis.

[76] M. Welk, J. Weickert, and G. Steidl. A four-pixel scheme for singular differential equations. In *Scale Space and PDE Methods in Computer Vision*, volume 3459 of *Lecture Notes in Computer Science*, pages 585–597, Springer, Berlin, Heidelberg, New York, 2005.

[77] H. White. Parametric statistical estimation using artifical neural networks. In *Mathematical perspectives on neural networks*, pages 719–775. L. Erlbaum Associates, Hilldale, NJ, 1996.

[78] G. Winkler, V. Aurich, K. Hahn, and A. Martin. Noise reduction in images: Some recent edge-preserving methods. *Pattern Recognition and Image Analysis*, 9:749–766, 1999.

[79] E. Zeidler, editor. *Nonlinear Functional Analysis and Applications I: Fixed-Point Theorems.* Springer-Verlag, New York, Inc., New York, NY, USA, 1986.

7

Nonlinear Analysis of Multi-Dimensional Signals: Local Adaptive Estimation of Complex Motion and Orientation Patterns

Christoph S. Garbe[1], Kai Krajsek[2], Pavel Pavlov[1], Björn Andres[1], Matthias Mühlich[2,4], Ingo Stuke[6], Cicero Mota[2,5], Martin Böhme[5], Martin Haker[5], Tobias Schuchert[3], Hanno Scharr[3], Til Aach[4], Erhardt Barth[5], Rudolf Mester[2], and Bernd Jähne[1]

[1] Interdisciplinary Center for Scientific Computing (IWR), University of Heidelberg, D-69120 Heidelberg, Germany
{Christoph.Garbe,pavel.pavlov,Bjoern.Andres}@iwr.uni-heidelberg.de
Bernd.Jaehne@iwr.uni-heidelberg.de

[2] Visual Sensorics and Information Processing Lab (VSI), Goethe University Frankfurt, D-60054 Frankfurt/M., Germany
krajsek@vsi.cs.uni-frankfurt.de, cicmota@gmail.com,
mester@iap.uni-frankfurt.de

[3] Institute for Chemistry and Dynamics of the Geosphere, Institute 3: Phytosphere, Research Center Jülich GmbH, D-52425 Jülich, Germany
t.schuchert@fz-juelich.de, h.scharr@fz-juelich.de

[4] Institute of Imaging & Computer Vision, RWTH Aachen University, D-52056 Aachen, Germany
mm@LfB.RWTH-Aachen.de, Til.Aach@LfB.RWTH-Aachen.de

[5] University of Lübeck, Institute for Neuro- and Bioinformatics, D-23538 Lübeck, Germany
{boehme,haker,barth}@inb.uni-luebeck.de

[6] University of Lübeck, Institute for Signal Processing, D-23538 Lübeck, Germany
Ingo.Stuke@t-online.de

7.1 Introduction

We consider the general task of accurately detecting and quantifying orientations in n-dimensional signals $\boldsymbol{s}$. The main emphasis will be placed on the estimation of motion, which can be thought of as orientation in spatio-temporal signals. Associated problems such as the optimization of matched kernels for deriving isotropic and highly accurate gradients from the signals, optimal integration of local models, and local model selection will also be addressed.

Many apparently different approaches to get a quantitative hold on motion have been proposed and explored, e.g., using the brightness constancy constraint [47, 89], the structure tensor [43, 49], blockwise matching or correlation [9, 10], quadrature filters [22, 45, 51], steerable filters and other filter-based approaches [18, 110], projection on polynomial subspaces [20], autocovariance based analysis [67], and many variants of these approaches. As mentioned previously, the estimation of motion is closely related to the estimation of orientation or linear symmetry in 2 D images [11, 39] and the computation of curvature [125, 132] in 2 D or higher dimensional spaces.

It is relatively well understood, how these various approaches can be employed in the case of simple motion patterns, which can – at least on a local scale – be approximated by a single rigid translational motion. This applies also to moderate amounts of noise, especially if the noise process is stationary, additive, and its amplitude distribution is unimodal.

However, most of the aforementioned approaches show severe limitations in case of complex motion patterns or strong disturbances, which are characteristic for real-life image data. Problematic situations occur for instance in case of motion superposition (due to transparency or specular reflection), temporal and/or spatial motion discontinuities, and spatio-temporal flow effects (relaxation, diffusion, etc.). It is this area, where the current challenges in motion analysis are found, in which significant contributions will be made throughout this chapter. Improved algorithms will necessarily be locally adaptive, nonlinear and based on modern signal processing tools, such as Total Least Squares, anisotropic diffusion, and Markov Random fields, or extend classical signal theory such as those presented by [22, 122].

An overview of the problem of multiple motions has been given in [12] and robust methods for multiple motions have been proposed. The problem of two motions has been first solved by Shizawa and Mase [108, 109]. Their approach is based on Fourier methods and on solving a six-dimensional eigensystem that limits the method to only two motions. Here we will show how to avoid such an eigensystem in case of one motion resulting in a simpler and faster solution for multiple motions. Important contributions in characterizing the spectral properties of multiple motions have been made [133]. In dealing with the problem of multiple motions, the useful and intuitive notions of "nulling filters" and "layers" have been introduced [16, 126]. Their approach is more general in that it treats the separation of motions into layers, but is also limited to the use of a discrete set of possible motions and a probabilistic procedure for finding the most likely motions out of the set.

The problems mentioned before are dealing with characteristics of the signal itself; but additionally we find several kinds of disturbances in real life image data. For instance: rain, snow or low illumination conditions cause different types of noise. Besides that, outdoor scenes often are afflicted by illumination changes, or specular reflections. Additionally, many scientific or medical applications acquire images at the limit of what is physically and technically feasible. This often introduces strong disturbances such as heavy signal-dependent noise.

As will be outlined, the estimation of complex orientation from image data represents an inverse problem. Paramount to solving this problem are adequate models. In Sect. 7.2 a number of such extended models will be presented, ranging from those incorporating brightness changes along given orientations over those incorporating multiple orientations to those deducing scene depth in motion models. The models constrain orientation to image intensities. Still, generally more unknowns are sought than can be fully recovered from these presented constraint equations. In Sect. 7.3 a number of approaches will be presented that make it feasible to derive the sought parameters by introducing additional constraints. Refined estimators will be presented that take statistics into account to perform maximum likelihood estimates. The presented algorithms are based on differential orientation models, relying on an accurate extraction of gradients from image intensities. In Sect. 7.4 schemes for computing these gradients from optimized filters will be presented. Closely related to estimating the orientation parameter of the introduced models is the task of selecting the correct model, given a noise level by which the image data is corrupted. Two such approaches are presented in Sect. 7.5. The correct model that can be retrieved from the image data will also depend on the signal structures. These structures can be categorized by their intrinsic dimension, a concept which is also introduced in this section. The inverse problem of estimating model parameters is performed by an optimal regularization. In Sect. 7.6 different regularization schemes are presented, that preserve anisotropies in the image signal. These approaches are also used for suppressing noise prior to performing the estimation, leading to optimum results. The problem of orientation estimation in multiple subspaces is a prominent one, that has significant impact, both on modeling and estimation, but for applications as well. Algorithms tackling this problem are detailed in Sect. 7.7. The developed algorithms make novel applications feasible. In Sect. 7.8 some exemplary applications are presented stemming from different fields such as environmental-, geo-, and life sciences. This chapter closes with some concluding remarks in Sect. 7.9.

7.2 Modeling Multi-Dimensional Signals

7.2.1 Motion and Subspace Models

We regard n-dimensional signals $s(\boldsymbol{x})$ defined over a region Ω, e.g., images and image sequences. Motions (translations) and orientations correspond to linear d-dimensional subspaces E of Ω with $1 \leq d < n$, such that

$$s(\boldsymbol{x}) = s(\boldsymbol{x} + k\boldsymbol{u}) \quad \forall k \in \mathbb{R} \text{ and } \forall\, \boldsymbol{x}, \boldsymbol{x} + k\boldsymbol{u} \in \Omega \text{ and } \boldsymbol{u} \in E. \tag{7.1}$$

Often one needs to (i) detect the existence of such a subspace E and (ii) estimate the parameter vector $k\boldsymbol{u}$, which corresponds (not always, see Sect. 7.3.2) to the direction of motion in the regarded volume. The values of s can be scalar as in gray-level images or vector valued (denoted $\boldsymbol{s}$) as in color or multispectral images. The estimation is often based on the fact that constancy of the signal in a certain direction in Ω, such as it is reflected in (7.1) implies linear differential constraints such as the classical *brightness constancy constraint equation (BCCE)*

$$\frac{\partial s}{\partial \boldsymbol{u}} = 0 \text{ for all } \boldsymbol{u} \in E\,. \tag{7.2}$$

This is the simplest special case of general partial differential equations which result from applying a suitable differential operator $\alpha(\boldsymbol{u})$ on the signal:

$$\alpha(\boldsymbol{u}) \circ s = 0 \tag{7.3}$$

and we will learn about more sophisticated operators $\alpha(\boldsymbol{u})$ later in this contribution. Assuming the constancy of a moving brightness pattern, motions can be interpreted as local orientations in spatio-temporal signals ($n = 3, d = 1$). Many motion models are based on Taylor expansions of (7.1) (see e.g. [24]). Writing (7.1) with time t as individual parameter, we obtain

$$s(\boldsymbol{x}(t), t) = s(\boldsymbol{x} + \boldsymbol{u}(x,t)\Delta t, t + \Delta t)\,, \tag{7.4}$$

where s is interpreted as the constant brightness signal produced by a spatial point $\boldsymbol{x}(t)$ moving in time t. First order approximation of (7.4) yields

$$\frac{\mathrm{d}s}{\mathrm{d}t} = 0 \quad \Leftrightarrow \quad \nabla^T_{(\boldsymbol{x})} s \cdot \boldsymbol{u} + \frac{\partial s}{\partial t} = 0 \quad \Leftrightarrow \quad \nabla^T_{(\boldsymbol{x},t)} s \cdot (\boldsymbol{u}^T, 1)^T = 0\,, \tag{7.5}$$

where ∇ is the gradient operator with respect to parameters given as indices and the general differential operator $\alpha(\boldsymbol{u})$ from (7.3) takes the form $\alpha(\boldsymbol{u}) := \nabla^T_{(\boldsymbol{x})}\boldsymbol{u} + \frac{\partial}{\partial t}$. Being based on derivatives, such models are called *differential models.* One may further model the motion field $\boldsymbol{u}(x,t)$ locally by applying a Taylor expansion

$$\nabla^T_{\boldsymbol{x}} s(\boldsymbol{u} + \boldsymbol{A}\Delta\boldsymbol{x}) + \frac{\partial s}{\partial t} = 0 \tag{7.6}$$

where the matrix $\boldsymbol{A} = \nabla_{\boldsymbol{x}} \boldsymbol{u}^T$ contains the spatial derivatives of $\boldsymbol{u}$, and $\Delta\boldsymbol{x} = \boldsymbol{x}_0 - \boldsymbol{x}$ are local coordinates. This is called an *affine motion model.* These and other motion models, i.e. parametrizations of $\boldsymbol{u}$, can be found e.g. in [24].

7.2.2 Multiple Motions and Orientations

In case of multiple motions and orientations, we are dealing with a (linear, multiplicative, or occluding) superposition of subspaces as defined in (7.1). In case of two subspaces, and additive superposition, one has

$$s(\boldsymbol{x}) = s_1(\boldsymbol{x} + k\boldsymbol{u}_1) + s_2(\boldsymbol{x} + k\boldsymbol{u}_2)\,. \tag{7.7}$$

The model for multiplicative superposition is simply

$$s(\boldsymbol{x}) = s_1(\boldsymbol{x} + k\boldsymbol{u}_1) \cdot s_2(\boldsymbol{x} + k\boldsymbol{u}_2)\,, \tag{7.8}$$

which can be transformed into (7.7) by taking the logarithm. The model for occluded superposition is

$$s(\boldsymbol{x}) = \chi s_1(\boldsymbol{x} + k\boldsymbol{u}_1) + (1 - \chi) s_2(\boldsymbol{x} + k\boldsymbol{u}_2)\,, \tag{7.9}$$

where $\chi(\boldsymbol{x})$ is the characteristic function that defines the occlusion. As we shall see in Sect. 7.7, the constraint in (7.2), and some novel mathematical tricks, can be used to detect and estimate multiple motions and orientations.

7.2.3 Dynamic 3 D Reconstruction

Since the early work of [48] optical flow has been used for disparity and therefore 3 D structure estimation. For disparity estimation time t in (7.4) is replaced by camera displacement r. Therefore time and camera displacement may be considered equivalent as "time-like" parameters. Combining equations with N time-like parameters sets the dimension d of subspace E to $d = N$ because brightness constancy applies for each of these parameters. We therefore get a 3-dimensional solution space when using time t and two camera displacement directions r_1 and r_2. Following (7.2), we achieve 3 D position and 3 D motion depending on the motion field either of the object or the camera. Using (7.4), we can determine motion of an object whereas replacing time t in this equation with camera position r, called structure from motion, yields disparity. Combining these estimations yields a higher dimensional solution space and the problem of determining the parameters in this space (see [97, 99]). Still assuming constant brightness now in time and in camera displacement direction we get according to (7.5)

$$\frac{\partial s}{\partial u} = 0 \quad \Leftrightarrow \quad \nabla^T s\boldsymbol{u} = 0 \quad \text{with} \quad \boldsymbol{u} = (\mathrm{d}x, \mathrm{d}y, \mathrm{d}r_1, \mathrm{d}r_2, \mathrm{d}t)^T\,. \tag{7.10}$$

Parameters like optical flow (u_1, u_2) or disparity ν are then obtained by combination of subspace solutions, e.g.,

$$u_1 = \left.\frac{\mathrm{d}x}{\mathrm{d}t}\right|_{\mathrm{d}r_1,\mathrm{d}r_2=0}, u_2 = \left.\frac{\mathrm{d}y}{\mathrm{d}t}\right|_{\mathrm{d}r_1,\mathrm{d}r_2=0}, \nu_1 = \left.\frac{\mathrm{d}x}{\mathrm{d}r_1}\right|_{\mathrm{d}t,\mathrm{d}r_2=0}, \nu_2 = \left.\frac{\mathrm{d}y}{\mathrm{d}r_2}\right|_{\mathrm{d}t,\mathrm{d}r_1=0} \tag{7.11}$$

where ν_1 and ν_2 are dependent estimates of ν which can be transformed into two independent estimates of ν (see Sect. 7.3.5). Further modeling the data by applying a surface patch model extends the above orientation model of (7.10). In the same way, standard optical flow can be extended to affine optical flow (see (7.5) and (7.6)).

$$\nabla_{\boldsymbol{x}}^T s(\boldsymbol{u} + \boldsymbol{A}\Delta\boldsymbol{x}) + \sum_{i=1}^{2} \frac{\partial s}{\partial r_i}\mathrm{d}r_i + \frac{\partial s}{\partial t}\mathrm{d}t = 0 \tag{7.12}$$

where u contains parameters for motion and disparity, matrix $\boldsymbol{A}$ parameters for depth motion and surface slopes, and $\Delta\boldsymbol{x} = \boldsymbol{x}_0 - \boldsymbol{x}$ are local coordinates. A detailed derivation of the model is proposed in [97] for special cases and in [99] for the general model.

Comparison of 3 D Reconstruction Models

In [97] a detailed error analysis of special cases of the model, i.e. 1 D camera grid without normals with and without z-motion, 2 D camera grid without any motion with and without normals, is presented. The analysis shows that with higher z-motion the errors due to linearization increase and that estimating the surface normals within this framework reduces systematic errors. Further an error analysis was done for the full model in [99]. There comparisons between (i) 1 D camera grid with and without normals and (ii) 1 D camera grid with the 2 D camera grid are shown. The full model performed well or better for all parameters and the additional parameters do not lead to instabilities. Also the 2 D model is more robust with respect to the aperture problem. All these analyses are performed using the estimation framework presented below (Secs. 7.3.1 and 7.3.5).

7.2.4 Brightness Changes

The basic approach of motion estimation requires brightness constancy along motion trajectories, as in the model (7.5), (7.6), (7.10) and (7.12) for single motions and in (7.7) for multiple motions. In gradient based optical flow techniques, brightness changes can be modeled by partial differential equations in a similar manner as in [41]

$$\alpha(\boldsymbol{u}) \circ \boldsymbol{s} = \frac{\mathrm{d}}{\mathrm{d}t} h(s_0, t, \boldsymbol{\beta}) , \tag{7.13}$$

where the brightness change may depend on the initial gray value s_0, the time t and a parameter vector $\boldsymbol{\beta}$. It is modeled by the function $h(s_0, t, \beta)$. Physically motivated brightness changes include exponential ($h(t) \propto \exp(\beta t)$) and diffusive ($h(t) \propto \beta \Delta s$ with the spatial Laplacian Δs) processes. Also, the simple linear brightness change ($h(t) \propto \beta t$) can be used quite advantageously if an accurate model of the actual brightness change model is unknown. In these standard cases, β is a term independent of time t.

In the next sections, specialized application driven brightness change model, as well as incorporations of these brightness changes to dynamic 3 D scene reconstruction and multiple motions will be introduced.

Brightness Changes for Fluid Flows

The measurement of fluid flows is an emerging field for optical flow computation. In a number of such applications, a tracer is visualized with modern digital cameras. Due to the projective nature of the imaging process, the tracer is integrated across a velocity profile. For a number of fluid flow configuration, the velocity profile can be approximated to leading order by

$$u(x_3) = A \cdot x_3^n,$$

where A is a term independent of the coordinate direction of integration x_3 and time t. Integration across such a profile leads to an intensity change, modeled by the differential equation [27]

$$\frac{\mathrm{d}s}{\mathrm{d}t} = u_1 \frac{\partial s}{\partial x} + u_2 \frac{\partial s}{\partial y} + \frac{\partial s}{\partial t} = -\frac{1}{n \cdot t}\, s\,. \tag{7.14}$$

This equation presents a generalization of the results obtained for Couette flow (shear driven flow, $n = 1$) and Poiseuille flow (pressure driven flow, $n = 2$). These brightness change models take into account effects such as Taylor dispersion and have been applied successfully to microfluidics [34] or in shear driven flows at the air–water interface [32].

Brightness Changes in Dynamic 3 D Reconstruction

Brightness changes can be incorporated into dynamic 3 D scene reconstruction suitably applying (7.13). One obtains the following constraint equation

$$\frac{\partial s}{\partial x}\,\mathrm{d}x + \frac{\partial s}{\partial y}\,\mathrm{d}y + \frac{\partial s}{\partial r}\,\mathrm{d}r + \frac{\partial s}{\partial t}\,\mathrm{d}t = s\frac{\partial h}{\partial t}\,\mathrm{d}t. \tag{7.15}$$

Brightness changes due to changing viewing direction and bidirectional reflectance distribution function (BRDF, see e.g. [40]) may be modeled only temporally, but changes due to inhomogeneous illumination need additional spatial modeling. A suitable brightness change function $h(\Delta X, \Delta Y, t)$ has been derived by Taylor expansion of changing BRDF influence and illumination intensity [106].

$$h(\Delta X, \Delta Y, t) \approx h(t, \boldsymbol{a}) := \sum_{i=1}^{2} (a_i + a_{i,x}\Delta X + a_{i,y}\Delta Y)\, t^i \tag{7.16}$$

with illumination parameter vector $\boldsymbol{a}$.

Comparison of Physics Based Brightness Variation Models for 3 D Reconstruction Model

A systematic error analysis using sinusoidal patterns in [106] demonstrates that modeling spatial variations of the BRDF, as shown in (7.16), improves estimation results. Figure 7.1 presents the improvement of the estimations on a reconstructed cube moving with U_Z=2 mm/frame while illuminated by a

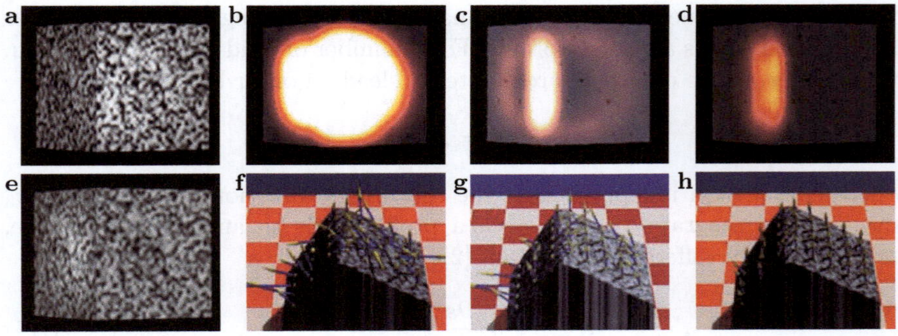

Fig. 7.1 Motion estimation of cube moving towards camera with spot light moving around cube center. (**a**) and (**e**): first and last image taken with central camera. (**b**)–(**d**): color coded model errors (projected on contrast reduced cube) for models without (**b**), constant temporal (**c**), and spatially varying temporal brightness change (**d**). Below the model errors, scaled motion estimates for the models are depicted, respectively (**f**)–(**h**)

rotating spot light. Figure 7.1 (a) and (b) show the first and the last image of the cube sequence. In Fig. 7.1 (b)–(d) the numerical error, i.e. the largest of the three eigenvalues of the structure tensor is depicted as color overlay on the central input image. Finally Fig. 7.1 (f)–(h) highlight the estimation of the 3 D scene flow for the three different models, i.e. constant, spatially constant but changing in time and both spatially and temporally changing BRDF.

Brightness Changes in Multiple Motions and Orientations

The multiple-motions model can be extended to account for brightness changes [103]. As with the operator $\alpha(\boldsymbol{u})$ for brightness constancy, one may define operators for additive, multiplicative or diffusive brightness changes. E.g. for multiplicative brightness changes (7.7) becomes

$$s(\boldsymbol{x},t) = s_1(\boldsymbol{x} - \boldsymbol{u}_1 t)k_1(t) + s_2(\boldsymbol{x} - \boldsymbol{u}_2 t)k_2(t)\,, \tag{7.17}$$

where k_1 and k_2 are scalar functions. It can be nullified via $\beta(\boldsymbol{u}_1, c_1) \circ \beta(\boldsymbol{u}_2, c_2) \circ s = 0$ using

$$\beta(\boldsymbol{u}, c) := u_x\partial_x + u_y\partial_y + \partial_t - c \tag{7.18}$$

if $k_1(t) \propto \exp(c_1 t)$ and $k_2(t) \propto \exp(c_2 t)$. As in the constant brightness case (see Sect. 7.2.2) the constraint equation is linear in mixed and therefore nonlinear parameters. A detailed analysis reveals that parameter estimation and disentangling motion parameters can be done as in the constant brightness case, but disentangling brightness parameters is done by solving a real instead of a complex polynomial [103].

7.3 Estimation of Local Orientation and Motion

7.3.1 The Structure Tensor Approach

All of the motion models presented in Sect. 7.2 linearly relate optical flow parameters, brightness changes and image signals. These constraints can be formulated generally as an inner product of a data vector $\boldsymbol{d}$ and a parameter vector p, resulting in

$$\boldsymbol{d}^{\top}\boldsymbol{p} = \alpha(\boldsymbol{u}) \circ s - \frac{\mathrm{d}}{\mathrm{d}t} h(s_0, t, \boldsymbol{\beta}) = 0\,, \tag{7.19}$$

where $\boldsymbol{d}$ contains image spatio-temporal derivatives and depending on the actual constraint, the gray values s themselves. The parameter vector $\boldsymbol{p}$ consists of the image velocities $\boldsymbol{u}$ as well as additional parameters such as those of brightness change $\boldsymbol{\beta}$ or those of higher order motion [8, 23, 25].

Equation (7.19) provides one constraint for several unknowns we seek to estimate. An additional constraint is that of local constancy of parameters. To this end, the constraints of type (7.19) can be constructed for each pixel in a spatio-temporal neighborhood, leading to a linear system of equations

$$\boldsymbol{D} \cdot \boldsymbol{p} = \boldsymbol{0}\,, \tag{7.20}$$

where $\boldsymbol{D} = [\boldsymbol{d}_1, \ldots, \boldsymbol{d}]^{\top}$. Assuming identical isotropic Gaussian noise (i.i.d.) in all measurements, the maximum likelihood estimate for the unknown parameter vector is given by the total least squares (TLS) solution [25, 90, 92, 123, 127]. The total least squares (TLS) method seeks to minimize $||\boldsymbol{D} \cdot \boldsymbol{p}||^2$, subject to the constraint that $\boldsymbol{p}^{\top}\boldsymbol{p} = 1$ to avoid the trivial solution. Usually the rows of $\boldsymbol{D}$ are weighted according to their distance from the central pixel by Gaussian weights w with standard deviation ρ. The structure tensor $\boldsymbol{J}_\rho$ then results from $||\boldsymbol{D} \cdot \boldsymbol{p}||^2$

$$||\boldsymbol{D} \cdot \boldsymbol{p}||^2 = \boldsymbol{p}^T \boldsymbol{D}^T \boldsymbol{W}_\rho \boldsymbol{D} \boldsymbol{p} =: \boldsymbol{p}^T \boldsymbol{J}_\rho \boldsymbol{p}\,, \tag{7.21}$$

where $\boldsymbol{W}_\rho$ is a diagonal matrix containing the Gaussian weights w. This formulation yields a solution, $\hat{\boldsymbol{p}}$, given by the right singular direction associated with the smallest singular value of row weighted $\boldsymbol{D}$ [86, 123] or the respective eigenvectors of $\boldsymbol{J}_\rho$. The sought parameter vector $\boldsymbol{p}$ is found by normalizing the last component of $\hat{\boldsymbol{p}}$ to unity [123]. The algorithmic aspects of TLS parameter estimation have been explored in some detail [86, 87, 123].

7.3.2 Beyond the Differential Approach: The Generalized Structure Tensor

The attempt to express brightness constancy along the motion trajectory by a first-order partial differential equation, that is: by using the BCCE of (7.2), is not the unique and not the most expressive way of specifying a relation between the entity that is sought (the motion vector $\boldsymbol{u}$) and the signal that can be observed. The BCCE describes the situation for a *continuous signal*, and it does not explicitly consider the different error terms that are caused by observation noise, spatio-temporal pixel aperture, and by the necessary discretization of the problem. Beyond that, the formulation in terms of derivatives or gradients does not lend itself so much for the development of motion estimation procedures that take into account the spectral characteristics of the image signal and the spectral characteristics of the noise.

Assuming brightness constancy along the motion trajectory, *all* higher order directional derivatives vanish in the motion direction:

$$\frac{\partial s}{\partial \boldsymbol{u}} \stackrel{!}{=} 0 \quad \bigcap \quad \frac{\partial^2 s}{\partial \boldsymbol{u}^2} \stackrel{!}{=} 0 \quad \bigcap \quad \cdots \tag{7.22}$$

A condition which is less stringent than (7.22), but nevertheless comprises as much as possible from these multitude of conditions in a single linear equation can be obtained by summing up the constraints:

$$\alpha(\boldsymbol{u}) \circ s \quad = \quad \alpha_1 \frac{\partial s}{\partial \boldsymbol{u}} + \alpha_2 \frac{\partial^2 s}{\partial \boldsymbol{u}^2} + \alpha_3 \frac{\partial^3 s}{\partial \boldsymbol{u}^3} \quad \stackrel{!}{=} 0 \tag{7.23}$$

The middle part of this equation is a generator for a very rich class of filter operators, parameterized by direction vector $\boldsymbol{u}$:

$$h(\boldsymbol{x} \,|\, \boldsymbol{u}) * s(\boldsymbol{x}) \stackrel{!}{=} 0 \, .$$

This means that all linear operators that do not let an ideal oriented signal $s(\boldsymbol{x})$ pass,[1] have the structure of (7.23). Since all oriented signals have power spectra that are concentrated on lines or planes in the Fourier domain, we can denote these filters as *oriented nulling filters.*

Like in the case of the normal BCCE, this equation will be satisfied almost never for a real image signal. Thus, we end up with optimization criteria like

$$\int_{\boldsymbol{x}} w(\boldsymbol{x}) \cdot |h(\boldsymbol{x} \,|\, \boldsymbol{u}) * s(\boldsymbol{x})|^2 \, \mathrm{d}\boldsymbol{x} \quad \longrightarrow \quad \min \tag{7.24}$$

where $h(\boldsymbol{x} \,|\, \boldsymbol{u})$ comprises the combination of directional derivatives of different order, and an optional *pre-filter* $p(\boldsymbol{x})$. This means: the *frequency-weighted* and *localized* directional variation of the signal is minimized in the direction of motion.

In standard differential motion estimation schemes, so-called *pre-filters* are used to compensate for the varying precision of the (discrete) derivative filters

[1] *Multiple motions:* The constraint (7.54) generalized to

$$h(\boldsymbol{x} \,|\, \boldsymbol{u}_1) * h(\boldsymbol{x} \,|\, \boldsymbol{u}_2) * s(\boldsymbol{x}) \stackrel{!}{=} 0 \, .$$

for different frequencies, and in particular to obtain isotropic performance of gradient estimates. Standard pre-filter design assumes that the Fourier spectra of the input signal and the noise are both white. For real signals, this is clearly not the case, as can be seen by inspecting the spatio-temporal autocovariance function of video signals. Furthermore, the possibly scenario-dependent distribution of motion vectors significantly controls the temporal part of the autocovariance function of the signal (see [57, 66]). This means that such pre-filters should be designed in consideration of the actual autocovariance function.

We will now generalize the concept of the *structure tensor* in consideration of (7.23), building upon a wider interpretation of the directional derivative operator.

We proceed by restating the relation between directional derivatives and steerable filters, which have been explored e.g. in [19, 26, 110]. The partial derivative in a direction specified by a unit vector $\boldsymbol{e}_r \in \mathbb{R}^3$ parameterized via spherical angles $\theta = (\theta_1, \theta_2)$ as

$$\boldsymbol{e}_r = (a_1(\theta),\, a_2(\theta),\, a_3(\theta))$$

is given by

$$\frac{\partial}{\partial \boldsymbol{e}_r} s(\boldsymbol{x}) \;=\; \boldsymbol{e}_r^T \cdot \boldsymbol{g}(x) \;=\; \boldsymbol{e}_r^T \cdot \nabla s(\boldsymbol{x}) \;=\; \sum_{i=1}^{3} a_i(\theta) \cdot \frac{\partial s(\boldsymbol{x})}{\partial x_i}\,. \qquad (7.25)$$

Following the reasoning on pre-filters presented in Sect. 7.3.2, we may insert a prefilter $p(\boldsymbol{x})$ (see e.g. [98, 110])

$$\frac{\partial}{\partial \boldsymbol{e}_r}(s(\boldsymbol{x}) * p(\boldsymbol{x})) \;=\; \left(\sum_{i=1}^{3} a_i(\theta_1, \theta_2) \cdot \left(\frac{\partial}{\partial x_i} p(\boldsymbol{x}) \right) \right) * s(\boldsymbol{x})\,. \qquad (7.26)$$

For $p(\boldsymbol{x})$ there are, therefore, many more functions under consideration than only a simple Gaussian kernel.[2] We can design $p(\boldsymbol{x})$ in a way that optimizes the signal/noise ratio at the output of the prefilter; this is (again) the *Wiener-type prefilter approach* [57, 68]. On the other hand, we may generalize the structure of the analysis scheme described by (7.26) and arrive at a *generalized class of structure tensors*, as will be shown in the following.

Steerable Oriented Signal Energy Determination

We abstract now from derivative filters and regard a family of steerable filter operators which can be written in the form [26]

$$h(\boldsymbol{x} \,|\, \theta) = \sum_{i=1}^{N} a_i(\theta) \cdot b_i(\boldsymbol{x})\,.$$

[2] In general, a binomial filter does much better than a sampled (and truncated) Gaussian.

Since the original signal is *sheared* instead of being rotated by motion, it is appropriate to design $h(\boldsymbol{x} \mid \theta)$ accordingly, however, we will not deal here with details of such *shearable filters*. The symbol θ stands for a general parameter (or parameter vector) that controls the direction in which the filter operator is being steered. The $b_i(\boldsymbol{x})$ are basis functions, $a_i(\theta)$ and $b_i(\boldsymbol{x})$ are subject to certain conditions discussed in [26]. This operator will now be applied to an input signal $s(\boldsymbol{x})$:

$$h(\boldsymbol{x} \mid \theta) * s(\boldsymbol{x}) = \sum_{i=1}^{N} a_i(\theta) \cdot (b_i(\boldsymbol{x}) * s(\boldsymbol{x})) \,.$$

As before, the local energy of the resulting signal will be computed. The localization of the computation is again ensured by the weight function $w(\boldsymbol{x})$:

$$Q(\theta) = \int_{\boldsymbol{x}} w(\boldsymbol{x}) \cdot (h(\boldsymbol{x} \mid \theta) * s(\boldsymbol{x}))^2 \, \mathrm{d}\boldsymbol{x}$$

A closer look reveals (using $g_i(\boldsymbol{x}) \equiv s(\boldsymbol{x}) * b_i(\boldsymbol{x})$):

$$\begin{aligned}
(h(\boldsymbol{x} \mid \theta) * s(\boldsymbol{x}))^2 &= \left(\sum_{i=1}^{N} a_i(\theta) \cdot (b_i(\boldsymbol{x}) * s(\boldsymbol{x})) \right)^2 = \left(\sum_{i=1}^{N} a_i(\theta) \cdot g_i(\boldsymbol{x}) \right)^2 \\
&= \sum_{i=1}^{N} \sum_{k=1}^{N} a_i(\theta) \cdot a_k(\theta) \cdot g_i(\boldsymbol{x}) \cdot g_k(\boldsymbol{x}) \,.
\end{aligned}$$

If now a local integration is performed across this squared signal, we obtain:

$$Q(\theta) = \sum_{i=1}^{N} \sum_{k=1}^{N} a_i(\theta) \cdot a_k(\theta) \int_{\boldsymbol{x}} w(\boldsymbol{x}) \cdot g_i(\boldsymbol{x}) \cdot g_k(\boldsymbol{x}) \, \mathrm{d}\boldsymbol{x} \,.$$

With the shorthand notation

$$J_{ik} \stackrel{\text{def}}{=} \int_{\boldsymbol{x}} w(\boldsymbol{x}) \cdot g_i(\boldsymbol{x}) \cdot g_k(\boldsymbol{x}) \, \mathrm{d}\boldsymbol{x}$$

we obtain $$Q(\theta) = \sum_{i=1}^{N} \sum_{k=1}^{N} a_i(\theta) \cdot a_k(\theta) \cdot J_{ik}$$

This is a quadratic form

$$Q(\theta) = \begin{pmatrix} a_1(\theta) \\ \vdots \\ a_N(\theta) \end{pmatrix}^T \begin{pmatrix} J_{11} & \cdots & J_{1N} \\ \vdots & \ddots & \vdots \\ J_{N1} & \cdots & J_{NN} \end{pmatrix} \begin{pmatrix} a_1(\theta) \\ \vdots \\ a_N(\theta) \end{pmatrix} = \boldsymbol{a}^T(\theta) \cdot \mathbf{J} \cdot \boldsymbol{a}(\theta)$$

$$\text{with} \quad \boldsymbol{a}(\theta) \stackrel{\text{def}}{=} \begin{pmatrix} a_1(\theta) \\ \vdots \\ a_N(\theta) \end{pmatrix} \quad \text{and} \quad \mathbf{J} \stackrel{\text{def}}{=} \begin{pmatrix} J_{11} & \cdots & J_{1N} \\ \vdots & \ddots & \vdots \\ J_{N1} & \cdots & J_{NN} \end{pmatrix} .$$

In the standard structure tensor approach, $N = 3$, and $h(\boldsymbol{x} \,|\, \theta)$ is the first order directional derivative which can be represented by a steerable set of $N = 3$ filters (each of them representing the directional derivative in one of the principal directions of space-time). It is not very surprising that in this case $\boldsymbol{a}(\theta)$ is a unit vector in $\mathbb{R}^3$, and the determination of the argument θ which minimizes $Q(\theta)$ boils down to a simple eigensystem problem, as given already in Sect. 7.3.1.

For synthesizing and steering a more general filter operator $h(\boldsymbol{x} \,|\, \theta)$, we know that the basis functions $b_i(\boldsymbol{x})$ should be polar-separable harmonic functions. The coefficient functions $a_i(\theta)$ will then be trigonometric functions of different (harmonic) frequencies [134], and the optimization problem will not be so simple to solve, though well-behaved. The design of the localization function $w(\boldsymbol{x})$ and the generalization of the directional derivative can be adapted to the signal and noise power spectra, respectively [57, 68]. Within this framework, a wide class of orientation selective steerable filters can be used to find principal orientations, if necessary they can equipped with a much more pronounced selectivity, offering the potential for higher accuracy.

7.3.3 A Mixed OLS–TLS Estimator

Local estimators of motion pool a constraint equation in a local neighborhood constructing an overdetermined system of equations. The parameters of the motion model can then be solved by ordinary least squares (OLS) [62] or by total least squares (TLS), resulting in the structure tensor approach. Using (OLS) techniques, the temporal derivatives are treated as erroneous observations and the spatial gradients as error free. This approach will lead to biases in the estimates, as all gradients are generally obscured by noise [44]. Under these circumstances the use of a total least squares (TLS) method [124] is the estimator of choice [85]. A number of physically induced brightness changes as well as those caused by inhomogeneous illumination can be modeled quite accurately by a source term in the constraint equation. Additionally does the computation of surface motion from range data lead to the same type of constraints [112]. The equation of motion is given by

$$\alpha(\boldsymbol{u}) \circ \boldsymbol{s} - 1 \cdot c = 0 \,, \tag{7.27}$$

where c is a constant, modeling the local brightness change linearly.

The data matrix of such a model for the TLS estimator contains a column of exactly known elements (the elements $D_{i,1} = -1$ for $i \in \{1, \ldots, n\}$ where n is the number of pixel in the local neighborhood) thus inducing a strong bias in the estimation. This bias can be efficiently eliminated by mixing the OLS and TLS estimator as presented by [28].

The data matrix $\boldsymbol{D}$ can be split into two submatrices $\boldsymbol{D} = [\boldsymbol{D}_1, \boldsymbol{D}_2]$, where $\boldsymbol{D}_1$ contains the p_1 exactly known observations. A QR factorization of $\boldsymbol{D}$ is performed, leading to

$$(\boldsymbol{D}_1, \boldsymbol{D}_2) = \boldsymbol{Q} \begin{pmatrix} \boldsymbol{R}_{11} & \boldsymbol{R}_{12} \\ \boldsymbol{0} & \boldsymbol{R}_{22} \end{pmatrix},$$

with $\boldsymbol{Q}$ being orthogonal and $\boldsymbol{R}_{11}$ upper triangular. The QR factorization is justified because the singular vectors and singular values of a matrix are not changed by multiplying it by an orthogonal matrix [37].

The solution for the sub system of equations $\boldsymbol{R}_{22}\boldsymbol{p}_2 = 0$ is computed in a TLS sense, which boils down to a singular value analysis of the data matrix $\boldsymbol{R}_{22}$ [124].

With the known estimate of $\boldsymbol{p}_2$ the system of equations $\boldsymbol{R}_{11}\boldsymbol{p}_1 + \boldsymbol{R}_{12}\boldsymbol{p}_2 = 0$ is solved for $\boldsymbol{p}_1$ by back-substitution. A comparative analysis has shown that the error in the mixed OLS–TLS estimates can be reduced by a factor of three as compared to standard TLS [28].

7.3.4 Simultaneous Estimation of Local and Global Parameters

Local estimation schemes, like all estimation schemes presented so far, e.g., the structure tensor method (see Sect. 7.3.1) or mixed OLS–TLS scheme (see Sect. 7.3.3), can be implemented efficiently in terms of storage needed and CPU time used. This is due to local formulation of the models and their parameters, because then all estimations can be done separately for each pixel neighborhood. In other words the model equation matrix is a block diagonal matrix with one block per pixel and one block after the other is processed. This is no longer true if global parameters have to be estimated as well. They introduce additional full rows in the model matrix, thus coupling all blocks. Thus the optimization problem occurring in the estimation process has to be treated as a large scale problem and can only be solved for practical applications if the problem structure is carefully exploited. In [17] an OLS estimation method is presented for simultaneous estimation of local and global parameters which full exploits the structure of the estimation matrix. It has comparable complexity and memory requirements as pure local methods. The numerical solution method makes use of the so called *Sherman–Morrison–Formula* [38], which allows to efficiently obtain the inverse of an easily invertible matrix (the block diagonal matrix) when it is modified by a low rank matrix (the few full rows).

Ordinary least squares (OLS) means the minimization problem is posed as

$$\min_{\boldsymbol{x}} \|\boldsymbol{A}\boldsymbol{x} - \boldsymbol{b}\|_2^2 \quad \Rightarrow \quad \bar{\boldsymbol{x}} = (A^T A)^{-1} A^T b\,, \tag{7.28}$$

where $\boldsymbol{x}$ are the sought for parameters, $\boldsymbol{A}$ and $\boldsymbol{b}$ are defined by the model. The solution vector $\bar{\boldsymbol{x}}$ is given by the Moore-Penrose pseudo-inverse $(A^T A)^{-1} A^T$ if $A^T A$ is invertible. The matrix A has the following block structure

$$A = \left[\begin{array}{ccc|c} B_1 & & & V_1 \\ & \ddots & & \vdots \\ & & B_N & V_N \end{array}\right] = \left[B \middle| V\right] \quad \text{and thus} \quad A^T A = \left[\begin{array}{c|c} B^T B & B^T V \\ \hline V^T B & V^T V \end{array}\right]$$

with $n_\Omega \times N_{\text{lp}}$-blocks B_i and $n_\Omega \times N_{\text{gp}}$-blocks V_i. Finally, the squared matrix can be decomposed as

$$A^T A = M + \text{RSR}^T \text{ with } M = \left[\begin{array}{c|c} B^T B & 0 \\ \hline 0 & V^T V \end{array}\right], R = \left[\begin{array}{c|c} B^T V & 0 \\ \hline 0 & \mathbf{I} \end{array}\right], S = \left[\begin{array}{c|c} 0 & \mathbf{I} \\ \hline \mathbf{I} & 0 \end{array}\right],$$

with matrix M block diagonal and matrix R low rank, $2N_{\text{gp}}$, so that the Sherman–Morrison–Woodbury formula gives an efficiently computable inverse:

$$(A^T A)^{-1} = (\mathbf{I} - M^{-1} R (S^{-1} + R^T M^{-1} R)^{-1} R^T) M^{-1} .$$

In addition to the matrix blocks $B_i^T B_i$ and $\sum_{i=1}^N V_i^T V_i$ of M one, therefore, only has to invert one further $(2N_{\text{gp}}) \times (2N_{\text{gp}})$ matrix, $(S^{-1} + R^T M^{-1} R)$, and all remaining calculations for computation of $\bar{x} = (A^T A)^{-1} A^T b$ can be performed as matrix vector products. As the inversion of the matrix blocks $B_i^T B_i$ is by far the most time consuming step in the computations of the algorithm, the computational burden is comparable to that of an OLS estimation *without* global parameters.

For an illustrative example of combined estimation of local optical flow and global camera gain as well as for further details on the numerical solution we refer to [17].

7.3.5 Simultaneous Estimation of 3 D Position and 3 D Motion

Modeling dynamic data acquired with a 2 D camera grid, the solution space is a three-dimensional subspace ($n = 5$, $d = 3$, see Sects. 7.2.1 and 7.2.3). Using the weighted TLS estimation scheme presented in Sect. 7.3.1, it is spanned by the three right singular vectors corresponding to the three smallest singular values of the data matrix $\boldsymbol{D}$. From these singular vectors the sought for parameters are derived by linear combination such that all but one component of $\mathrm{d}r_1$, $\mathrm{d}r_2$, $\mathrm{d}t$ vanish. The parameters for disparity surface slopes and depth motion occur twice in the model (see (7.11)). Their estimates cannot be combined straightforward because they are not independent in the original coordinate system. In order to decouple these measurements, we first estimated their error covariance matrix as proposed in [91]. After diagonalizing this matrix via a rotation in r_1–r_2-space (for disparity and surface slopes) or x–y-space (for u_z), we achieve independent measurements which can be combined respecting their individual errors (see [97]). A more detailed description of decoupling the motion estimates is presented in [99].

7.3.6 Motion Estimation with the Energy Operator

For continuous signals $\boldsymbol{s} = \boldsymbol{s}(\boldsymbol{x})$ the energy operator E is defined as in [71]

$$E(\boldsymbol{s}) := D(\boldsymbol{s})^2 - \boldsymbol{s} \cdot D^2(\boldsymbol{s}) \ . \tag{7.29}$$

Here D denotes an abstract derivative or pseudo-derivative operator.

Implementation for Optic Flow Estimation

We implemented the energy operator in three dimensions for optic flow estimation based on three-dimensional derivative filters in the spatio-temporal domain in a straightforward manner using (7.29) by replacing D^2 with the Hessian of the sequence. We used the modified Sobel filters as described in [101] and [98]. For comparison between different implementations, the Hessian of the spatio-temporal image data was computed either by twice applying first order derivative filters or second order derivative masks.

Numerical Experiments

In our experiments we computed the energy operator on original resolution. We conducted measurements of the average angular error for a synthetic sequence without noise, for synthetic sequences with noise and a real world test sequence acquired by a camera.

Since we estimated image sequences with ground truth, we compared the best results of a total least squares local approach for the energy operator and the structure tensor. For results, see Table 7.1.

Table 7.1 Results for optic flow estimation with the energy tensor and comparison with the structure tensor

Sequence	Derivative filter	Optimal integration scale	Average angular error
Sinus pattern	**Structure tensor**	0.24	4.589143
	energy operator by ...		
	first order derivative	3.54	10.2803
	second order derivative	3.43	11.59
Marble	**Structure tensor**	3.2	3.509298
	energy operator by ...		
	first order derivative	2.59	3.215193
	second order derivative	6.02	4.498214
Street	**Structure tensor**	1.57	4.589143
	energy operator by ...		
	first order derivative	6.1	10.251023
	second order derivative	5.4	9.240537

On the Effect of the Bandpass Filtering: Filter Bandwidth Versus Wavelet Scale

We investigated the dependency of the average angular error as a function of the bandwidth and wavelet scale simultaneously. Here we mean the bandwidth of the bandpass filter or the spread of the wavelet, used to filter the input image sequence. As a result it turns out that there is an optimal point in the bandwidth-scale plane which minimizes the error measure, see also Figs. 7.2 and 7.3.

Despite the preprocessing of the image sequence, our experiments showed that there is a need of post-integration for the energy operator to achieve optimal average angular error of the estimated flow fields. For orientation estimation with the energy operator, [21, page 498] reported similar results and applied a Gaussian post-filtering with $\sigma = 1$ and a smoothing window size of 7×7.

The accuracy gain for the real world Marbled Block sequence due to this post-filtering is approximately 7.7%. This improvement is achieved under optimal parameter setting for both operators, the structure tensor and the energy operator. For the synthetic sequence with discontinuity and the Street sequence we measured higher accuracy with the structure tensor.

The energy operator requires additional computation steps, such as the calculation of the Hessian of the image data. For this reason and because of the measurements in our experiment we recommend the structure tensor for motion estimation in the scale space.

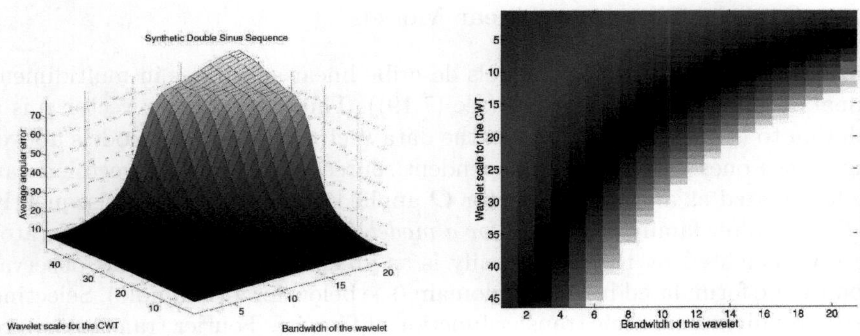

Fig. 7.2 The optimal average angular error as a function of the bandwidth and scale of the Mexican hat wavelet for a synthetic sinus pattern sequence with discontinuity. In the left image the range is set between 6 and 11 degrees, in order to show the region around the minimum

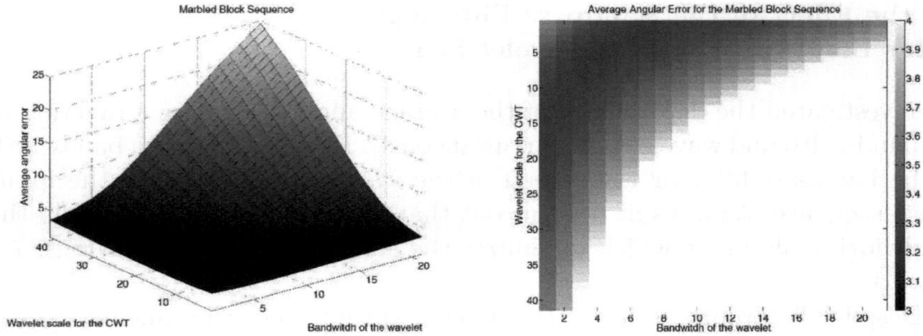

Fig. 7.3 The optimal average angular error as a function of the bandwidth and scale of the Mexican hat wavelet for the Marbled Block sequence. In the left image the range is set between 3 and 4 degrees, in order to show the region around the minimum

7.4 Filter Design

All models described above contain derivative filters of first order, e.g. (7.2), (7.5) or (7.11), second order (7.54) or higher orders, e.g., if brightness changes are due to diffusive processes or more than two orientations are present. They are discretized by finite differences using convolution kernels optimized with respect to the model assumptions, scales and/or noise present in the data.

7.4.1 Optimal Filters for Linear Models

As shown in Sect. 7.2 linear models describe linear subspaces in multidimensional data in the form $\boldsymbol{d}^T \boldsymbol{p} = 0$ (see (7.19)). Thus a parameter vector $\boldsymbol{p}$ is a solution to (7.19) if it is normal to the data vector $\boldsymbol{d}$. In all the models above, except the ones with a data independent source-term, the data vector $\boldsymbol{d}$ can be formulated as an operator vector $\boldsymbol{\mathcal{O}}$ applied to the signal s. Consequently a discrete filter family is *optimal for a model* if the *orientation* of the operator vector calculated by the filter family is as precise as possible. This observation can be formulated in Fourier domain (see below for an example). Selecting fixed size filter sets, their transfer functions (TFs, i.e. Fourier transforms) are known. The coefficients of the whole filter set may then be optimized simultaneously by adapting a TF optimization scheme first presented in [98]. There a weighted 2-norm of the difference of an ansatz function $f_a(\tilde{\boldsymbol{k}})$ and a reference function $f_r(\tilde{\boldsymbol{k}})$ is minimized

$$c(\boldsymbol{h}) = \sqrt{\int w^2(\tilde{\boldsymbol{k}}) \left(f_r(\tilde{\boldsymbol{k}}) - f_a(\tilde{\boldsymbol{k}}, \boldsymbol{h}) \right)^2 \mathrm{d}\tilde{\boldsymbol{k}}} \quad \text{with} \quad \int w^2(\tilde{\boldsymbol{k}}) \mathrm{d}\tilde{\boldsymbol{k}} = 1$$

with normalized wave-numbers $\tilde{\boldsymbol{k}}$, i.e. Nyquist wave-number equal to 1. The normalized weight function $w(\tilde{\boldsymbol{k}})$ allows to specify statistical importance of different wave vectors $\tilde{\boldsymbol{k}}$ and thus allows for scale selection.

Example: Filter Families for Two Transparent Motions

The operator vector $\boldsymbol{\mathcal{O}}$ for transparent motion can be extracted from (7.54)

$$\boldsymbol{\mathcal{O}} = [\partial_{xx}, \partial_{xy}, \partial_{yy}, \partial_{xt}, \partial_{yt}, \partial_{tt}]^T \tag{7.30}$$

and its TF is $\hat{\boldsymbol{\mathcal{O}}} = -\pi^2 \left[\tilde{k}_x^2\,,\, \tilde{k}_x\tilde{k}_y\,,\, \tilde{k}_y^2\,,\, \tilde{k}_x\tilde{k}_t\,,\, \tilde{k}_y\tilde{k}_t\,,\, \tilde{k}_t^2\right]^T$. The reference function, therefore, is $\hat{\boldsymbol{\mathcal{O}}}$ normalized by its length

$$f_r(\tilde{k}) = \frac{[\tilde{k}_x^2\,,\, \tilde{k}_x\tilde{k}_y\,,\, \tilde{k}_y^2\,,\, \tilde{k}_x\tilde{k}_t\,,\, \tilde{k}_y\tilde{k}_t\,,\, \tilde{k}_t^2]^T}{\sqrt{\tilde{k}_x^4 + \tilde{k}_x^2\tilde{k}_y^2 + \tilde{k}_y^4 + \tilde{k}_x^2\tilde{k}_t^2 + \tilde{k}_y^2\tilde{k}_t^2 + \tilde{k}_t^4}}\,.$$

We discretize $\boldsymbol{\mathcal{O}}$ (from (7.30)) using fixed size separable kernels. The filter family consists of only four 1 D filters: a first order derivative D^1, a second order derivative D^2 and two smoothing kernels I^1 and I^2. Please note the upper indices. The 3 D filters are then $\partial_{xy} = \mathcal{D}_x^1 * \mathcal{D}_y^1 * \mathcal{I}_t^1$ and $\partial_{xx} = \mathcal{D}_x^2 * \mathcal{I}_y^2 * \mathcal{I}_t^2$, where $*$ denotes convolution and lower indices denote the application direction. All filters not introduced above, can be derived by suitably exchanging lower indices. All one-dimensional filters are constrained to numerical consistency order 2 (see [98]). We refer to [96] for further details. Using the transfer functions $\hat{\mathcal{D}}_{xx}$, $\hat{\mathcal{D}}_{xy}$, $\hat{\mathcal{D}}_{yy}$, $\hat{\mathcal{D}}_{xt}$, $\hat{\mathcal{D}}_{yt}$, and $\hat{\mathcal{D}}_{tt}$ of these filter kernels, we get the ansatz function

$$f_a(\tilde{k}) = \frac{[\hat{\mathcal{D}}_{xx}, \hat{\mathcal{D}}_{xy}, \hat{\mathcal{D}}_{yy}, \hat{\mathcal{D}}_{xt}, \hat{\mathcal{D}}_{yt}, \hat{\mathcal{D}}_{tt}]^T}{\sqrt{\hat{\mathcal{D}}_{xx}^2 + \hat{\mathcal{D}}_{xy}^2 + \hat{\mathcal{D}}_{yy}^2 + \hat{\mathcal{D}}_{xt}^2 + \hat{\mathcal{D}}_{yt}^2 + \hat{\mathcal{D}}_{tt}^2}}\,.$$

Example for $5 \times 5 \times 5$- filters the optimization results in the kernels ($c = 1.6\mathrm{e}-12$)

$$\begin{aligned}
I^1 &= [0.01504, 0.23301, 0.50390, 0.23301, 0.01504] \\
I^2 &= [0.01554, 0.23204, 0.50484, 0.23204, 0.01554] \\
D^1 &= [0.06368, 0.37263, 0, -0.37263, -0.063681] \\
D^2 &= [0.20786, 0.16854, -0.75282, 0.16854, 0.20786]
\end{aligned}$$

For larger kernels we refer to [96].

7.4.2 Steerable and Quadrature Filters

Quadrature filters have become an appropriate tool for computing the local phase and local energy of one-dimensional signals. In the following, we describe the relation between steerable filter and quadrature filter. For a detailed description on steerable filters we refer to [55].

The main idea of a quadrature filter is to apply two filters to a signal such that the sum of the square filter responses reflect the local energy of the signal. Furthermore, the local energy should be group invariant, i.e. the filter outputs should be invariant with respect to the deformation of the signal by the corresponding group. In order to achieve group invariance, we construct our quadrature filter from the basis of a unitary group representation. Groups with a unitary representation are compact groups and Abelian groups [130]. The even $\boldsymbol{h}_e$ and odd $\boldsymbol{h}_o$ components of the quadrature filter are constructed by a vector valued impulse response consisting of the basis functions of a unitary representation of dimension m_e and m_o, respectively.

$$\boldsymbol{h}_e = \begin{pmatrix} h_{e1}(\boldsymbol{x}) \\ h_{e2}(\boldsymbol{x}) \\ \vdots \\ h_{em_e}(\boldsymbol{x}) \end{pmatrix}, \quad \boldsymbol{h}_o = \begin{pmatrix} h_{o1}(\boldsymbol{x}) \\ h_{o2}(\boldsymbol{x}) \\ \vdots \\ h_{om_o}(\boldsymbol{x}) \end{pmatrix}.$$

It can be shown that all basis functions belonging to the same subspace attain the same parity. The filter responses of $\boldsymbol{h}_e$ and $\boldsymbol{h}_o$ are denoted as the filter channels $\boldsymbol{c}_e = s(\boldsymbol{x}) * \boldsymbol{h}_e(\boldsymbol{x})$ and $\boldsymbol{c}_o = s(\boldsymbol{x}) * \boldsymbol{h}_o(\boldsymbol{x})$, respectively. The square of the filter response of each channel are denoted as even and odd energies. Due to the unitary representation, both energies are invariant under the corresponding group action

$$E_s = (\mathbf{D}(g)\boldsymbol{c}_s)^T (\mathbf{D}(g)\boldsymbol{c}_s) = \boldsymbol{c}_s^T \boldsymbol{c}_s \,, \quad s \in \{e, o\} \,.$$

Note that the inner product is taken with respect to the invariant subspace, not with respect to the function space. The local energy of the signal is given by the sum of the even and odd energy. In the following we will examine the properties of the filter channels required to achieve a phase invariant local energy when applied to bandpass signals. In the ideal case, a simple[3] bandpass filtered signal consists of only one wave vector $\boldsymbol{k}_0$ and its Fourier transform[4] reads with the Dirac delta distribution $\delta(\boldsymbol{k})$

$$S(\boldsymbol{k}) = S_0 \delta(\boldsymbol{k} - \boldsymbol{k}_0) + S_0^* \delta(\boldsymbol{k} + \boldsymbol{k}_0) \,. \tag{7.31}$$

We start with examining the Fourier transform of the even and odd energies

$$E_s = \boldsymbol{c}_s^T \boldsymbol{c}_s = \sum_{j=1}^{m_s} (s(\boldsymbol{x}) * h_{sj}(\boldsymbol{x}))^2 \,. \tag{7.32}$$

Applying the convolution theorem to E_s reads

$$\mathcal{F}\{E_s\}(\boldsymbol{k}) = \sum_{j=1}^{m_s} (S(\boldsymbol{k}) H_{sj}(\boldsymbol{k})) * (S(\boldsymbol{k}) H_{sj}(\boldsymbol{k})) \,.$$

---

[3] simple signal: signal with intrinsic dimension one.

[4] Note that the Fourier transformed entities are labeled with capital letters.

Inserting the signal (7.31) in the equation above, computing the convolution and performing the inverse Fourier transformation results in a phase invariant local energy

$$E = 2|S_0|^2 \left(\sum_{j=1}^{m_e} |H_{ej}(\boldsymbol{k}_0)|^2 + \sum_{k=1}^{m_o} |H_{ok}(\boldsymbol{k}_0)|^2 \right) .$$

Thus, each steerable filter whose basis can be decomposed as described above is a quadrature filter which is invariant with respect to the corresponding group action.

7.4.3 Design and Application of Wiener-Optimized Filters and Average Masks

The filter masks described in Sect. 7.4 have been optimized for ideal noiseless signals. However, the fact that all real world images are with different extents corrupted by noise has thus been neglected. In the following we present how these filters have to be adapted in case of noisy signals. A detailed treatment can be found in [54].

The Signal and Noise Adapted Filter Approach

The signal and noise adapted (SNA)-filter approach is motivated by the fact that we can exchange the directional derivative filter $d_{\boldsymbol{r}}(\boldsymbol{x}_n)$ in the BCCE by any other steerable filter $h_{\boldsymbol{r}}(\boldsymbol{x}_n)$ which only nullifies the signal when applied in the direction of motion [69]. The shape of the frequency spectrum of any *rank 2* signal[5] is a plane $K_{\boldsymbol{r}}$ going through the origin of the Fourier space and its normal vector $\boldsymbol{n}$ points to the direction of motion $\boldsymbol{r}$ ([42], p. 316). Thus, the transfer function[6] $H_{\boldsymbol{r}}(\boldsymbol{f})$ has to be zero in that plane, but the shape of $H_{\boldsymbol{r}}(\boldsymbol{f})$ outside of plane $K_{\boldsymbol{r}}$ can be chosen freely as long as it is not zero at all. If the impulse response $h_{\boldsymbol{r}}(\boldsymbol{x}_n)$ shall be real-valued, the corresponding transfer function $H_{\boldsymbol{r}}(\boldsymbol{f})$ has to be real and symmetric or imaginary and antisymmetric or a linear combination thereof. The additional degrees of freedom to design the shape outside $K_{\boldsymbol{r}}$ make it possible to consider the spectral characteristics of the signal and the noise which are encoded in the second order statistical moments in the filter design. In the following section the derivation of an optimal filter is shown which is a special case of the more general framework presented in [84] and for the special case of motion estimation in [57].

General Model of the Observed Signal

The general idea of the SNA-filter proposed first in [70] is to combine Wiener's theory of optimal filtering with a desired ideal impulse response. The term

[5] The rank of a signal is defined by the rank of the corresponding structure tensor.
[6] The Fourier transforms of functions are denoted here by capital letters.

ideal, in this case, means that the filter is designed for noise free signal $s(\boldsymbol{x})$. But signals are always corrupted by noise. Our goal is now to adapt the ideal filter $h_{\boldsymbol{r}}(\boldsymbol{x})$ to more realistic situations where signal is corrupted by noise. We model the observed image signal z at position i, j, k in a spatio-temporal block of dimension $N \times N \times N$ by the sum of the ideal (noise free) signal s and a noise term v: $z(i,j,k) = s(i,j,k) + v(i,j,k)$. For the subsequent steps it is convenient to arrange the s, v, and z of the block in vectors $\boldsymbol{s} \in \mathbb{R}^M, \boldsymbol{v} \in \mathbb{R}^M$ and $\boldsymbol{z} \in \mathbb{R}^M$. The extraction of a single filter response value $\hat{g}$ can thus be written as the scalar product $\hat{g} = \boldsymbol{x}^T\boldsymbol{z}$ using a filter coefficient vector $\boldsymbol{x} \in \mathbb{R}^M$. The corresponding equation for the actual filter output $\hat{g}$ reads:

$$\hat{g} = \boldsymbol{x}^T\boldsymbol{z} = \boldsymbol{x}^T(\,\boldsymbol{s} + \boldsymbol{v}) = \boldsymbol{x}^T\,\boldsymbol{s} + \boldsymbol{x}^T\boldsymbol{v}\,. \tag{7.33}$$

Our task is to choose $\boldsymbol{x}^T$ in such a way that the filtered output $\hat{g}$ approximates, on an average, the desired output $g = \boldsymbol{h}^T\boldsymbol{s}$ for the error-free case as closely as possible. The next step is to define the statistical properties of the signal and the noise processes, respectively. Let the noise vector $\boldsymbol{v} \in \mathbb{R}^N$ be a zero-mean random vector with covariance matrix $\mathbb{E}\left[\boldsymbol{v}\boldsymbol{v}^T\right] = \mathbf{C}_v$ (which is in this case equal to its correlation matrix $\mathbf{R}_v$). Furthermore, we assume that the process which has generated the signal $\boldsymbol{s} \in \mathbb{R}^N$ can be described by the expectation $\mathbb{E}\left[\boldsymbol{s}\right] = \boldsymbol{m}_s$ of the signal vector, and an autocorrelation matrix $\mathbb{E}\left[\boldsymbol{s}\boldsymbol{s}^T\right] = \mathbf{R}_s$. Our last assumption is that noise and signal are uncorrelated $\mathbb{E}\left[\boldsymbol{s}\boldsymbol{v}^T\right] = 0$.

Designing the Optimized Filter

Knowing these first and second order statistical moments for both the noise as well as the signal allows the derivation of the optimum filter $\boldsymbol{x}$. For this purpose, we define the approximation error $e := \hat{g} - g$ between the ideal output g and the actual output $\hat{g}$. The expected squared error Q as a function of the vector $\boldsymbol{x}$ can be computed from the second order statistical moments:

$$\begin{aligned} Q(\boldsymbol{x}) &= \mathbb{E}\left[e^2\right] = \mathbb{E}\left[\hat{g}^2\right] - 2\mathbb{E}\left[g\hat{g}\right] + \mathbb{E}\left[g^2\right] \\ &= \boldsymbol{h}^T\mathbf{R}_s\boldsymbol{h} - 2\boldsymbol{x}^T\mathbf{R}_s\boldsymbol{h} + \boldsymbol{x}^T(\mathbf{R}_s^T + \mathbf{R}_v)\boldsymbol{x} \end{aligned}$$

We see that a minimum mean squared error (MMSE) estimator can now be designed. We set the derivative $\partial Q(\boldsymbol{x})/\partial\boldsymbol{x}$ to $\mathbf{0}$ and after solving for $\boldsymbol{x}$ we obtain

$$\boldsymbol{x} = \underbrace{(\mathbf{R}_s^T + \mathbf{R}_v)^{-1}\mathbf{R}_s}_{\mathbf{M}}\,\boldsymbol{h} = \mathbf{M}\boldsymbol{h}\,. \tag{7.34}$$

Thus, the desired SNA-filter is obtained by a matrix vector multiplication. The ideal filter designed for the ideal noise free case is multiplied with a matrix $\mathbf{M}$ composed out of the correlation matrices of signal and noise. In principle, this result is a direct consequence of the Gauss–Markov theorem.

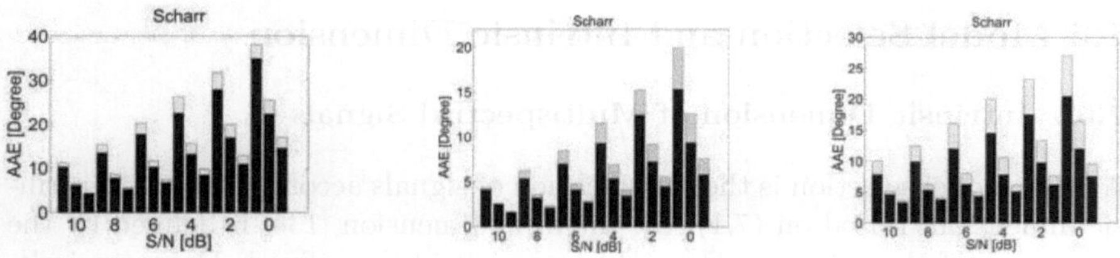

Fig. 7.4 Bar plots of the average angular error (AAE) vs. the signal to noise (S/N) ratio. For every S/N ratio the AAE for three different densities is depicted: From left to right: 86%, 64% and 43%. The gray bar denotes the AAE with the original, the black ones the AAE with the optimized filters. Note that the experiment is performed at rather low S/N ratio in range 10–0 dB

Experimental Results

In this section, we present examples which show the performance of our optimization method. For the test we use three image sequences, together with the true optical flow: "Yosemite" (without clouds) "diverging tree" and "translating tree", sequences.[7] The optical flow has been estimated with the tensor based method described in Sect. 7.3.1. For all experiments, an averaging volume of size $11 \times 11 \times 11$ and filters of size $5 \times 5 \times 5$ are applied. For the weighting function $w(\boldsymbol{x})$ we chose a sampled Gaussian function with width $\sigma = 8$ (in pixels) in all directions. For performance evaluation, the average angular error (AAE) [4] is computed. The AAE is computed by taking the average over 1000 trials with individual noise realization. In order to achieve a fair comparison between the different filters but also between different signal-to-noise ratios S/N, we compute all AAEs for three different but fixed densities[8] determined by applying the *total coherence measure* and the *spatial coherence measure* [42]. We optimized the described in Sect. 7.4 denoted as Scharr filter in the following, for every individual signal to noise ratio S/N in a range from 10 to 0 dB (for i.i.d. noise). We then applied both the original Scharr and its corresponding SNA-filters.

As expected and shown in Fig. 7.4, the SNA-filters yield a better performance than the original non-adapted filters in case the image sequence being corrupted by noise for all types of ideal filters. The performance of the Scharr filter is increased by optimizing it to the corresponding images sequence. We can conclude that for these cases the optimum shape of the filter is mainly determined by the signal and noise characteristics, whereas for higher signal to noise ratios the systematical optimization plays a greater role.

---

[7] The diverging and translating tree sequence has been taken from Barron's web-site and the Yosemite sequence from *http://www.cs.brown.edu/people/black/images.html*

[8] The density of an optical flow field is defined as the percentage of estimated flow vectors which have been used for computing the AAE with respect to all estimated flow vectors.

7.5 Model Selection and Intrinsic Dimension

7.5.1 Intrinsic Dimension of Multispectral Signals

A basic model selection is the classification of signals according to their intrinsic dimension. Based on (7.1), the intrinsic dimension [135] is defined by the dimension of the subspace E to which the signal is confined. More precisely, the intrinsic dimension of an n-dimensional signal s is $n - d$ if s satisfies the constraint in (7.1) [74, 81]. Therefore, when estimating subspaces, it is essential to know the intrinsic dimension of the signal.

Furthermore, the intrinsic dimension is relevant to image and video coding due to the predominance of regions with intrinsic dimension 0 and 1 in natural images [136], and the fact that images and videos are fully determined by the regions with intrinsic dimension 2, i.e., the whole image information is contained in the 2 D regions [5, 72].

The intrinsic dimension of scalar images can be estimated with differential methods based, for example, on the structure tensor, the Hessian, and the energy tensor. All these methods have been generalized to vector-valued (multispectral) images in [81]. A further extension is based on the concept of fractional intrinsic dimension for combinations of subspaces [77], e.g., multiple-motion layers. More general approaches for estimating the intrinsic dimension are based on the compensation principle [135] and the Volterra–Wiener theory of nonlinear systems [60].

7.5.2 Model Selection by Residual Analysis

In local optical flow estimation under parametric model assumptions, the parameters of optical flow models are allowed to depend on non-trivial subsets of the spatiotemporal volume. The exploitation of the full potential of this approach involves the problem of selecting appropriate motion models for each of these subsets. While a simple optical flow model fails to approximate data of higher intrinsic complexity under low noise conditions, a complex model is prone to over-fitting in the presence of noise. Various information criteria (e.g. AIC [2], BIC [107]) have been proposed that penalize model complexity in order to avoid over-fitting. In the context of motion estimation, the model selection problem has been discussed by Wechsler et al. [128] as well as by Gheissari et al. [36]. Gheissari et al. point out that "[...] the available information theoretic model selection criteria are based on the assumptions that noise is very small and the data size is large enough" and that this assumption is often violated in computer vision applications [36]. Hence, they suggest to consider the constraint surfaces of parametric models as thin plates and to penalize the strain energy of these plates according to a physical model. As this penalization incorporates only second order derivatives of the model surfaces, it cannot be used to distinguish different linear models. Moreover, if information on the distribution of noise is available from camera calibration measurements or noise estimation, probabilistic model selection criteria that incorporate this information should be employed. Residual analysis fills the gap between information theoretic penalization and heuristic surface modeling. Following the general idea of Cootes et al. [15], we suggest in [3] to assess parametric optical flow models by measuring the discrepancy between the empirical distribution of regression residuals and the pdf predicted from theory. The additive Errors-in-Variables (EIV) model claims the existence of a true signal $\tau : \Omega \to \mathbb{R}$ and, for all $\boldsymbol{x} \in \Omega$, a random variable $\varepsilon(\boldsymbol{x})$ (noise) such that

$$\forall \boldsymbol{x} \in \Omega : s(\boldsymbol{x}) = \tau(\boldsymbol{x}) + \varepsilon(\boldsymbol{x}) . \tag{7.35}$$

Optical flow estimation under the assumption of the BCCE is performed on the partial derivatives of the signal which are approximated by linear shift invariant operators. The overlap of the finite impulse response masks of these operators in the computation of derivatives at nearby pixels introduces correlation to the entries of the data matrix D_s and data term $\boldsymbol{d}_s$ used in TLS estimation. As these entries are linear in the derivatives, they can be decomposed with respect to (7.35) into

$$D_s(\boldsymbol{x}) = D_\tau(\boldsymbol{x}) + D_\varepsilon(\boldsymbol{x}) \quad \text{and} \quad \boldsymbol{d}_s(\boldsymbol{x}) = \boldsymbol{d}_\tau(\boldsymbol{x}) + \boldsymbol{d}_\varepsilon(\boldsymbol{x}) . \tag{7.36}$$

Equilibration as proposed by Mühlich [82] is used to derive from the covariance matrices of the vectors $\text{vec}([D_s(\boldsymbol{x}), \boldsymbol{d}_s(\boldsymbol{x})])$ (column-wise vectorization of the matrix $[D_s(\boldsymbol{x}), \boldsymbol{b}_s(\boldsymbol{x})]$) square equilibration matrices $W_L(\boldsymbol{x})$ and $W_R(\boldsymbol{x})$ to estimate $\hat{\boldsymbol{p}}(\boldsymbol{x})$ by TLS on the data $W_L(\boldsymbol{x})[D_s(\boldsymbol{x}), \boldsymbol{d}_s(\boldsymbol{x})]W_R^T(\boldsymbol{x})$ instead of $[D_s(\boldsymbol{x}), \boldsymbol{d}_s(\boldsymbol{x})]$. $W_R^T(\boldsymbol{x})\hat{\boldsymbol{p}}(\boldsymbol{x})$ is then taken as an estimate of the initial problem. If the distribution of noise in the signal is known, regression residuals can be tested to be in accordance with the theoretically expected distribution. Given ETLS estimates $\hat{\boldsymbol{p}} : \Omega \to \mathbb{R}^k$ (k being the number of model parameters), the residuals are given by the mapping $\hat{\boldsymbol{r}} : \Omega \to \mathbb{R}^m$ (m being the number of pixels in the neighborhood for which a consistent model is assumed) such that

$$\hat{\boldsymbol{r}} := W_L[D_s, \boldsymbol{d}_s]W_R^T \begin{pmatrix} \hat{\boldsymbol{p}} \\ -1 \end{pmatrix} . \tag{7.37}$$

In principle, the theoretical pdf of these residuals is determined by the joint pdf of the entries of D_s and $\boldsymbol{d}_s$. The latter is obtained from the EIV model, the motion models, and the derivative operators. However, there is a direct influence to the residual pdf by the factor $[D_s, \boldsymbol{d}_s]$ as well as an indirect influence by the pdf of the estimates $\hat{\boldsymbol{p}}$. In the following, we assume $\hat{\boldsymbol{p}}$ to be deterministic. Then, the residuals (7.37), expressed as

$$\forall \boldsymbol{x} \in \Omega : \quad \hat{\boldsymbol{r}} = \underbrace{\left(\begin{pmatrix} \hat{\boldsymbol{p}} \\ -1 \end{pmatrix}^T W_R \otimes W_L \right)}_{=: R} \text{vec}([D_s, \boldsymbol{d}_s]) ,$$

are obtained from the deterministic linear mapping defined by the matrix R, applied to the vector $\text{vec}([D_s, \boldsymbol{d}_s])$ of which the covariance matrix C is known. The covariance matrices of the residual vectors are therefore given by $C_r := \mathrm{cov}(\hat{\boldsymbol{r}}) = RCR^T$. From the Cholesky factorization $LL^T = C_r$ follows that $\hat{\boldsymbol{r}}' := L^{-1}\hat{\boldsymbol{r}}$ is decorrelated, i.e.,

$$\mathrm{cov}(\hat{\boldsymbol{r}}') = 1\!\mathrm{l}_m \,, \tag{7.38}$$

while $\mathbb{E}(\hat{\boldsymbol{r}}') = L^{-1}R\mathbb{E}(\mathrm{vec}([D_s, \boldsymbol{d}_s]))$. From (7.36) follows $\mathbb{E}(\mathrm{vec}([D_s, \boldsymbol{d}_s])) = \mathrm{vec}([D_\tau, \boldsymbol{d}_\tau]) + \mathbb{E}(\mathrm{vec}([D_\varepsilon, \boldsymbol{d}_\varepsilon]))$. Under the assumption that the entries of $[D_\varepsilon, \boldsymbol{d}_\varepsilon]$ have zero mean, it follows

$$\mathbb{E}(\hat{\boldsymbol{r}}') = L^{-1}W_L[D_\tau, \boldsymbol{d}_\tau]W_R^T \begin{pmatrix} \hat{\boldsymbol{p}} \\ -1 \end{pmatrix} . \tag{7.39}$$

In practice, it depends on the appropriateness of the parametric model as well as on the empirical distribution of noise whether or not

$$[D_\tau, \boldsymbol{d}_\tau]W_R^T \begin{pmatrix} \hat{\boldsymbol{p}} \\ -1 \end{pmatrix} = 0 \tag{7.40}$$

holds, in which case (7.39) implies that

$$\mathbb{E}(\hat{\boldsymbol{r}}') = 0 \,. \tag{7.41}$$

If, in addition, noise is i.i.d. according to a normal distribution with known variance then, it follows from (7.38) and (7.41) that the entries of the decorrelated residual vector $\hat{\boldsymbol{r}}'$ from ETLS estimation form a set of independent standard normally distributed random variables. We therefore suggest to test this set of residuals to be standard normally distributed. Deviations from the standard normal distribution are then taken as indications of inappropriateness of the motion model. Testing for this deviation is performed by the Kolmogorov–Smirnov test, Pearson's χ^2 test, the Anderson–Darling test as well as the absolute difference of the vectors of the first j moments of the empirical and theoretical distribution. In order to specifically test for properties of the model selector, we generated a variety of sequences from given two-dimensional displacement fields by warping of an initial frame. Grayvalue structure on multiple scales was introduced to this frame in order to avoid the aperture problem. Zero mean Gaussian noise was added to the sequences. Results of model selection from the models LC (locally constant), LSS (local shear/stretch), LRD (local rotation/divergence), LAF (local affine) and LPL (local planar) are shown in Fig. 7.5 for a sequence featuring motion patterns of different parametric form (top) as well as for a simulated continuous current (bottom). From the different shading in Fig. 7.5b, it can be seen that model selection is in accordance with the true displacement field. Motion patterns are identified correctly. The incidental choice of overly complex models is explained by the fact that a higher order model with the additional parameters correctly estimated as zero cannot be distinguished from the simpler model by means of residual analysis. The most complex model is correctly selected at motion discontinuities.

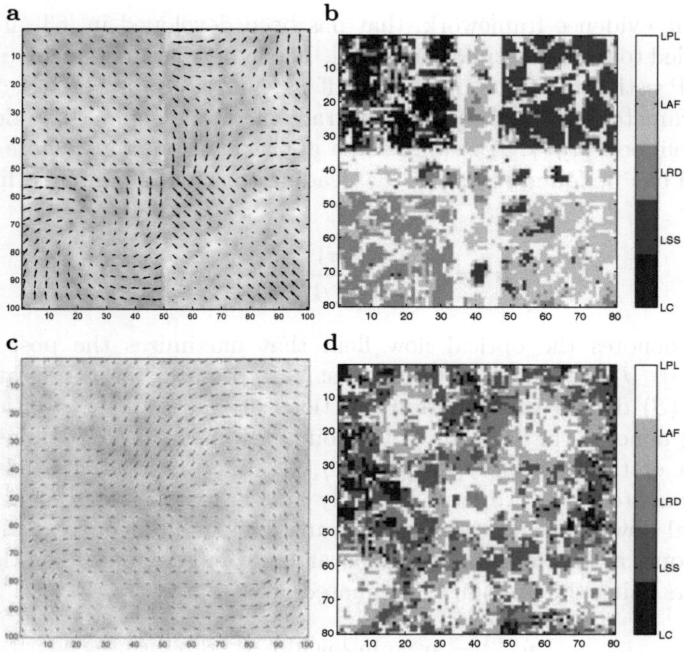

Fig. 7.5 Model selection from 11 × 11 × 3 motion neighborhoods of simulated sequences at 0.5% noise-to-signal amplitude ratio by comparison of 5 moments of the residual distribution. (**a**) displacement field of the types sequence, (**b**) according model selection, (**c**) displacement field of the current sequence, (**d**) according model selection

7.5.3 Bayesian Model Selection

In the following we consider energy functionals of the form

$$E(\boldsymbol{u}) = \int \gamma_1 \psi_1(\alpha_1(\boldsymbol{u})s) + \gamma_2 \psi_2(\alpha_2(\boldsymbol{u})s) + \beta \psi_3(|\nabla \boldsymbol{u}|^2)\, \mathrm{d}\Omega\,, \qquad (7.42)$$

where Ω is the image domain over which integration takes place. Here, $\psi_i(\alpha_i(\boldsymbol{u})s)$, $i = 1, 2$ are different data terms and we define $|\nabla \boldsymbol{u}|^2 = |\nabla u_1|^2 + |\nabla u_2|^2$. The goal is to determine all hyper-parameters $\boldsymbol{\gamma} = (\gamma_1, \gamma_2), \beta$ directly from the data, i.e., the relative weights of the data term as well as the regularization term are to be estimated simultaneously with the optical flow. For a detailed description of the method we refer to [53].

Thus, different data models are selected for the given data set. For estimating the hyper-parameter we explore the well known relation between variational and probabilistic methods, i.e., each (discrete approximated) energy functional can be interpreted as the energy of the posterior pdf $p(\boldsymbol{u}|\boldsymbol{g}, \boldsymbol{\gamma}, \beta)$ of the optical flow. The main idea to estimate the hyper-parameters are to explore the evidence framework, that has been developed in [63] and firstly been applied to motion estimation in [59]. The hyper-parameters are estimated by a MAP estimate of the posterior pdf $p(\gamma, \beta|\boldsymbol{g}) \propto p(\boldsymbol{g}|\gamma, \beta)p(\gamma, \beta)$ of the hyper-parameters given the observed gradient of the signal. The derivation of the likelihood $p(\boldsymbol{g}|\gamma, \beta)$ require some approximation steps whose detailed derivation can be found in [53, 56, 58] leading to the approximated likelihood function

$$p(\boldsymbol{g}|\gamma, \beta, \hat{\boldsymbol{u}}) = \frac{(2\pi)^N}{Z_L(\gamma) Z_p(\beta) \det \mathbf{Q}^{\frac{1}{2}}} \exp\left(-\hat{J}\right) ,$$

where $\hat{\boldsymbol{u}}$ denotes the optical flow field that maximizes the posterior pdf $p(\boldsymbol{u}|\boldsymbol{g}, \gamma, \beta)$, $\hat{J}$ the energy of the joint pdf $p(\boldsymbol{g}, \boldsymbol{u}|\gamma, \beta)$ taken at $\hat{\boldsymbol{u}}$ and $Z_L(\gamma), Z_p(\beta)$ denote the partition functions of Gaussian distributions. The matrix $\mathbf{Q}$ denotes the Hessian of the joint pdf energy $J(\boldsymbol{u}, \boldsymbol{g})$ taken at the maximum of the posterior pdf $p(\boldsymbol{u}|\boldsymbol{g}, \gamma, \beta)$. Since $\hat{\boldsymbol{u}}$ itself depends on the hyper-parameters γ, β we have to apply an iterative scheme for estimating the optical flow field and the hyper-parameters simultaneously, i.e., we estimate the optical flow for fixed hyper-parameters and estimate then the hyper-parameters using the previously estimated optical flow.

$$\begin{aligned} \boldsymbol{u}^{k+1} &= \arg\min_{\boldsymbol{u}} \left\{ p(\boldsymbol{u}|\boldsymbol{g}, \hat{\gamma}^k, \beta^k) \right\} \\ \gamma^{k+1} &= \arg\max_{\gamma} \left\{ p(\gamma, \beta^k|\boldsymbol{g}) \right\} \\ \beta^{k+1} &= \arg\max_{\beta} \left\{ p(\gamma^k, \beta|\boldsymbol{g}) \right\} . \end{aligned}$$

This procedure is repeated until convergence (Figure 7.6 shows some experimental results).

7.6 Anisotropic Regularization for Orientation Estimation

So far, we focused on modeling, model selection and estimation of parameters in multidimensional signals. A major source of estimation inaccuracy is due to the data being noisy, or corrupted, and thus not fulfilling the constraint equations selected. As long as the data belongs to the same population or distribution, one can reduce the influence of noise by smoothing, i.e., grouping measurements belonging together. For single orientation or single motion data this can be done by anisotropic diffusion with a diffusion tensor. A suitable diffusion process is described in Sect. 7.6.1. As parameter selection is non-trivial, a learning approach interpreting the data as a Markovian random field may be advantageous (see Sect. 7.6.2). This approach does not end in anisotropic diffusion in the same way as it is usually applied for data denoising. Anisotropic diffusion with a diffusion tensor can be derived from a cost functional where

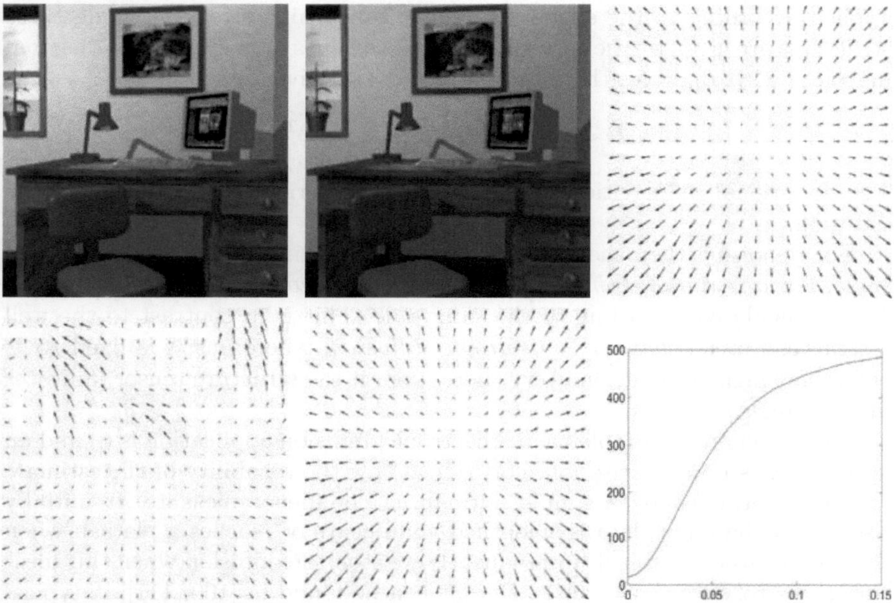

Fig. 7.6 Upper figures (from left to right): first frame of the "Office" sequence; second frame of the "Office sequence" with a brightness decay; estimated flow field using $\gamma_1 = 1$ and $\gamma_2 = 0$; Lower figures (from left to right): estimated flow field using $\gamma_1 = 1$ and $\gamma_2 = 0$; estimated flow field using the Bayesian model selection approach for; ratio of both estimated likelihood hyper-parameters γ_2/γ_1 vs. the gradient of the brightness change

the *expectation* of the motion constraint (7.5). This is not only of theoretic interest and opens a connection to stochastic partial differential equations, but also allows to construct diffusion-like denoising schemes for other (linear) models, e.g., two transparently overlaid motions (see Sect. 7.6.3).

7.6.1 Flow Adapted Anisotropic Diffusion

Anisotropic diffusion filtering evolves the acquired, noisy initial multi-dimensional signal $s(\boldsymbol{x}, 0)$ via an evolution equation:

$$\frac{\partial s}{\partial t} = \nabla \cdot (D \nabla s) \,. \tag{7.43}$$

Here D is the diffusion tensor, a positive definite symmetric matrix and $s(\boldsymbol{x}, t)$ is the evolving signal. Here t is diffusion time not to be confused with the time coordinate x_3 if s is an image sequence. As we will see in Sect. 7.6.3 this diffusion is appropriate for signals with up to single orientation only.

The diffusion tensor D usually applied in anisotropic diffusion uses the same eigenvectors e_i as the structure tensor J_ρ (see (7.21)) constructed for single orientation (7.2) or single motion constraint (7.5). Thus smoothing is applied according to the signal structures. Smoothing strengths along these structures are given by the corresponding eigenvalues λ_i of D. Given a diagonal matrix L with $L_{ii} = \lambda_i$, the diffusion tensor D is given by

$$D = (\boldsymbol{e}_1, \boldsymbol{e}_2, \boldsymbol{e}_3)\, L\, (\boldsymbol{e}_1, \boldsymbol{e}_2, \boldsymbol{e}_3)^T \,. \tag{7.44}$$

The directional diffusivities $\lambda_i, i \in \{1, \ldots, n\}$ determine the behavior of the diffusion.

For denoising they shall be large for small eigenvalues μ_i of J_ρ and vice versa. For *orientation-enhancing anisotropic diffusion* introduced in [100], all λ_i are calculated according to the same rule. This is in contrast to the well established *edge-enhancing anisotropic diffusion*, where one λ is fixed to 1, enforcing single orientation everywhere, even if the structure tensor indicates model violation.

One of the major problems in anisotropic diffusion application is to find an appropriate stopping criterion. For optical flow the reliability of the estimate can be determined by a simple normalized confidence measure. It initially rises when the input data is smoothed and decays or reaches a plateau when the data is over-smoothed. Consequently we stop the diffusion when this measure stops to raise significantly. Details can be found in [100, 111]. A typical smoothing result is shown in Fig. 7.7.

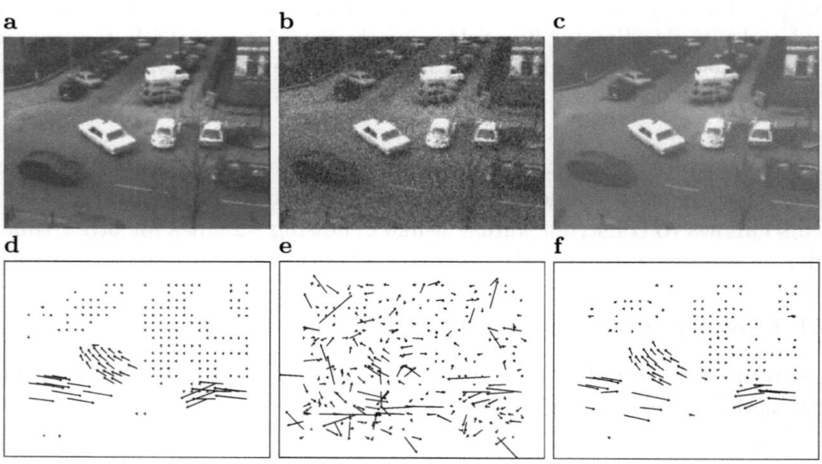

Fig. 7.7 Hamburg taxi sequence: **(a)** one frame in the original sequence, **(b)** the same frame with noise (std. dev. $\sigma = 20$) added and **(c)** reconstructed frame. **(d)** Velocity field on the original data, **(e)** on the noisy data and **(f)** on the reconstructed data

7.6.2 Anisotropic Diffusion as Maximum Likelihood Estimator

A crucial ingredient to successfully use anisotropic diffusion for image denoising is the appropriate selection of directional diffusivities λ_i (see (7.44)). In [102] it turned out that isotropic nonlinear diffusion and anisotropic diffusion correspond to isotropic and directional statistical models, respectively. Having training data at hand, as, e.g., noisy images f_i and noise-free images g_i derived in different operation modes of a focused-ion-beam tool (see Fig. 7.8), image statistics are computable as histograms (see Fig. 7.9).

The problem of recovering the image g from f can then be posed as the maximization of

$$p(g|f) \propto \prod_i \left(p(f_i|g_i) \prod_{j=1}^{J} p(\boldsymbol{n}_j \nabla g_i) \right) \qquad (7.45)$$

where $p(g|f)$ approximates the posterior probability of the image g conditioned on the observed, noisy, image f. The likelihood term $p(f_i|g_i)$, at every pixel, i, is defined by noise statistics. The spatial prior term exploits a Markov Random Field assumption [35] which defines the prior in terms of local neighborhood properties. Here it is defined in terms of the spatial derivatives, ∇g_i, at a pixel i, in J different directions $\boldsymbol{n}_j$, and uses learned image statistics to assess the prior probability. If ∇g_i is computed using neighborhood differences (as in Fig. 7.9), then (7.45) can be viewed as a standard Markov Random Field (MRF) formulation of the regularization problem [35]. Calculating such a spatial prior as histograms of the eigenvalues μ of the structure tensor $\boldsymbol{J}_\rho$ (see (7.21)), results in anisotropic diffusion. Exploiting this relationship provides a principled way of formulating anisotropic diffusion problems and results in a fully automatic algorithm in which all parameters are learned from training data. The resulting anisotropic diffusion algorithm has many of the benefits of Bayesian approaches along with a well-behaved numerical discretization. For further details on this approach and its performance we refer to [102].

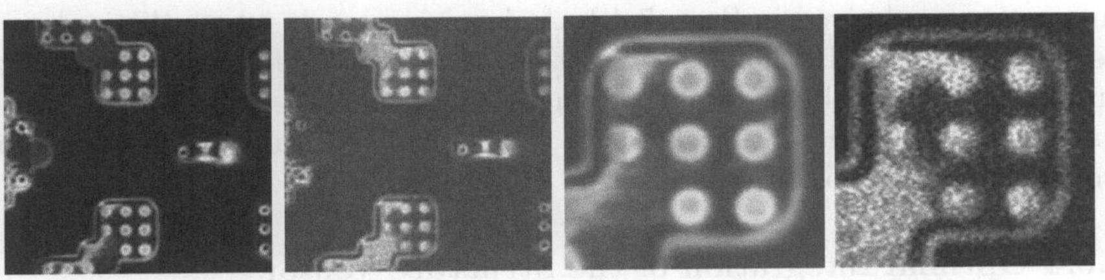

Fig. 7.8 Noise-free and noisy silicon chip images acquired by a focused-ion-beam tool. From left to right: high quality image, lower quality image, sub-regions of high and low quality images

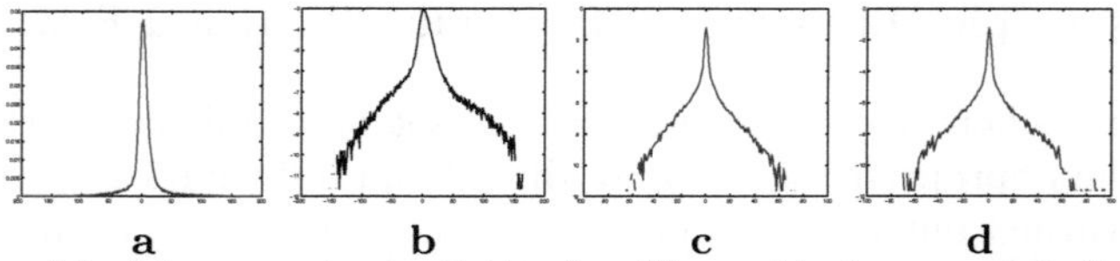

Fig. 7.9 Empirical image noise statistics for silicon chip images. (a) distribution of image noise $(f_i - g_i)$. (b) log of image noise distribution. (c) and (d) log probability of horizontal and vertical image derivatives

7.6.3 Anisotropic Diffusion for Multiple Motions

Standard diffusion may be derived from a cost function

$$E(g) = \int_{\Omega} (g-f)^2 + \alpha |\nabla g|^2 \mathrm{d}x \,. \tag{7.46}$$

The first term in (7.46) is usually called *data term*, corresponding to the posterior probability term in (7.45). Modeled purely quadratic is equivalent to $p(g|f)$ being zero-mean Gaussian. The second term is the *smoothness term*, corresponding to the prior probability, modeled by (the positive part of) a zero-mean Gaussian of $|\nabla g|$. An extension of this constraint has been proposed by Mumford and Shah [88]. Its connection to (not tensor driven) anisotropic diffusion can be found in [105].

In [95] a cost function penalizing violation of a linear model was introduced. Using e.g. (7.5), one gets

$$E(g) = \int_{\Omega} (g-f)^2 \mathrm{d}x + \alpha \int_{\Omega} < (\nabla^T g \boldsymbol{u})^2 > \mathrm{d}x \tag{7.47}$$

Minimizing (7.47) by iteratively fulfilling the minimality condition given by calculous of variations yields exactly anisotropic diffusion with a diffusion tensor $mD =< \boldsymbol{u}\boldsymbol{u}^T >$. In [95] it is shown that this tensor can be approximated by a diffusion tensor as constructed in (7.44). This observation allows to construct diffusion-like reconstruction schemes for linear models. Plugging the 2 D equivalent of (7.54) into (7.47) and minimizing it as before yields such a scheme, which has been implemented using optimized filters for transparent motion as derived in Sect. 7.4.1. A denoising result on synthetic data is depicted in Fig. 7.10. One observes that standard edge-enhancing diffusion produced artifacts by enforcing a single orientation model, while enforcing the novel double orientation diffusion yields results visually indistinguishable from the original noise-free image.

7.6.4 Optimal Integration of the Structure Tensor

In the following we generalized the signal and noise adapted filter approach discussed in 7.4.3 such that it is able to preserve edges (edge preserving Wiener (EPW) filter) and generalize it from scalar valued signals to tensor valued signals. For a detailed description we refer to [52].

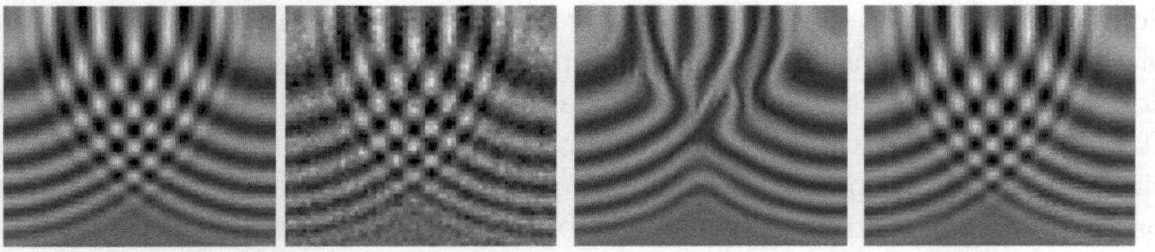

Fig. 7.10 Denoising of transparently overlaid ring pattern. From left to right: original image, image with noise added, reconstruction via edge-enhancing single orientation anisotropic diffusion, reconstruction via double orientation diffusion

The Scalar Valued Edge Preserving Wiener Filter

The estimation of a true underlying image value s_j at position j from a linear but not shift invariant filtering of the observable image $\boldsymbol{z}$ can be written in the form $\hat{s}_j = \boldsymbol{m}_j^T \boldsymbol{z}$. Our task is to choose $\boldsymbol{m}_j$ in such a way that the filtered output $\hat{s}_j$ approximates, on an average, the desired output s_j for the error-free case as closely as possible in the least mean squares sense. Therefore, it is necessary to model the statistical properties of the signal and the noise processes, respectively. Let the noise vector $\boldsymbol{v} \in \mathbb{R}^N$ be a zero-mean random vector with covariance matrix $\mathbf{C}_{\boldsymbol{v}} \in \mathbb{R}^{N\times N}$ (which is in this case equal to its correlation matrix $\mathbf{R}_{\boldsymbol{v}}$). Furthermore, we assume that the process that generates the signal $\boldsymbol{s} \in \mathbb{R}^N$ can be described by the expectation $\boldsymbol{w}_s = \mathbb{E}\left[\boldsymbol{s}\right]$ of the signal vector, and its autocorrelation matrix $\mathbf{R}_{\boldsymbol{s}}$. Furthermore, let $\mathbf{R}_{\boldsymbol{s}s_j} \in \mathbb{R}^{1\times N}$ denote the correlation matrix between the image value s_j and the whole image $\boldsymbol{s}$. The filter $\boldsymbol{m}_j$ is then determined by minimizing the mean squared error between the estimated signal value and the actual one

$$\boldsymbol{m}_j = \arg\min_{\tilde{\boldsymbol{m}}_j} \left\{ \mathbb{E}\left[||\tilde{\boldsymbol{m}}_j^T \boldsymbol{z} - s_j||^2 \right] \right\} . \tag{7.48}$$

Knowing the second order statistical moments for both the noise and signal as well as the observation matrix, the Gauss–Markov theorem delivers the optimal filter (for a detailed derivative of mean squared error based filters see e.g. [50])

$$\boldsymbol{m}_j = \left(\mathbf{K}_j \mathbf{R}_{\boldsymbol{s}} \mathbf{K}_j^T + \mathbf{R}_{\boldsymbol{v}} \right)^{-1} \mathbf{K}_j \mathbf{R}_{\boldsymbol{s}s_j} . \tag{7.49}$$

In following, we discuss the extension of this concept to matrix valued data.

The Edge Preserving Tensor Valued Wiener Filter

As already mentioned in the introduction, most important tensors for image processing are square positive (semi-)definite matrices denoted by $P(n)$ in the following where n is the size of the matrix. This set of tensors does not form a subspace of the tensor vector space. For example, multiplying a positive definite matrix by -1 yields a negative definite matrix and hence leads out of the set $P(n)$. Thus, applying image processing techniques to $P(n)$ requires additional care since even simple linear operations might destroy the basic structure of the data. In [118] the proposed nonlinear diffusion scheme is shown to preserve positiveness of the processed tensor field. An equivalent proof based on discrete filtering can be found in [129] which uses the fact that the proposed diffusion filters are convex filters. This is also the basis for the design of our tensor valued EPW-filter, i.e., we design the tensor valued EPW-filter as a convex filter. A map $\mathbf{F} : \mathbb{R}^N \to \mathbb{R}^N$ is denoted as a convex filter (see e.g. [131]) if for each $\boldsymbol{z} \in \mathbb{R}^N$ there are weights $w_{ij}(\boldsymbol{z})$ with

$$(\mathbf{F}\boldsymbol{z})_k = \sum_{t=1}^{N} w_{kt}(\boldsymbol{z}) z_t, \quad w_{tk}(\boldsymbol{z}) \geq 0 \ \forall k, \ \sum_{t=1}^{N} w_{kt}(\boldsymbol{z}) = 1 \,. \tag{7.50}$$

If each component of the tensor-field is processed with the same convex filter, it is simple to prove the positiveness of the processed tensor field. This implies that we have to model each matrix component by the same process and thus use the same statistical model as in the scalar case for each matrix element. We have to design a filter mask whose sum is equal one and where each element is non-negative. The first requirement can easily be obtained by a proper normalization. The second requirement is not guaranteed by (7.48). In order to keep each element non-negative, further constraints are introduced to the optimization procedure

$$\boldsymbol{m}_j = \arg\min_{\tilde{\boldsymbol{m}}_j} \left\{ \mathbb{E}\left[||\tilde{\boldsymbol{m}}_j^T \boldsymbol{z} - s_j||^2 \right] \right\} \quad \text{such that} \quad (\boldsymbol{m}_j)_k \geq 0 \,. \tag{7.51}$$

In contrast to (7.48), a closed form solution does not exist for the non-negative least squares problem and numerical methods (Chap. 23, p. 161 in [61]) need to be applied. Experimental results are shown in Fig. 7.11.

7.6.5 Tensor Field Integration by Nonlinear Diffusion

Here we present a method for adaptive integration of tensor fields with respect to motion estimation. The smoothing is fulfilled by a suitable nonlinear diffusion strategy, which is then applied on the manifold of the matrices with same eigenvalues, see [119].

As already discussed in [119, 120, 121], there is a general smoothing on corners, when we use the nonlinear diffusion based on total variation flow. In order to circumvent this drawback [13], defines a coherence dependent map, which stops the diffusion near corners. We propose the formalism of curvature preserving partial differential equations (PDE's) to avoid oversmoothing on corners. This is just a result of the analysis in [119, 120, 121]. It is a direct approach following the formalism for the heat flow equation, constrained on a curve. The adaptive, curve dependent metrics drives the diffusion according to

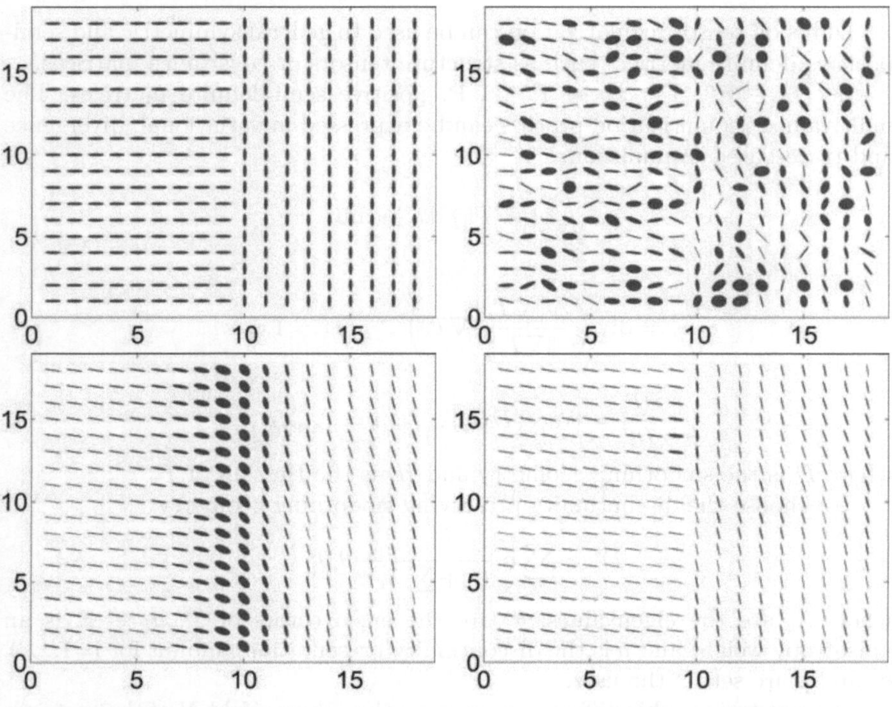

Fig. 7.11 Upper left: original tensor field; **upper right**: left tensor field corrupted by additive Gaussian noise ($\sigma_v = 0.3$ on each matrix element); **lower left**: processed tensor field by our EPW-filter with $\beta_1, \beta_2 = 0$; **lower right**: EPW-filter with $\beta_1 = 6$, $\beta_2 = 1$

the desired behavior by itself and avoids dependency of the smoothed tensor field on the derivatives of the steering geometry.

The next theoretical part of this section is to choose a proper integration scheme, which constraints the diffusion flow on the manifold of the matrices with the same spectrum. We represent different flows and conduct experiments with the isospectral flow. In optic flow and motion estimation not only the computation of the flow field is important, but also the confidence of the estimated flow. Most of the confidence measures rely on the eigenvalues of the second order tensor in the estimated pixel. The isospectral flow leaves the eigenvalues of second order tensors untouched. This is an intrinsic property of the flow, at least analytically. This means, the confidence measure is preserved locally. That's why we decided to employ an integration scheme, based on the isospectral flow. Additionally it should be mentioned, that the isospectral flow represents a good trade-off between performance and computational costs [119, p. 149].

Diffusion tensor regularization can be used to general symmetric and semi-positive definite matrices such as structure tensors or covariance matrices.

Let $\Omega \in \mathbb{R}^n$. $T : \Omega \to P^{n\times n}$, P: positive semi-definite matrices. The multivalued regularization process can be expressed in variational, divergence and trace-based formulation.

$$\int_{\Omega} \phi(\|\nabla T\|)\, \mathrm{d}x \to \min$$

$$\frac{\partial T_i}{\partial t} = \operatorname{div}\left(\frac{\phi\prime\|\nabla T_i\|}{\|\nabla T_i\|}\nabla T_i\right), \qquad i = 1, ..., n,$$

$$\frac{\partial T_i}{\partial t} = \operatorname{trace}(DH_i), \qquad i = 1, ..., n,$$

where D is the smoothing geometry and H_i is the Hessian of T_i.

We choose the discontinuity-preserving smoothing geometry

$$D := \sum a_i \frac{1}{(1+\sum \lambda_i)^p} \theta_i \theta_i^T,$$

where λ_i are the eigenvalues, θ_i are the eigenvectors of T, $a_i \in \mathbb{R}$ is an anisotropy weight and p is the discontinuity preservation amount for i=1,...,3. a_i and p are set by the user.

We constrained this diffusion process on the submanifold N of the matrices with the given set of eigenvalues by using the isospectral flow

$$\frac{\partial T}{\partial t} = [T, [T, -\mathcal{L} + \mathcal{L}^T]],$$

where L is the matrix, corresponding to an unconstrained Lagrangian, describing the regularization process.

A suitable integration scheme is

$$\begin{aligned} T_{t+dt} &:= A_t(x)^T\, T_t(x)\, A_t(x) \\ A_t(x) &:= \mathrm{e}^{-dt[\mathcal{L}(x)\mathcal{L},T]}\,. \end{aligned} \tag{7.52}$$

Numerical Experiments

We measured the performance of both techniques on a synthetic sequence without noise, containing a moving sinus pattern with discontinuity.

We achieved a small accuracy gain, compared to the isotropic smoothing, for synthetic sequences without noise by using the isospectral flow. The accuracy gain vanishes after adding Gaussian noise to the synthetic sequence, thus the method is noise sensitive. The tuning of parameters is crucial and it's difficult to find a trade-off between preserving of discontinuities and noise sensitivity of the diffusion flow. There is a thin line for the proper parameter setting, at least in our implementation of the isospectral flow by matrix exponential for the integration scheme (7.52).

7.6.6 Adaptive Anisotropic Filtering

Section 7.6 emphasized the importance of reducing the noise inherent in the data for modeling, model selection, and estimation of parameters in multidimensional signals. Standard procedures smooth the data by averaging measurements $s(\boldsymbol{x})$ over neighborhoods in the domain of s. Apparently, size and shape of the neighborhoods are crucial to the performance of such procedures.

The method of Adaptive Anisotropic Filtering, based on Adaptive Weights Smoothing [93], determines size and shape of such neighborhoods making two structural assumptions on s. First, in a neighborhood of local flow homogeneity, s takes the form of (7.1) (with $\mathrm{d} = 1, n = 2$) and can thus be approximated by a univariate function $f(u) \approx f(\nabla^T s \cdot \boldsymbol{x})$. A second assumption of local linearity regarding f leads to an approximation of s of the form

$$s(\boldsymbol{x}) \approx a + c\beta^T(\boldsymbol{x} - \boldsymbol{x}_0), \tag{7.53}$$

where β denotes a vector parallel to the gradient of s.

Noise reduction in the data with respect to the above formulation can be achieved by estimating the parameters a for each data point. This can be accomplished in an iterative procedure where an initial estimate of β determines an elliptic neighborhood which is used to estimate the parameters a and c. In a subsequent step the parameters a and c are used to improve the estimate of β and so on. In each iteration the elliptic neighborhoods are increased until further extension of the neighborhoods will consider data that significantly deviate from the estimated model with respect to a given threshold.

7.7 Estimation of Multiple Subspaces

7.7.1 Introduction to Multiple Subspaces

Multiple superimposed subspaces as defined in Sect. 7.2.2 can be estimated by extending the constraint in (7.2) based on the observation that all directional derivatives within the different subspaces must equal zero. Therefore, the concatenated, directional derivatives of s must equal zero, a condition that leads to a nonlinear constraint on the components of the $\boldsymbol{u}_i$. This constraint can be linearized by introducing the mixed-motion or mixed-orientation parameters (MMPs and MOPs), which then need to be separated such as to yield the $\boldsymbol{u}_i$´s.

For the case of scalar image sequences with two additively superimposed motions, such a constraint was first used in [108]. After successively applying $\alpha(\boldsymbol{u}_1)$ and $\alpha(\boldsymbol{u}_2)$ to (7.7), the resulting constraint $\alpha(u)\alpha(v)f = 0$, is linear in the MMPs, i.e., we obtain a model that is linear in its non-linear parameters. The nonlinear parameters can then be estimated using linear estimation techniques for the MMPs, which are then separated by solving a complex polynomial as shown below.

Here we summarize the comprehensive results obtained for multiple motions in videos and multiple orientations in images. The final goal, however, is the estimation of any combination of any number of subspaces with arbitrary intrinsic dimensions and signal dimension. First steps toward a comprehensive classification of such combinations have been done in [75, 76, 77]. In [83, 117] the problem of estimating multiple orientations in n-dimensional signals has been solved.

7.7.2 Multiple Motions

Analytical Solutions for Multiple Motions

First consider two additive motion layers s_1, s_2 as defined by (7.7). The constraint for the velocities becomes

$$\begin{aligned}\alpha(\boldsymbol{u}_1) \circ \alpha(\boldsymbol{u}_2) \circ s = c_{xx}s_{xx} + c_{xy}s_{xy} + c_{xt}s_{xt} \\ +c_{yy}s_{yy} + c_{yt}s_{yt} + c_{tt}s_{tt} = 0\,,\end{aligned} \tag{7.54}$$

$$\begin{aligned}c_{xx} = u_{1,x}u_{2,x}\,,\; c_{xy} = (u_{1,x}u_{2,y} + u_{1,y}u_{2,x})\,,\; c_{xt} = (u_{1,x} + u_{2,x})\,, \\ c_{yy} = u_{1,y}u_{2,y}\,,\; c_{yt} = (u_{1,y} + u_{2,y})\,,\; c_{tt} = 1\,.\end{aligned} \tag{7.55}$$

Equation (7.54) is nonlinear in the motion parameters themselves, but linear in the MMPs, which, therefore, can be estimated by standard linear techniques, e.g. [73, 114].

MMP Decomposition with Complex Polynomials

In [73] a general solution for decomposing an arbitrary number of superimposed motions has been proposed. Here we sketch the idea for the case of two motions. The interpretation of motion vectors as complex numbers $\boldsymbol{u} = u_x + iu_y$ enables us to find the motion parameters as the roots of the complex polynomial

$$Q(z) = (z - \boldsymbol{u}_1)(z - \boldsymbol{u}_2) = z^2 - (c_{xt} - ic_{yt})z + (c_{xx} - c_{yy} + ic_{xy}), \tag{7.56}$$

whose coefficients are expressed in terms of the MMPs. The generalization of this approach to N overlaid motion layers is straightforward.

Solutions for Multiple Motions Based on Regularization

A major benefit of the above approach to multiple motions is that it involves a linearization of the problem such that it becomes mathematically equivalent to the problem of estimating only one motion. As a consequence, the regularization methods used for single motions can be applied. In [114, 115], the well-known algorithm for single-motion estimation proposed by Horn and

Schunck has been extended to the case of multiple motions by the use of the following regularization term:

$$N = (\partial_x c_{xx})^2 + (\partial_y c_{xx})^2 + (\partial_x c_{yy})^2 + (\partial_y c_{yy})^2 + (\partial_x c_{xy})^2 + (\partial_y c_{xy})^2 + (\partial_x c_{xt})^2 + (\partial_y c_{xt})^2 + (\partial_x c_{yt})^2 + (\partial_y c_{yt})^2 .$$

The MMPs c are obtained as the values that minimize the above term together with the squared optical-flow term in (7.54), i.e.

$$\int\!\!\int (\alpha(\boldsymbol{u}_1) \circ \alpha(\boldsymbol{u}_2) \circ s)^2 + \lambda^2 N \, \mathrm{d}\Omega .$$

λ is the regularization parameter and Ω the image plane over which we integrate. Note that working on the MMPs has the great advantage that we obtain an Euler-Lagrange system of differential equations that is linear, which would not be the case, when working directly on the motion vectors themselves.

Block-Matching for Multiple Motions

The framework for multiple motions has been extended such as to include block-matching techniques for estimating an arbitrary number of overlaid motions [113]. To estimate N motions, $N+1$ images of the sequence are needed. A regularized version of the algorithm has also been derived based on Markov Random Fields [116], and the increased robustness has been demonstrated.

Separations of Motion Layers

A benefit of multiple-motion estimation is that the parameters of the multiple motions can be used to separate the motion layers. This problem has been solved in the Fourier domain, where the inherent singularities can be better understood and interpolated than in the spatio-temporal domain [79, 114].

Occluded Motions

For occluded motions as defined by (7.9), one obtains the constraint

$$\alpha(\boldsymbol{u}_1) \circ \alpha(\boldsymbol{u}_2) \circ s = -\alpha(\boldsymbol{u}_1) \circ \left[\chi(\boldsymbol{x} - t\boldsymbol{u}_1)\right] (\boldsymbol{u}_1 - \boldsymbol{u}_2) \cdot \nabla s_2(\boldsymbol{x} - t\boldsymbol{u}_2) . \quad (7.57)$$

In [80], it has been shown that

$$\alpha(\boldsymbol{u}_1) \circ \alpha(\boldsymbol{u}_2) \circ s = q(\boldsymbol{x}, t, \boldsymbol{u}_1, \boldsymbol{u}_2) \delta(B(\boldsymbol{x} - t\boldsymbol{u}_1)) , \quad (7.58)$$

where

$$q(\boldsymbol{x}, t, \boldsymbol{u}_1, \boldsymbol{u}_2) = -(\boldsymbol{u}_1 - \boldsymbol{u}_2) \cdot \boldsymbol{N}(\boldsymbol{x} - t\boldsymbol{u}_1) (\boldsymbol{u}_1 - \boldsymbol{u}_2) \cdot \nabla s_2(\boldsymbol{x} - t\boldsymbol{u}_2) . \quad (7.59)$$

$B(\boldsymbol{x}) = 0$ determines the occluding boundary of $\boldsymbol{N}(\boldsymbol{x}) = \nabla B(\boldsymbol{x})$ is the unit normal to the boundary. Equation (7.58) shows that the occlusion distortion is (i) restricted to the occluding boundary, (ii) minimal when the normal to the boundary is orthogonal to the relative motion (the difference between fore- and background motions) and maximal when the two vectors are aligned, (iii) proportional to the intensity gradient of the background pattern. By a Fourier analysis of occluded motions, it has been revealed that the decay of the distortion is hyperbolic for both straight and curved boundaries and the exact expression for the distortion term has been derived for the case of straight boundaries [80]. Based on these results, a hierarchical motion-estimation algorithm has been designed that obtains very good estimates at the occluding boundary by avoiding the there localized distortion [6, 7, 78, 80].

7.7.3 Multiple Orientations

Orientation Estimation in Tensor Form

If m orientations in n-variate signal are to be found, this problem can either be written using the mixed orientation parameters (MOP) vector or, alternatively, in tensor notation. The latter form was presented in [83], including the generalization to p-dimensional signals, for instance color or multi-spectral images, which will not be handled here.

In order to express orientation estimation in tensor form, we first define the sought entity, the orientation tensor, as outer product of all individual orientation unit vectors $\boldsymbol{u}_i$:

$$\mathcal{U} = \boldsymbol{u}_1 \otimes \cdots \otimes \boldsymbol{u}_n \, .$$

With the tensor scalar product

$$\langle \mathcal{O}, \mathcal{U} \rangle := \sum_{k_1,\ldots,k_m=1}^{n} (\mathcal{O})_{k_1\cdots k_m} (\mathcal{U})_{k_1\cdots k_m} \, , \tag{7.60}$$

we can define the data constraint as

$$\langle \mathcal{O}, \mathcal{U} \rangle = 0 \tag{7.61}$$

where

$$(\mathcal{O})_{k_1\cdots k_m} = \prod_{i=1}^{m} \frac{\partial s}{\partial x_{k_i}}$$

is the data tensor generated for the occluding orientations model. In a similar way, the additive superposition of multiple orientations leads to

$$\langle \mathcal{T}, \mathcal{U} \rangle = 0 \quad \text{with} \quad (\mathcal{T})_{k_1\cdots k_m} = \frac{\partial^m s}{\partial x_{k_1} \cdots \partial x_{k_m}} \, . \tag{7.62}$$

Symmetry Properties of the Data Tensors

The commutativity in the definitions of (7.61) and (7.62) is the key to the understanding of multiple orientations. The data tensors are invariant against any arbitrary permutation of indices and therefore have some very pronounced symmetry properties.

For $m = 2$, the data tensors $\mathcal{O}$ and $\mathcal{T}$ are symmetric $n \times n$ matrices, but for higher m, concepts from matrix algebra will not suffice and a tensor notation becomes necessary. We therefore define the *space of fully symmetric m-tensors* which are invariant to *any arbitrary* permutation of indices – and not just to some special permutations only. We define

$$\mathbb{R}_{\oplus}^{n\times\cdots\times n} = \left\{ \mathcal{T} \in \mathbb{R}^{n\times\cdots\times n} \middle| (\mathcal{T})_{i_1\cdots i_m} = (\mathcal{T})_{P(i_1\cdots i_m)} \right\}$$

with $P(i_1 \cdots i_m)$ denoting any arbitrary permutation of the indices $i_1 \cdots i_m$.

For both models – occluding or additive superposition – the resulting data tensors are fully symmetric and from this symmetry property follows an import consequence: The sought orientation tensor cannot be estimated uniquely, but only up any arbitrary permutation of indices. One cannot distinguish between a "first" orientation, "second" orientations and so on, all we can compute a *set* of m orientations.

Fortunately, this problem can be resolved by restricting the sought orientation tensor to those tensors which are invariant to index permutations, i.e. to $\mathbb{R}_{\oplus}^{n\times\cdots\times n}$. Within this set, the solution becomes uniquely determined – at least in non-degenerate cases.

Estimation of the Orientation Tensor

Stacking the independent elements of data tensor $\mathcal{O}$ (resp. $\mathcal{T}$) and of the symmetrized (and therefore uniquely determined) orientation tensor $\mathcal{U}$ generates two vectors which can be understood as generalization of the double-orientation constraint handled in [1]. See [83] for details.

In many applications, for instance in feature extraction for tracking, this is already sufficient. If, on the other hand, a decomposition into the underlying individual orientations is necessary, then an additional problem arises: the orientation tensor estimated so far is overparameterized because "true" orientation tensors must belong to the set

$$\mathbb{R}_{\circledast}^{n\times\cdots\times n} = \left\{ \sum_{P(i_1\cdots i_m)} \boldsymbol{u}_{i_1} \otimes \cdots \otimes \boldsymbol{u}_{i_m} \middle| \boldsymbol{u}_{i_1}, \ldots, \boldsymbol{u}_{i_m} \in \mathbb{R}^n \setminus \{\mathbf{0}\} \right\}, \quad (7.63)$$

which we will denote as set of *minimal fully symmetric tensors*. They are constructed by summing up all possible permutations of outer products and obviously form a subset of $\mathbb{R}_{\oplus}^{n\times\cdots\times n}$.

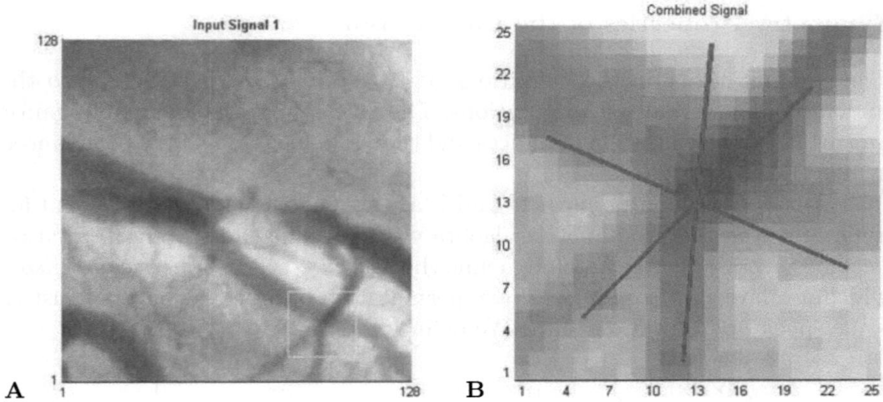

Fig. 7.12 Modeling blood vessels in X-ray images often requires *three* orientations. (**A**): Input image. (**B**): Region of interest (light square in (A)) and detected orientations

In [83], methods for estimating an arbitrary number of orientations in images (arbitrary m for $n = 2$) and for double orientation estimation ($m = 2$) in general n-variate signals are presented. The second problem boils down to the approximation of a general $n \times n$ matrix by a rank-2 matrix and the first problem, the estimation of multiple orientations in images, can be formulated as finding the roots of a degree m real polynomial.

Especially the estimation of more than two orientations in images has many applications in image processing and computer vision, for instance in industrial quality control or medical image processing. Figure 7.12 shows an X-ray image of superposed blood vessels and the estimated orientations.

7.8 Applications

7.8.1 Transport Processes

The transport of energy, mass and momentum is one ubiquitous phenomenon in our world, spanning all branches of science. It is one condition for the continued existence of complex dynamic systems. As such, only through this transport living organisms can coordinate and keep upright their metabolism and other processes of life. Plants, for example, have developed complex vascular systems (the xylem) to transport nutrients and water from the roots to the rest of the plant. At the same time, all living tissue is continuously supplied with energy from the leaves, in form of the organic products of photosynthesis.

In plant physiology, it is a longstanding question how water and nutrients are transported in the xylem of the plant's leaf and which external factors influence it. While bulk dependencies are known, a detailed analysis has eluded research due to inadequate measurement techniques. The transport processes in leaves are especially important since it is known that they can be regulated by plants according to environmental forcing. Also, due to these regulatory mechanisms, the plant can cope with cuts in the leaf and still supply areas affected by these cuts with xylem sap by other pathways. Still, very little is known about these mechanisms. To shed light on the transport of the xylem inside leafs the advanced techniques presented in this chapter have been employed on thermographic image sequences of heated water parcels in plant leaves [33]. The experimental set-up as well as a comparative measurement to ground truth is presented in Fig. 7.13. Through these measurements, quantitative measurements could be made in different parts of the leaf in dependence of external environmental forcings acting on the leaf [33].

On much smaller scales, a current trend in chemical and biochemical analytics as well as in medical diagnostics is the move to microfluidic mixers and "lab-on-a-chip" applications. Huge surface to volume ratios are achievable by micro channels with highly controlled boundary conditions, leading to more efficient reaction kinetics with less by-products. Even on these minute structures, the transport of energy, mass and momentum as well as the measurements thereof is vital for a better understanding of the processes involved. In these flow regimes, molecular tagging velocimetry (MTV) is an alternative approach to the standard technique of micro particle imaging velocimetry (μPIV) for measuring fluid flows. In MTV, a pattern is written to the fluid with an UV laser to uncage dyes and thus making them fluorescent. Although this approach has been used since the late eighties of the last century, measuring fluid flow with this technique has had one significant drawback: Due to the flow profile, the uncaged dyes would appear to diffuse in a process termed Taylor dispersion. Two frames of a typical image sequence can be seen in Fig. 7.14. This dispersive process leads to significant uncertainties in the measurements, as it is difficult to correct for this dispersion. Due to Taylor dispersion, the use of MTV is very limited. Nevertheless, it represents the only technique available for situations in which particles cannot be used for visualizing the flow, such as packed columns. Motion models such as those

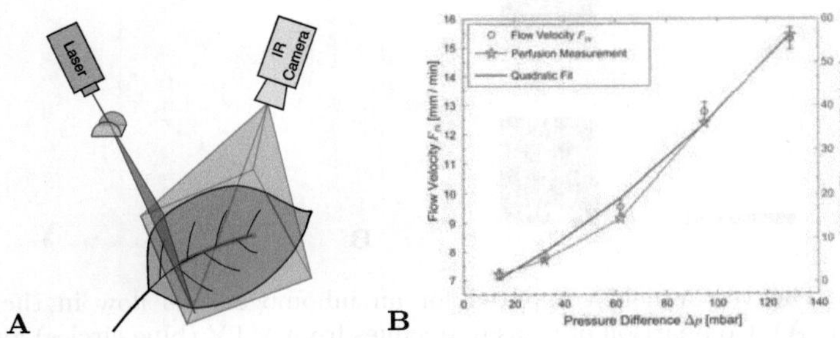

Fig. 7.13 The set-up for measuring the xylem flow in plant leaves with active thermography in (**A**) and the results of a ground truth measurement in (**B**)

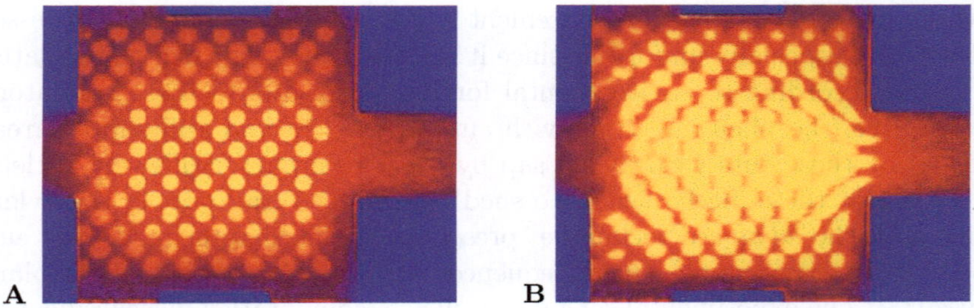

Fig. 7.14 In (**A**) and (**B**) two frames of a microfluidic image sequences are shown. The implication of Taylor dispersion can clearly be observed. Structures seem to diffuse in the direction of fluid flow

described in Sect. 7.2.4 were used to make feasible a highly accurate technique for measuring microfluidic flows based on MTV [31, 34, 94]. A comparison of MTV, µPIV and ground truth can be seen in Fig. 7.15.

In environmental sciences, it is the transport of the same quantities, energy, momentum and mass, that is the driving force in most weather conditions on short scales and in climatic variability on longer time periods. It plays the dominant role on aquatic processes in the oceans such as ocean currents and the thermohaline circulation, the meridional overturning circulation (MOC). The two compartments of atmosphere and ocean are in contact at the ocean surface where energy, mass and momentum is exchanged between them. Due to experimental difficulties of measuring inside the boundary layer, extending less than one millimeter into the water body which is undulated by waves of several centimeters heights. Applying the advanced techniques for the estimation of complex motion presented in this chapter has lead to significant advances in the field of air-sea interactions. Estimating the total derivative of temperature T with respect to time $\mathrm{d}T/\mathrm{d}t = \alpha(\boldsymbol{u}) \circ T = c$ from thermographic

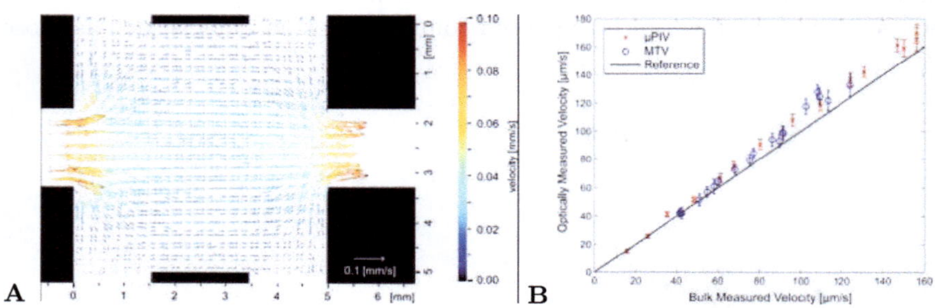

Fig. 7.15 The vector field computed for an inhomogeneous flow in the mixing chamber in (**A**). Comparison of measured values from MTV (blue circles) and µPIV (red crosses) compared to ground truth measurement obtained from a flow meter (solid black line) in (**B**)

Fig. 7.16 Growing pine needle. 1st (left) and 100th (right) image of the sequence, the frames indicate the 256×128 sheath region

image sequences of the air–water interface has made it possible to accurately estimate the net heat flux as well as the transfer velocity of heat [29, 30, 104]. The viscous shear stress could be deduced from active thermography [32]. This will make it possible to perform process studies, relating the transport of heat with that of momentum in the same footprint spatially and temporally highly resolved.

7.8.2 Growing Plant Leaves and Roots

In Sect. 7.4.1 optimal filters for transparent motion estimation have been derived. As demonstrated in [96] using these filters reduce systematical errors in transparent motion estimation. Motion estimation of a growing pine needle (see Fig. 7.16) shows the effect of using different filter families. The sheath region consists of a transparent layer, becoming more and more opaque the closer to the base to the needle. Motion underneath this layer shall be measured in order to quantify local growth. In the current data sets only rigid motion is visible, indicating that the growth zone is completely hidden in the opaque region at the base of the needle. Figure 7.17 depicts that using simple

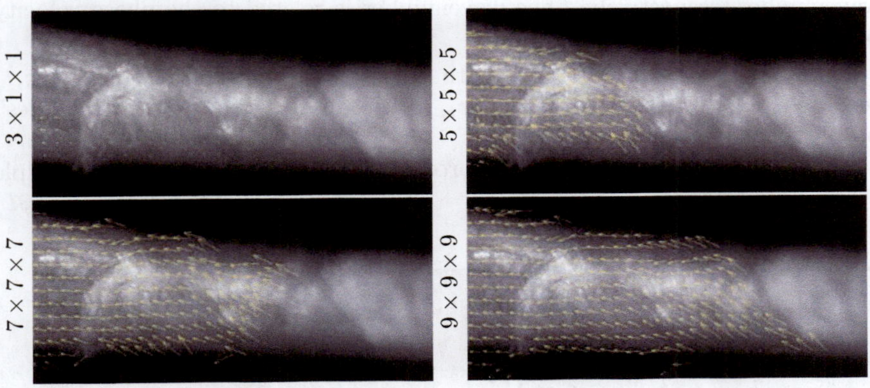

Fig. 7.17 Growing pine needle. From left to right and top to bottom: motion estimation results on the sheath region using $3 \times 1 \times 1$, $5 \times 5 \times 5$, $7 \times 7 \times 7$, and $9 \times 9 \times 9$-filters. Vectors longer than 4 pixels/frame are cut off. Vectors are scaled by a factor of 4

central distances $3 \times 1 \times 1$-filters, reliable results can be achieved nowhere in the transparent region (same holds for $3 \times 3 \times 3$-filters). The larger the optimized filters, the larger and visibly more accurate the motion fields become.

The 3 D model proposed in Sect. 7.2.3 has been integrated in a screening setup established at ICG 3, Jülich. Depth and surfaces slopes of a plant (kalanchoe) reconstructed from an input sequence of 49 images, i.e., 7×7 camera positions, demonstrate the accuracy of the model. Figure 7.18 shows the central image of the input sequence and of depth estimates rendered by *povray* [14]. Furthermore, Fig. 7.18 shows a close-up view to compare a leaf with the estimated parameters in detail. Reconstructed depth and surfaces slopes match the original quite well.

7.8.3 Analysis of Seismic Signals

The analysis of seismic surface waves can provide valuable information about the subsurface structure, which is of interest for engineering purposes and seismic hazard assessment (see e.g. [46]).

A seismic signal typically contains several superposed propagation modes, each with its own dispersion and attenuation characteristics. A typical propagation model is

$$S^t(x, f) = \sum_{l=1}^{L} \mathrm{e}^{-x(\alpha_l(f) + \mathbf{I}k_l(f))} R_l(f)\,, \qquad (7.64)$$

where $S^t(x, f)$ is the spectrum of the signal recorded at distance x from the source, L is the number of modes, $R_l(f)$ is the spectrum of the source event for each mode, and $\alpha_l(f)$ and $k_l(f)$ are the frequency-dependent attenuation and wavenumber, respectively. (The wavenumber is related to the phase velocity by the expression $k_l(f) = \frac{2\pi f}{c_l(f)}$).

The problem is now to estimate $\alpha_l(f)$ and $k_l(f)$ from signals $s(x, t)$ recorded at stations $x \in \{x_1, \dots, x_m\}$, we call this the "Multiple Modes" problem.

Note that the Multiple Motions problem is a specialization of the Multiple Modes problem where we have $k_l(f) = 2\pi \frac{f}{c_l}$ and $\alpha_l(f) \equiv 0$ for all $l = 1, \dots, L$,

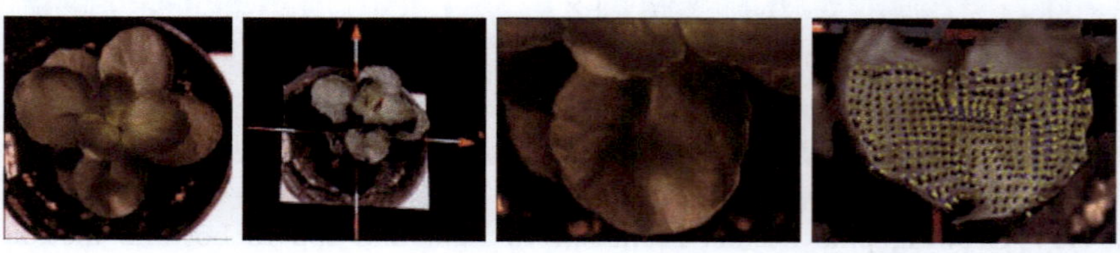

Fig. 7.18 Reconstruction of plant using 49 camera positions. From left to right: central image of input sequence, rendered depth estimates, close-up view on leaf and estimated depth with surface slopes

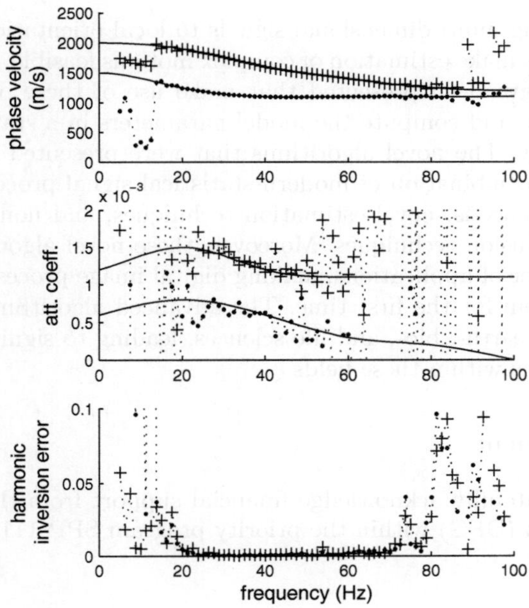

Fig. 7.19 Estimation results obtained on a signal containing two superposed propagation modes; noise level $\sigma = 0.001$, 20 recording stations

i.e., all frequency components propagate with the same phase velocity c_l and there is no attenuation.

There are several existing approaches to solving the Multiple Modes problem, e.g., by estimating a wavelet operator [46]. We have developed a solution that is based on the technique of harmonic inversion by filter diagonalization [64, 65]. This technique successfully separates modes that overlap in both time and frequency, but it is quite sensitive to noise (Fig. 7.19 shows a sample result). However, our interest in the technique stems from the fact that, like the Multiple Motions technique, it is based on solving an eigenproblem. Based on this observation, we conjecture that the ideas used in the solution of the Multiple Motions problem can be generalized to yield a solution for the Multiple Modes problem. However, we have not been able to find such a solution so far.

7.9 Conclusions

In this chapter the results of a fruitful and highly collaborative research initiative have been presented. A number of previously untackled and long-standing problems of estimating local and complex orientations were addressed in a multi-disciplinary effort. We have presented novel formulations for constraint equations, linking multi-dimensional signals to local orientations. These constraint equations make estimation of complex motions feasible. Also, a number of algorithms have been presented that make use of these orientation constraint equations and compute the model parameters in a statistically sound and efficient way. The novel algorithms that were presented in this chapter result from the combination of modern statistical signal processing, differential geometric analysis, novel estimation techniques, and nonlinear adaptive filtering and diffusion techniques. Moreover, these novel algorithms were applied to a number of applications, making digital image processing feasible to a number of them for the first time. The advanced algorithms were used in environmental-, earth-, bio-, and life sciences, leading to significant advances and contributions within these fields.

Acknowledgement

The authors gratefully acknowledge financial support from the German Science Foundation (DFG) within the priority program SPP 1114.

References

[1] T. Aach, C. Mota, I. Stuke, M. Mühlich, and E. Barth. Analysis of superimposed oriented patterns. *IEEE Transactions on Image Processing*, 15(12):3690–3700, 2006.

[2] H. Akaike. A new look at statistical model identification. *IEEE Transactions on Automatic Control*, 19:716–723, 1974.

[3] B. Andres, F. Hamprecht, and C. S. Garbe. Selection of local optical flow models by means of residual analysis. In F. Hamprecht, C. Schnörr, and B. Jähne, editors, *Pattern Recognition*, volume 4713, pages 72–81, 2007 Springer.

[4] J. L. Barron, D. J. Fleet, and S. S. Beauchemin. Performance of optical flow techniques. *International Journal of Computer Vision*, 12:43–77, 1994.

[5] E. Barth, T. Caelli, and C. Zetzsche. Image encoding, labeling, and reconstruction from differential geometry. *CVGIP: Graphical Models and Image Processing*, 55(6):428–46, November 1993.

[6] E. Barth, I. Stuke, and C. Mota. Analysis of motion and curvature in image sequences. In *Proceedings IEEE Southwest Symposium Image Analysis and Interpretation*, pages 206–10, Santa Fe, NM, April 7–9, 2002. IEEE Computer Press.

[7] E. Barth, I. Stuke, T. Aach, and C. Mota. Spatio-temporal motion estimation for transparency and occlusion. In *Proceedings IEEE International Conference Image Processing*, volume III, pages 69–72, Barcelona, Spain, September 14–17, 2003. IEEE Signal Processing Soc.

[8] J. Bergen, P. Anandan, K. Hanna, and R. Hingorani. Hierarchical model-based motion estimation. In *Proceedings of the European Conference on Computer Vision*, Lecture Notes in Computer Vision, pages 237–252, Berlin, 1992. Springer.

[9] M. Bierling. Displacement estimation by hierarchical blockmatching. In *SPIE Vol. 1001 Visual Communications and Image Processing*, pages 942–951, Cambridge, MA, November 1988.

[10] M. Bierling and R. Thoma. Motion compensating field interpolation using a hierarchically structured displacement estimator. *Signal Processing*, 11:387–404, 1986.

[11] J. Bigün and G. H. Granlund. Optimal orientation detection of linear symmetry. In *ICCV*, pages 433–438, London, UK, 1987. IEEE Computer Society Press.

[12] M. J. Black and P. Anandan. The robust estimation of multiple motions: Parametric and piecewise-smooth flow fields. *Computer Vision and Image Understanding*, 63(1):75–104, January 1996.

[13] T. Brox, J. Weickert, B. Burgeth, and P. Mrázek. Nonlinear structure tensors. *Image and Vision Computing*, 24(1):41–55, January 2006.

[14] C. Cason. Persistence of vision ray tracer (POV-Ray), version 3.6, Windows, 2005.

[15] T. F. Cootes, N. Thacker, and C. J. Taylor. Automatic model selection by modelling the distribution of residuals. In *ECCV 2002, LNCS 2353*, pages 621–635, Berlin, 2002. Springer.

[16] T. Darrel and E. Simoncelli. "Nulling" filters and the separation of transparent motions. In T. Darrel and E. Simoncelli, editors, *CVPR*, pages 738–739, New York City, NY, 1993.

[17] M. Diehl, R. Küsters, and H. Scharr. Simultaneous estimation of local and global parameters in image sequences. In *Forth Workshop Dynamic Perception*, Bochum, Germany, 14–15 November 2002.

[18] M. Elad, P. Teo, and Y. Hel-Or. Optimal filters for gradient-based motion estimation. In *Proceedings of the International Conference on Computer Vision (ICCV'99)*, 1999.

[19] H. Farid and E. P. Simoncelli. Optimally rotation-equivariant directional derivative kernels. In *7th International Conference on Computer Analysis of Images and Patterns*, Kiel, Germany, 1997.

[20] G. Farnebäck. *Spatial Domain Methods for Orientation and Velocity Estimation*. Lic. Thesis, Linköping University, Linköping, Sweden, March 1999.

[21] M. Felsberg and E. Jonsson. Energy tensors: Quadratic, phase invariant image operators. In W. Kropatsch, R. Sablatnig, and A. Hanbury, editors, *Pattern Recognition*, pages 493–500, Berlin, Heidelberg, 2005. Springer.

[22] M. Felsberg and G. Sommer. The multidimensional isotropic generalization of quadrature filters in geometric algebra. In Proc. Int. Workshop on Algebraic Frames for the Perception-Action

Cycle, Kiel (2000), G. Sommer and Y. Zeevi, Eds., *Lecture Notes in Computer Science*, Springer-Verlag, Heidelberg. accepted. http://citeseer.ist.psu.edu/felsberg00multidimensional.html

[23] D. J. Fleet, M. J. Black, Y. Yacoob, and A. D. Jepson. Design and use of linear models for image motion analysis. *International Journal of Computer Vision*, 36:171–193, 2000.

[24] D. J. Fleet and Y. Weiss. Optical flow estimation. In N. Paragios, Y. Chen, and O. Faugeras, editors, *Mathematical Models in Computer Vision: The Handbook*. Berlin, 2005. Springer.

[25] L. Florac, W. Niessen, and M. Nielsen. The intrinsic structure of optical flow incorporating measurement duality. *International Journal of Computer Vision*, 27:263–286, 1998.

[26] W. T. Freeman and E. H. Adelson. The design and use of steerable filters. *IEEE Transactions on Pattern Analysis and Machine Intelligence*, 13(9):891–906, 1991.

[27] C. S. Garbe. Fluid flow estimation through integration of physical flow configurations. In F. Hamprecht, C. Schnörr, and B. Jähne, editors, *Pattern Recognition*, Heidelberg, 2007. Springer. Accepted.

[28] C. S. Garbe, H. Spies, and B. Jähne. Mixed OLS-TLS for the estimation of dynamic processes with a linear source term. In L. Van Gool, editor, *Pattern Recognition*, volume LNCS 2449 of *Lecture Notes in Computer Science*, pages 463–471, Zurich, CH, 2002. Springer-Verlag.

[29] C. S. Garbe, H. Spies, and B. Jähne. Estimation of surface flow and net heat flux from infrared image sequences. *Journal of Mathematical Imaging and Vision*, 19(3):159–174, 2003.

[30] C. S. Garbe, U. Schimpf, and B. Jähne. A surface renewal model to analyze infrared image sequences of the ocean surface for the study of air-sea heat and gas exchange. *Journal of Geophysical Research*, 109 (C08S15):1–18, 2004.

[31] C. S. Garbe, K. Roetmann, and B. Jähne. An optical flow based technique for the non-invasive measurement of microfluidic flows. In *12th International Symposium on Flow Visualization*, pages 1–10, Göttingen, Germany, 2006.

[32] C. S. Garbe, K. Degreif, and B. Jähne. Estimating the viscous shear stress at the water surface from active thermography. In C. S. Garbe, R. A. Handler, and B. Jähne, editors, *Transport at the Air Sea Interface - Measurements, Models and Parametrizations*, pages 223–239. Berlin, Heidelberg, 2007. Springer-Verlag.

[33] C. S. Garbe, R. Pieruschka, and U. Schurr. Thermographic measurements of xylem flow in plant leaves. *New Phytologist*, 2007. Submitted.

[34] C. S. Garbe, K. Roetmann, V. Beushausen, and B. Jähne. An optical flow MTV based technique for measuring microfluidic flow in the presence of diffusion and Taylor dispersion. *Experiments in Fluids*, 2007. Accepted.

[35] S. Geman and D. Geman. Stochastic relaxation, Gibbs distribution, and the Bayesian restoration of images. *Pattern Analysis and Machine Intelligence*, 6(6):721–741, 1984.

[36] N. Gheissari, A. Bab-Hadiashar, and D. Suter. Parametric model-based motion segmentation using surface selection criterion. *Computer Vision and Image Understanding*, 102:214–226, 2006.

[37] G. H. Golub and C. F. van Loan. *Matrix Computations*. Baltimore and London, 3rd edition, 1996. The Johns Hopkins University Press.

[38] G. H. Golub and C. F. van Loan. *Matrix Computations*. Baltimore and London, 3rd edition, 1996. Johns Hopkins University Press.

[39] G. H. Granlund and H. Knutsson. *Signal Processing for Computer Vision*. Dordrecht, The Netherlands, 1995. Kluwer Academic.

[40] H. Haussecker. Radiation. In Jähne, B., Haussecker, H., and Geißler, P., editors, *Handbook of Computer Vision and Applications*, volume 1, chapter 2, pages 7–35. San Diego, CA, 1999. Academic Press.

[41] H. Haussecker and D. J. Fleet. Computing optical flow with physical models of brightness variation. *IEEE PAMI*, 23(6):661–673, June 2001.

[42] H. Haussecker and H. Spies. Motion. In *Handbook of Computer Vision and Applications*. San Diego, CA, 1999. Academic Press.

[43] H. Haussecker and H. Spies. Motion. In Jähne, B., Haussecker, H., and Geißler, P., editors, *Handbook of Computer Vision and Applications*, volume 2, chapter 13. San Diego, CA, 1999. Academic Press.

[44] H. Haussecker, C. Garbe, et al. A total least squares for low-level analysis of dynamic scenes and processes. In *DAGM*, pages 240–249, Bonn, Germany, 1999. Springer.

[45] D.J. Heeger. Optical flow from spatiotemporal filters. *IJCV*, 1:279–302, 1988.

[46] M. Holschneider, M. S. Diallo, M. Kulesh, M. Ohrnberger, E. Lück, and F. Scherbaum. Characterization of dispersive surface waves using continuous wavelet transforms. *Geophysical Journal International*, 163: 463–478, 2005.

[47] B. K. P. Horn and B. Schunk. Determining optical flow. *Artificial Intelligence*, 17:185–204, 1981.

[48] B.K.P. Horn and B.G. Schunck. Determining optical flow. *Artificial Intelligence*, 17:185–204, 1981.

[49] B. Jähne. *Spatio-Temporal Image Processing*. Lecture Notes in Computer Science. Berlin, 1993. Springer Verlag.

[50] S. M. Kay. *Fundamentals of Statistical Signal Processing, Volume I: Estimation Theory*. 1993. Prentice Hall PTR.

[51] H. Knutsson and G.H. Granlund. Texture analysis using two-dimensional quadrature filters. In *IEEE Workshop Computer Architecture for Pattern Analysis and Image Data Base Management*, Pasadena, CA, 1983.

[52] K. Krajsek and R. Mester. The edge preserving wiener filter for scalar and tensor valued images. In *Pattern Recognition, 28th DAGM-Symposium*, pages 91–100, Berlin, Germany, September 2006. Springer.
[53] K. Krajsek and R. Mester. Bayesian model selection for optical flow estimation. In *Pattern Recognition, 29th DAGM-Symposium*, pages 142–151, Heidelberg, Germany, September 2007. Springer.
[54] K. Krajsek and R. Mester. Signal and noise adapted filters for differential motion estimation. In *Pattern Recognition, 27th DAGM-Symposium*, pages 476–484, Vienna, Austria, September 2005. Springer.
[55] K. Krajsek and R. Mester. A unified theory for steerable and quadrature filters. In A. Ranchordas, H. Araújo and B. Encarna✓cão, edl-itors, *VISAPP 2006: Proceedings of the First International Conference on Computer Vision Theory and Applications*, 2 Volumes, Setúbal, Portugal, February 25–28, 2006, INSTICC, pages 48–55, 2006. ISBN 972-8865-40-6.
[56] K. Krajsek and R. Mester. A maximum likelihood estimator for choosing the regularization parameters in global optical flow methods. In *IEEE International Conference on Image Processing*, pages 1081–1084 Atlanta, USA, October 2006. ISBN 1-4244-0481-9.
[57] K. Krajsek and R. Mester. Wiener-optimized discrete filters for differential motion estimation. In B. Jähne, E. Barth, R. Mester, and H. Scharr, editors, *First International Workshop on Complex Motion, Günzburg, Germany, October 2004*, volume 3417 of *Lecture Notes in Computer Science*, Berlin, 2005. Springer Verlag.
[58] K. Krajsek and R. Mester. Marginalized maximum a posteriori hyperparameter estimation for global optical flow techniques. In *Bayesian Inference and Maximum Entropy Methods in Science and Engineering*, AIP Conference Proceedings, Visul Sensorics and Information Proceeding Lab, Institute for Computer Science, J.W. Goethe University, Frankfurt am Main, Germany, Volume 872, pages 311–318, November 2006.
[59] K. Krajsek and R. Mester. On the equivalence of variational and statistical differential motion estimation. In *Southwest Symposium on Image Analysis and Interpretation*, pages 11–15, Denver, CO/U.S.A., March 2006. ISBN 1-4244-0069-4.
[60] G. Krieger, C. Zetzsche, and E. Barth. Nonlinear image operators for the detection of local intrinsic dimensionality. In *Proceedings of the IEEE Workshop Nonlinear Signal and Image Processing*, pages 182–185, 1995.
[61] C.L. Lawson and R.J. Hanson. *Solving Least-Squares Problems.* Englewood Cliffs, NJ, 1974. Prentice-Hall.
[62] B. Lucas and T. Kanade. An iterative image registration technique with an application to stereo vision. In *DARPA Image Understanding Workshop*, pages 121–130, 1981.
[63] D. J. C. MacKay. Bayesian interpolation. *Neural Computation*, 4(3): 415–447, 1992.

[64] V. A. Mandelshtam. On harmonic inversion of cross-correlation functions by the filter diagonalization method. *Journal of Theoretical and Computational Chemistry*, 2(4):1–9, 2003.

[65] V. A. Mandelshtam and H. S. Taylor. Harmonic inversion of time signals and its applications. *Journal of Chemical Physics*, 107(17):6756–6769, 1997.

[66] R. Mester. Some steps towards a unified motion estimation procedure. In *Proceedings of the 45th IEEE MidWest Symposium on Circuits and Systems (MWSCAS)*, August 2002.

[67] R. Mester. Orientation estimation: conventional techniques and a new approach. In *Proceedings of the European Signal Processing Conference (EUSIPCO2000)*, Tampere, FI, September. 2000.

[68] R. Mester. A new view at differential and tensor-based motion estimation schemes. In B. Michaelis, editor, *Pattern Recognition 2003*, Lecture Notes in Computer Science, Magdeburg, Germany, September 2003. Springer Verlag.

[69] R. Mester. On the mathematical structure of direction and motion estimation. In *Workshop on Physics in Signal and Image Processing*, Grenoble, France, January 2003.

[70] R. Mester. A system-theoretical view on local motion estimation. In *Proceedings of the IEEE SouthWest Symposium on Image Analysis and Interpretation*, Santa Fé, NM, April 2002. IEEE Computer Society.

[71] S. K. Mitra, H. Li, I.-S. Lin, and T.-H. Yu. A new class of nonlinear filters for image enhancement. In *International Conference on Acoustics, Speech, and Signal Processing, 1991. ICASSP-91., 1991*, volume 4, pages 2525–2528, Toronto, Ontario, Canada, April 1991. IEEE.

[72] C. Mota and E. Barth. On the uniqueness of curvature features. In G. Baratoff and H. Neumann, editors, *Dynamische Perzeption*, volume 9 of *Proceedings in Artificial Intelligence*, pages 175–178, Köln, 2000. Infix Verlag. ISBN 3 89838 020 3.

[73] C. Mota, I. Stuke, and E. Barth. Analytic solutions for multiple motions. In *Proceedings IEEE International Conference Image Processing*, volume II, pages 917–20, Thessaloniki, Greece, October 7–10, 2001. IEEE Signal Processing Soc.

[74] C. Mota, T. Aach, I. Stuke, and E. Barth. Estimation of multiple orientations in multi-dimensional signals. In *IEEE International Conference Image Processing*, pages 2665–8, Singapore, October 24–27, 2004.

[75] C. Mota, M. Door, I. Stuke, and E. Barth. Categorization of transparent-motion patterns using the projective plane. Technical report, Preprint series of the DFG priority program 1114 "Mathematical methods for time series analysis and digital image processing", November 2004.

[76] C. Mota, M. Dorr, I. Stuke, and E. Barth. Analysis and synthesis of motion patterns using the projective plane. In Bernice E. Ragowitz and Thrasyvoulos N. Pappas, editors, *Human*

Vision and Electronic Imaging Conference IX, Volume 5292 of Proceedings of SPIE, pages 174–181, 2004. http://www.inb.uni-luebeck.de/publications/pdfs/MoDoStBa046.pdf

[77] C. Mota, M. Dorr, I. Stuke, and E. Barth. Categorization of transparent-motion patterns using the projective plane. *International Journal of Computer & Information Science*, 5(2):129–140, June 2004.

[78] C. Mota, I. Stuke, T. Aach, and E. Barth. Spatial and spectral analysis of occluded motions. Technical report, Preprint series of the DFG priority program 1114 "Mathematical methods for time series analysis and digital image processing", November 2004.

[79] C. Mota, I. Stuke, T. Aach, and E. Barth. Divide-and-conquer strategies for estimating multiple transparent motions. In B. Jähne, E. Barth, R. Mester, and H. Scharr, editors, *Complex Motion, 1. International Workshop, Günzburg, October 2004*, volume 3417 of *Lecture Notes in Computer Science*, Berlin, 2005. Springer Verlag.

[80] C. Mota, I. Stuke, T. Aach, and E. Barth. Spatial and spectral analysis of occluded motions. *Signal Processing: Image Communication. Elsevier Science*, 20–6:529–536, 2005.

[81] C. Mota, I. Stuke, and E. Barth. The intrinsic dimension of multispectral images. In *MICCAI Workshop on Biophotonics Imaging for Diagnostics and Treatment*, pages 93–100, 2006.

[82] M. Mühlich. *Estimation in Projective Spaces and Application in Computer Vision*. PhD thesis, Johann Wolfgang Goethe Universität Frankfurt am Main, 2005.

[83] M. Mühlich and T. Aach. A theory for multiple orientation estimation. In H. Bischof and A. Leonardis, editors, *Proceedings European Conference on Computer Vision 2006*, number 3952 in LNCS, pages (II) 69–82, 2006. Springer.

[84] M. Mühlich and R. Mester. A statistical unification of image interpolation, error concealment, and source-adapted filter design. In *Proceedings of the Sixth IEEE Southwest Symposium on Image Analysis and Interpretation*, Lake Tahoe, NV/U.S.A., March 2004.

[85] M. Mühlich and R. Mester. The role of total least squares in motion analysis. In *ECCV*, pages 305–321, Freiburg, Germany, 1998.

[86] M. Mühlich and R. Mester. The role of total least squares in motion analysis. In *Proceedings European Conference on Computer Vision ECCV 1998*, Lecture Notes on Computer Science, pages 305–321, 1998.

[87] M. Mühlich and R. Mester. Subspace methods and equilibration in computer vision. Technical Report XP-TR-C-21, Institute for Applied Physics, Goethe-Universitaet, Frankfurt, Germany, November 1999.

[88] D. Mumford and J. Shah. Optimal approximations by piecewise smooth functions and associated variational problems. *Communications Pure and Applied Mathematics XLII*, pages 577–685, 1989.

[89] H. H. Nagel and W. Enkelmann. An investigation of smoothness constraints for the estimation of dispalcement vector fields from image

sequences. *Pattern Analysis and Machine Intelligence*, 8(5):565–593, September 1986.

[90] O. Nestares, D. J. Fleet, and D. Heeger. Likelihood functions and confidence bounds for total-least-squares problems. In *CVPR'00*, volume 1, 2000.

[91] O. Nestares, D.J. Fleet, and D.J. Heeger. Likelihood functions and confidence bounds for total-least-squares problems. In *IEEE Conference on Computer Vision and Pattern Recognition*, pages 523–530, Hilton Head, SC, Vol. I, 2000.

[92] N. Ohta. Optical flow detection using a general noise model. *IEICE Transactions on Information and Systems*, E79-D(7):951–957, July 1996.

[93] J. Polzehl and V. Spokoiny. Adaptive weights smoothing with applications to image restoration. *Journal of the Royal Statistical Society B*, 62(2):335–354, 2000.

[94] K. Roetmann, C. S. Garbe, W. Schmunk, and V. Beushausen. Micro-flow analysis by molecular tagging velocimetry and planar raman-scattering. In *Proceedings of the 12th International Symposium on Flow Visualization*, Göttingen, Germany, 2006.

[95] H. Scharr. Diffusion-like reconstruction schemes from linear data models. In *Pattern Recognition 2006*, Lecture Notes in Computer Science 4174, pages 51–60, Berlin, 2006. Springer Verlag.

[96] H. Scharr. Optimal second order derivative filter families for transparent motion estimation. In *EUSIPCO, 2007*, pages 302–306, Poznan, Poland, 2007. EURASIP. ISBN 978-83-921340-2-2.

[97] H. Scharr. Towards a multi-camera generalization of brightness constancy. In B. Jähne, E. Barth, R. Mester, and H. Scharr, editors, *Complex Motion, 1. International Workshop, Günzburg, October 2004*, volume 3417 of *Lecture Notes in Computer Science*, Berlin, 2005. Springer Verlag.

[98] H. Scharr. *Optimal Operators in Digital Image Processing.* PhD thesis, Interdisciplinary Center for Scientific Computing, Univ. of Heidelberg, 2000.

[99] H. Scharr and T. Schuchert. Simultaeous estimation of depth, motion and slopes using a camera grid. In T. Aach L. Kobbelt, T. Kuhlen, and R. Westermann, editors, *Vision Modeling and Visualization 2006*, pages 81–88, Aachen, Berlin, November 22–24, 2006. Aka. ISBN 3-89838-081-5.

[100] H. Scharr and H. Spies. Accurate optical flow in noisy image sequences using flow adapted anisotropic diffusion. *Signal Processing: Image Communication*, 20(6):537–553, 2005.

[101] H. Scharr, S. Körkel, and B. Jähne. Numerische isotropieoptimierung von FIR-filtern mittels querglättung. In E. Paulus and F. M. Wahl, editors, *Mustererkennung 1997*, pages 367–374, 1997. Springer. ISBN 3-540-63426-6.

[102] H. Scharr, M.J. Black, and H.W. Haussecker. Image statistics and anisotropic diffusion. In *International Conference on Computer Vision, ICCV 2003*, pages 840–847, Nice, France, 2003. IEEE Computer Society. ISBN 0-7695-1950-4.
[103] H. Scharr, I. Stuke, C. Mota, and E. Barth. Estimation of transparent motions with physical models for additional brightness variation. In *13th European Signal Processing Conference, EUSIPCO*, pages 1–8, 2005. EURASIP. ISBN 975-00188-0-X.
[104] U. Schimpf, C. S. Garbe, and B. Jähne. Investigation of transport processes across the sea surface microlayer by infrared imagery. *Journal of Geophysical Research*, 109(C08S13), 2004.
[105] C. Schnörr. A study of a convex variational diffusion approach for image segmentation and feature extraction. *Journal of Mathematical Imaging and Vision*, 8(3):271–292, 1998.
[106] T. Schuchert and H. Scharr. Simultaneous estimation of surface motion, depth and slopes under changing illumination. In *Pattern Recognition 2007*, Lecture Notes in Computer Science 4713, pages 184–193, Springer Verlag, 2007. ISBN 103-540-74933-0.
[107] G. Schwarz. Estimating the dimension of a model. *Annals of Statistics*, 6(461):464, 1978.
[108] M. Shizawa and K. Mase. Simultaneous multiple optical flow estimation. In *IEEE Conference Computer Vision and Pattern Recognition*, I, pages 274–278, Atlantic City, NJ, June 1990. IEEE Computer Press.
[109] M. Shizawa and K. Mase. Principle of superposition: A common computational framework for analysis of multiple motion. In *Proceedings of the IEEE Workshop on Visual Motion*, pages 164–172, Princeton, NJ, 1991. IEEE Computer Society Press.
[110] E. P. Simoncelli. Design of multi-dimensional derivative filters. In Proceedings 1994 *International Conference on Image Processing*, Volume 1, pages 790–794, Austin, TX, USA, November 13–16 1994. IEEE Computer Society. ISBN 0-8186-6950-0.
[111] H. Spies and H. Scharr. Accurate optical flow in noisy image sequences. In *Proceedings of the Eighth International Conference on Computer Vision (ICCV-01)*, Volume I, Vancouver, British Columbia, Canada, July 7–14, 2001. IEEE Computer Society
[112] H. Spies, O. Beringer, et al. Analyzing particle movements at soil interfaces. In B. Jähne, H. Haussecker, and P. Geißler, editors, *Handbook on Computer Vision and Applications*, volume 3, pages 699–718. San Diego, CA, 1999. Academic Press.
[113] I. Stuke, T. Aach, E. Barth, and C. Mota. Estimation of multiple motions by block matching. In W. Dosch and R. Y. Lee, editors, *Proceedings ACIS 4th International Conference Software Engineering, Artificial Intelligence, Networking and Parallel/Distributed Computing*, pages 358–62, Lübeck, Germany, October 16–18, 2003.

[114] I. Stuke, T. Aach, C. Mota, and E. Barth. Estimation of multiple motions: regularization and performance evaluation. In B. Vasudev, T. R. Hsing, A. G. Tescher, and T. Ebrahimi, editors, *Image and Video Communications and Processing 2003*, volume 5022 of *Proceedings of SPIE*, pages 75–86, May 2003.

[115] I. Stuke, T. Aach, C. Mota, and E. Barth. Linear and regularized solutions for multiple motion. In *Proceedings IEEE International Conference Acoustics, Speech and Signal Processing*, volume III, pages 157–60, Hong Kong, April 2003. IEEE Signal Processing Soc.

[116] I. Stuke, T. Aach, E. Barth, and C. Mota. Estimation of multiple motions using block-matching and Markov random fields. In S Panchanathan and B Vasudev, editors, *Visual Communications and Image Processing 2004, IS&T/SPIE 16th Annual Symposium Electronic Imaging*, San Jose, CA, January 18–22, 2004.

[117] I. Stuke, E. Barth, and C. Mota. Estimation of multiple orientations and multiple motions in multi-dimensional signals. In *IEEE XIX Brazilian Symposium on Computer Graphics and Image Processing (SIBGRAPI'06)*, pages 341–348, 2006.

[118] B. Burgeth T. Brox, J. Weickert and P. Mrázek. Nonlinear structure tensors. Revised version of technical report No. 113, Saarland University, Saarbrücken, Germany, 2004.

[119] D. Tschumperle. *PDE's based regularization of multivalued images and applications.* PhD thesis, Université de Nice-Sophia, 2002.

[120] D. Tschumperle. Fast anisotropic smoothing of multi-valued images using curvature preserving pde's. Technical report, Equipe Image/GREYC, UMR CNRS 6072, 6 Bd du Maréchal Juin, 14050 Caen Cedex, France, 2005.

[121] D. Tschumperle and R. Deriche. Vector-valued image regularization with pdes: a common framework for different applications. *IEEE Transactions on Pattern Analysis and Machine Intelligence*, 27(4):506–517, 2005.

[122] M. Unser. Splines: A perfect fit for signal and image processing. *IEEE Signal Processing Magazine*, 16(6):22–38, November 1999.

[123] S. Van Huffel and J. Vandewalle. Analysis and properties of the generalized total least squares problem $\boldsymbol{Ax} \approx \boldsymbol{B}$ when some or all columns in A are subject to error. *SIAM Journal on Matrix Analysis and Applications*, 10(3):294–315, 1989.

[124] S. Van Huffel and J. Vandewalle. *The Total Least Squares Problem: Computational Aspects and Analysis.* Society for Industrial and Applied Mathematics, Philadelphia, PA 1991.

[125] N. J. Walkington. Algorithms for computing motion by mean curvature. *SIAM Journal on Mathematical Analysis*, 33(6):2215–2238, 1996. doi: http://dx.doi.org/10.1137/S0036142994262068.

[126] J.Y.A. Wang and E.H. Adelson. Spatio-temporal segmentation of video data. In *Proceedings of the SPIE: Image and Video Processing II, vol. 2182*, pages 120–131 *San Jose, February 1994*, 1994.

[127] J. Weber and J. Malik. Robust computation of optical flow in a multi-scale differential framework. *International Journal of Computer Vision*, 14(1):67–81, 1995.

[128] H. Wechsler, Z. Duric, F. Y. Li, and V. Cherkassky. Motion estimation using statistical learning theory. *Pattern Analysis and Machine Intelligence*, 26(4):466–478, 2004.

[129] J. Weickert and T. Brox. Diffusion and regularization of vector- and matrix-valued images. In M. Z. Nashed, O. Scherzer, editors, *Inverse Problems, Image Analysis, and Medical Imaging. Contemporary Mathematics*, pages 251–268, Providence, 2002. AMS.

[130] E.P. Wigner. *Group Theory and its Application to Quantum Mechanics of Atomic Spectra.* New York, 1959. Academic Press.

[131] G. Winkler. *Image Analysis, Random Fields and Markov Chain Monte Carlo Methods. A Mathematical Introduction.* Berlin, 2002. Springer.

[132] M. Worring and A.W.M. Smeulders. Digital curvature estimation. *CVGIP: Image Understanding*, 58(3):366–382, 1993. ISSN 1049-9660. doi: http://dx.doi.org/10.1006/ciun.1993.1048.

[133] W. Yu, K. Daniilidis, S. Beauchemin, and G. Sommer. Detection and characterization of multiple motion points. In *18th IEEE Conference on Computer Vision and Pattern Recognition*, vol. 1, pages 171–177, IEE Computer press Fort Collins, CO, 1999.

[134] W. Yu, K. Daniilidis, and G. Sommer. A new 3D orientation steerable filter. In *Proceedings DAGM 2000.* Berlin, September 2000. Springer.

[135] C. Zetzsche and E. Barth. Fundamental limits of linear filters in the visual processing of two-dimensional signals. *Vision Research*, 30: 1111–7, 1990.

[136] C. Zetzsche, E. Barth, and B. Wegmann. The importance of intrinsically two-dimensional image features in biological vision and picture coding. In Andrew B. Watson, editor, *Digital Images and Human Vision*, pages 109–38. MIT Press, October 1993.

Index

Understanding Complex Systems

Jirsa, V.K.; Kelso, J.A.S. (Eds.)
Coordination Dynamics: Issues and Trends
XIV, 272 p. 2004 [978-3-540-20323-0]

Kerner, B.S.
The Physics of Traffic:
Empirical Freeway Pattern Features,
Engineering Applications, and Theory
XXIII, 682 p. 2004 [978-3-540-20716-0]

Kleidon, A.; Lorenz, R.D. (Eds.),
Non-equilibrium Thermodynamics
and the Production of Entropy
XIX, 260 p. 2005 [978-3-540-22495-2]

Kocarev, L.; Vattay, G. (Eds.)
Complex Dynamics in Communication
Networks
X, 361 p. 2005 [978-3-540-24305-2]

McDaniel, R.R.Jr.; Driebe, D.J. (Eds.)
Uncertainty and Surprise in Complex Systems:
Questions on Working with the Unexpected
X, 200 p. 2005 [978-3-540-23773-0]

Ausloos, M.; Dirickx, M. (Eds.)
The Logistic Map and the Route to Chaos –
From the Beginnings to Modern Applications
XX, 413 p. 2006 [978-3-540-28366-9]

Kaneko, K.
Life: An Introduction to Complex Systems
Biology
XIV, 369 p. 2006 [978-3-540-32666-3]

Braha, D.; Minai, A.A.; Bar-Yam, Y. (Eds.)
Complex Engineered Systems – Science Meets
Technology
X, 384 p. 2006 [978-3-540-32831-5]

Fradkov, A.L.
Cybernetical Physics – From Control of Chaos
to Quantum Control
XII, 241 p. 2007 [978-3-540-46275-0]

Aziz-Alaoui, M.A.; Bertelle, C. (Eds.)
Emergent Properties in Natural
and Artificial Dynamical Systems
X, 280 p. 2006 [978-3-540-34822-1]

Baglio, S.; Bulsara, A. (Eds.)
Device Applications of Nonlinear Dynamics
XI, 259 p. 2006 [978-3-540-33877-2]

Jirsa, V.K.; McIntosh, A.R. (Eds.)
Handbook of Brain Connectivity
X, 528 p. 2007 [978-3-540-71462-0]

Krauskopf, B.; Osinga, H.M.;
Galan-Vioque, J. (Eds.)
Numerical Continuation Methods
for Dynamical Systems
IV, 412 p. 2007 [978-1-4020-6355-8]

Perlovsky, L.I.; Kozma, R. (Eds.)
Neurodynamics of Cognition and Consciousness
XI, 366 p. 2007 [978-3-540-73266-2]

Qudrat-Ullah, H.; Spector, J.M.; Davidsen, P. (Eds.)
Complex Decision Making – Theory and Practice
XII, 337 p. 2008 [978-3-540-73664-6]

beim Graben, P.; Zhou, C.; Thiel, M.;
Kurths, J. (Eds.)
Lectures in Supercomputational Neuroscience –
Dynamics in Complex Brain Networks
X, 378 p. 2008 [978-3-540-73158-0]

Fuchs, A.; Jirsa, V.K.K. (Eds.)
Coordination: Neural, Behavioral
and Social Dynamics
XIV, 348 p. 2008 [978-3-540-74476-4]

Dahlhaus, R.; Kurths, J.; Maass, P.;
Timmer, J. (Eds.)
Mathematical Methods in Signal Processing
and Digital Image Analysis
XIV, 292 p. 2008 [978-3-540-75631-6]

FSC
www.fsc.org
MIX
Papier aus verantwortungsvollen Quellen
Paper from responsible sources
FSC® C105338

If you have any concerns about our products,
you can contact us on
ProductSafety@springernature.com

In case Publisher is established outside the EU,
the EU authorized representative is:
Springer Nature Customer Service Center GmbH
Europaplatz 3, 69115 Heidelberg, Germany

Printed by Libri Plureos GmbH
in Hamburg, Germany